SCIENCE

SCIENCE

A HISTORY OF DISCOVERY
IN THE TWENTIETH CENTURY

TREVOR I. WILLIAMS

Oxford • New York
OXFORD UNIVERSITY PRESS
1990

Volume editor Robert Peberdy
Art editor Ayala Kingsley
Designers Frankie Macmillan, Tony de Saulles
Picture research manager Alison Renney
Picture research Diane Hamilton (USA), Charlotte Ward-Perkins
Project editor Peter Furtado

AN EQUINOX BOOK

Planned and produced by
Equinox (Oxford) Ltd
Musterlin House
Jordan Hill Road
Oxford, England
OX2 8DP

Copyright © Equinox (Oxford) Ltd
1990

Published in the United States of
America by
Oxford University Press, Inc.
200 Madison Avenue
New York NY 10016

Oxford is a trademark of Oxford
University Press

Library of Congress
Cataloging-in-Publication Data
Science: a history of discovery in the
twentieth century/edited by
Trevor I. Williams.
    p.   cm.
    "An Equinox book."
    Includes index.
    ISBN 0-19-520843-9
    1. Science–History–20th
century.   2. Technology–History–
20th century.   3. Science–Social
aspects.   4. Technology–Social
aspects. I. Williams. Trevor Illtyd.
Q125.S4317 1990
509'.04–dc20

ISBN 0-19-520843-9

Printing (last digit) 9 8 7 6 5 4 3 2 1

Printed in Singapore by
CS Graphics

CONTRIBUTORS

Clive Bagshaw   University of
Leicester, UK 146–147
James R. Bartholomew   Ohio State
University, Columbus, USA 200
Michael Bravo   Darwin College,
Cambridge, UK 114
P. Thomas Carroll   Rensselaer
Polytechnic Institute, Troy, USA 154
Dougal Dixon   Freelance science
writer, UK 190–191
John Emsley   King's College,
London, UK 84
Robert Fox   Linacre College,
Oxford, UK 223
Thomas F. Glick   Boston
University, USA 145
Anna Guagnini   Linacre College,
Oxford, UK 116
John Hodgson   *Biotechnology*,
London, UK 176, 212
Alexander Kohn   Orgenics Ltd,
Yavne, Israel 68–69

Kwang-Ting Liu   National Taiwan
University, Taipei, Taiwan
162–163
Zhores Medvedev   National
Institute for Medical Research,
London, UK 134–135
Iain Nicolson   Hatfield Polytechnic,
UK 52–53, 226–227
J.R. Ravetz   Council for Science
and Society, London, UK 104–105,
138–139
Anthony Storr   Oxford, UK 77
Christine Sutton   Department of
Nuclear Physics, Oxford, UK 86–87,
120–121
Peter Turvey   Science Museum,
London, UK 34–35, 172–173
Shiv Visvanathan   Centre for the
Study of Developing Societies,
Delhi, India 183
Sten Wahlström   Karlstad,
Sweden 49

Paul Weindling   Wellcome Unit for
the History of Medicine, Oxford,
UK 46–47, 110–111
Tom Wilkie   *The Independent*,
London, UK 136
A.B. Zahlan   Zahlan Consultants
Ltd, London, UK 206

ADVISORY EDITORS

Michael J. Clark   Wellcome
Centre, London; formerly of the
Centre de recherche en histoire
des sciences et des techniques,
Paris, France
Yehuda Elkana   Tel-Aviv
University, Israel
David Landes   Harvard University,
USA
Charles Webster   All Souls College,
Oxford, UK; formerly of the
Wellcome Unit for the History of
Medicine, Oxford

CONTENTS

PREFACE

In 1944 G.M. Trevelyan added a new dimension to the study of history with the publication of his *English Social History*. In the preface to this he asserted that "without social history, economic history is barren and political history is unintelligible". Today it can be argued that for a proper understanding of history a fourth dimension is necessary: the history of science and technology. Although these are widely acknowledged as major forces in the shaping of modern civilization, most historians have been reluctant to give them due weight in general studies. This book is designed to redress that imbalance by presenting, in a form understandable by the general reader, an account of how science and technology have developed and influenced society worldwide in the 20th century.

This period is significant in that the turn of the century saw an important change of emphasis. Until then science had been carried along mainly by the Scientific Revolution, epitomized by Newtonian physics, and technology by the engineering advances of the Industrial Revolution and its aftermath. But by 1900 mechanistic interpretations of the workings of nature no longer sufficed to explain a whole range of new phenomena, especially in the field of electricity. Long cherished beliefs – such as the laws of the conservation of mass and of energy – had to give way to relativity theory and quantum mechanics.

The 20th century is now in its last decade but the oldest citizens can personally recall immense changes in the pattern of life. They have adapted to the telephone, the cinema, radio and television; electric light and heating; the automobile, the aeroplane and – vicariously – space flight; refrigeration and frozen foods; mechanical and chemical agriculture; and the computer. There is no ideal way of marshaling such a complex mass of information and a pragmatic approach is unavoidable. The one I have chosen is essentially that of a gridiron: the text is organized in six horizontal chronological bands and three vertical strands corresponding to technology, the biological and medical sciences, and the physical sciences.

Currently there is much debate about the importance of dates in the teaching of history. While it is arguable that precise dates are not necessarily individually important, they do provide a guide to the order in which events occurred, which is certainly significant. For this reason six time charts have been included. To emphasize the role of the individual there are almost 200 short biographies. It must be recognized, however, that there is an increasing tendency toward anonymity; much original work is done within large corporations and government departments, and those concerned are not always publicly identified.

While the chronological and thematic framework – reinforced by liberal photographic illustration and artwork – provides a satisfactory way of displaying the main events and their interrelationships, material auxiliary to the main text has been included to give emphasis to certain features such as national and international organizations, and political and social implications. Many of these articles have been written by authors with special expertise in these fields: I am glad to acknowledge their contribution.

Trevor I. Williams   Oxford

INTRODUCTION

Traditionally, historians concerned themselves primarily with political and economic history. Today, not only is social history generally acknowledged as a discipline in its own right but the importance of a fourth dimension – the history of science, technology and medicine – is becoming increasingly recognized. The significance of this fourth dimension is the theme of this book. It is, nevertheless, a truism that more has happened in science during this century, now in sight of its close, than in the whole of previous history. The acceleration of new discoveries, and their growing complexity, makes it difficult to discern and present a wholly consistent picture of the course of events. The problem is compounded, of course, when we try to weave the scientific and technological thread into an historical tapestry which also has to incorporate strands of political, economic and social history, which are also susceptible to different interpretations. Nevertheless, although the picture that emerges must necessarily be obscure and blurred in many points of detail the broad outline can be quite clearly delineated.

The meaning of science and technology

Before attempting to construct this picture, however, some definitions are necessary – in particular, what do we mean by the terms science and technology respectively? Increasingly, technology is regarded as simply the application of science to practical ends, but this is an oversimplification. Many technologies – metal working, pottery and textiles, to mention only a few – were empirically developed thousands of years ago quite independently of ideas about the laws of nature, which are the concerns of science. Even the Industrial Revolution, which may be roughly dated 1760–1830, made little use of the growing corpus of scientific knowledge. More important was the mechanization of traditional processes and the use of water and steam power. There is a good deal of truth in the aphorism that science owes more to the steam engine – through the direction of attention to thermodynamic problems – than the steam engine does to science. But the 19th century saw a new trend. Increasingly technology advanced not empirically but through the application of science, both that revealed by academic scientists pursuing knowledge for its own sake and that revealed by what we would now call targeted research; that is, research conducted with the object of achieving a specific objective. But although much targeted research is complementary to the relevant technology it is by no means identical with it. The technology concerns itself with turning the acquired knowledge to a practical industrial end, whether it be a new chemical process or a new or improved machine. This is a matter for specialist engineers – mechanical, electrical, civil and chemical.

If science is thus increasingly the mainspring of technology, how then is science itself to be defined? That technology so substantially predates science in itself indicates a fundamental difference between them but does not preclude a steady convergence with the passage of time, technology becoming increasingly dependent on the application of scientific rather than empirical knowledge. This is indeed implicit in the commonly used phrase "science-based industry", which denotes for example, the chemical, electronic, communications, electrical, pharmaceutical and many other industries which exist only because they utilize the products of scientific research. However, before discussing this convergence it is constructive to attempt separate definitions of science and technology.

In the present context it is unprofitable to consider what science denoted in Classical times and, with one proviso, we shall restrict ourselves to science in the modern sense, stemming from the Scientific Revolution which began in Europe in the middle of the 16th century. The one proviso is that this Revolution was not wholly spontaneous and indigenous, but stemmed from the reintroduction of Classical learning into Europe through Arabic texts which incorporated not only the new discoveries of Islam but drew also on the scientific and mathematical concepts of the East, notably India. From China came four major discoveries: papermaking, printing from movable type, gunpowder and the mariner's compass.

The Revolution had a dual character. On the one hand it was intellectual; questioning established dogma and putting forward new interpretations of old ideas. On the other, it sought to advance knowledge not only by observation of natural phenomena but also by deliberately contrived experiments. It was in the event an unstoppable movement, but that is not to say that it did not meet considerable opposition. In the 17th century Galileo came within sight of burning by the Inquisition for daring to support the heliocentric theory of the Universe advanced by Copernicus in 1543 in his monumental *De revolutionibus orbium celestium* (*on the Revolutions of the Celestial Spheres*). In the mid-19th century Darwin's theory of evolution was fiercely attacked by the established Church on purely theological grounds; the scientific evidence was judged irrelevant. As recently as 1940, the Russian geneticist N.I. Vavilov was thrown into a concentration camp – where he died – because his ideas, scientifically unassailable, were incompatible with the dialectical materialism of Communism.

The essence of the new science was that it was objective, in that the validity of any reported experiment could be checked by any other investigator. It was also cumulative, each advance resting on already established factual evidence. Thus was built up an ever-increasing knowledge of the nature of the world around us. But the objectivity also left room, and need, for imagination. The mere collection of facts is in itself unrewarding: real progress comes with recognition that certain observations fall within a pattern, that they can all be accommodated within some wider theoretical concept. The validity or otherwise of the theory can be tested objectively on the basis of prediction based on it. If the predictions prove correct the case for the theory is strengthened; if they are wrong it must be modified or even abandoned. On this basis some theories are short-lived while others go from strength to strength. Nevertheless, in science it is an article of faith that no theory is unchallengeable, though naturally the more firmly established and comprehensive a theory the more well-founded must be the evidence advanced to refute it.

Therein lies the uniqueness of science as an intellectual discipline. It is a systematic and coherent organization of knowledge built up from generation to generation; it is objective and not subjective; it can at any time be put to the test anywhere in the world. But this very dispassionate objectivity

which is the strength of science has also been the basis of criticism of it. Science, it is said, does nothing to answer the fundamental philosophical arguments that have raged since the dawn of civilization – the nature of good and evil; the merits of different religious beliefs; the rights of man; the vagaries of human nature. This is true, but the answer is simple: scientists ask the question "How?"; it is for philosophers to ask "Why?" The scientific method has been extraordinarily successful in revealing the working of the physical world but there is no reason to suppose that it should be equally successful with metaphysics. When scientists step outside their rigid professional framework, as they do, they prove to be as disparate and passionate as the world at large in arguing such questions.

Among those who sought to relate science to philosophy was William Whewell, Master of Trinity College, Cambridge, England. It is to him that we owe the word scientist. In 1840 he

wrote: "We very much need a name to describe a cultivator of science in general. I should incline to call him a Scientist." Previously, such scholars had referred to themselves as natural philosophers, the word natural denoting not only their interest in natural phenomena but emphasizing that they did not concern themselves with such free-thinking as in past times might have embroiled them with the Church. The French similarly signaled a change of direction: their *savants* became known as *hommes de science*. Likewise, the *Wissenschaftler* appeared in Germany, though often with rather more catholic interests than his opposite numbers in France or Britain.

While the ultimate goal of science is to discover the laws that govern the world around us, the natural philosophers were not slow to perceive that the knowledge so gained could be turned to practical advantage. When it was founded in 1662 the Royal

▲ **A botany class at Tarpaulin Cove, Massachusetts, USA, 1895.**

Society of London specifically defined its responsibility as "promoting by the authority of experiments the sciences of natural things and of useful arts ... to the advantage of the human race." In France, where the Académie des Sciences was founded in 1666, the emphasis on practicality was even more emphatic, and the academicians were paid a royal pension to pursue their inquiries.

So we come, via science and scientists, to consideration of technology. Historically, as we have noted, this was distinct from applied science, if only because it existed long before science in its modern sense existed. Though there has been a remarkable convergence during the present century the two are still by no means synonymous.

That technology is not readily definable in precise terms reflects the fact that, like many such words, it has not only changed its meaning over the years but is still fluid. After a lapse of more than three centuries it is impossible to know exactly what the 40 original Fellows of the Royal Society had in mind when referring to the "useful arts" which were to be inspired by the application of scientific knowledge. From their early discussions it is clear, however, that they had particularly in mind the mechanization of industrial processes, which was already beginning and was to reach its full flowering with the beginning of the industrial revolution a century later. Thus their "useful arts" can be roughly equated with the technology of the day.

In the early 18th century the emphasis was still on engineering, as exemplified by Phillips' *Technology: a description of Arts especially the mechanical* (1706). But the great 18th-century work in this field was Denis Diderot's vast and comprehensive *Encyclopédie...* (Encyclopedia of science, arts and industry) published in France in 28 volumes (1751–72). Two of its volumes were devoted to fine illustration plates and, significantly, these were described as relating not only to science and the mechanical arts, but also to the "liberal arts". The latter included glass making, leather working, agriculture, baking, brewing and soap boiling.

In 1866 there appeared Charles Tomlinson's lesser, but still substantial, *Cyclopaedia of Useful Arts, Mechanical and Chemicals Manufactures, Mining and Engineering*. This brings us much closer to the modern concept of technology, a term embracing both the ancient and still largely empirical technologies and the application of science to the practical and industrial arts in the broadest sense. Nevertheless a distinction between the two was still perceived. Thus when the Science Museum was instituted in London in 1853 it came under the Privy Council's Science and Art Department with the National Museum of Science and Industry as its full title. Later, in 1882, the department was reconstituted as the Department of Applied Science and Technology.

Technical education

Thus at the turn of the century it was generally understood that there were distinctions between science, applied science and technology but where the boundaries should be drawn was by no means agreed. Equally, the practitioners in these fields were not well defined. It is an over-simplification to suppose that the pattern was the same throughout the Western world. In Britain, for example, scientists – with very few exceptions – were firmly entrenched in the "ivory towers" of academia, taking a positive pride in regarding science as a strictly intellectual pursuit, leaving practical applications and commercial development to – by inference – lesser mortals.

◄ Soviet electrification in the 1930s: the Dneproges hydroelectric dam.

Elsewhere, more enlightened views prevailed. In the United States the importance of technological universities was well understood. The internationally famous Massachusetts Institute of Technology was founded in 1861 and the Stevens Institute of Technology, New Jersey, in 1870. The Morrill Land Grant College Act of 1862, had been a powerful stimulus. This provided that each State should be granted land – 30,000 acres for each State and representative then in Congress – to support at least one college "to teach such branches of learning as are related to agriculture and the mechanic arts." In Germany, the older universities tended to show the same antipathy toward industry and commerce as their British counterparts and positive steps were taken to redress the balance with the Technische Hochschulen, mostly founded in the second half of the 19th century. They conferred their own degrees and had a status comparable to that of the older foundations: they included the Technical University of Munich (1868) and the Berlin Technische Hochschule (1879). At professional level these maintained strong links with German industry.

On the other side of the globe Japan, late in adopting industrialization, also recognized the importance of advanced technology, and this was certainly a factor in its defeat of Russia in 1905. The teaching of science and technology was emphasized in the five imperial universities founded between 1877 (Tokyo) and 1911 (Tohoku). Not until after the Revolution, in a succession of five-year plans, did Russia begin a massive program of higher education with emphasis on science and technology. This prompted the Soviet leader Lenin's famous dictum that "Electrification plus Soviet power equals Communism".

The creation and development of new science-based industries, and innovation in old ones, was possible only with increasing numbers of highly trained technical staff. In the main, however, they were subservient to nontechnical management. The new industries were often founded and controlled by powerful and ambitious men of vision who often had little or no formal education of any kind, let alone in science or technology. Among such were Andrew Carnegie, Thomas Edison, Henry Ford, and Sam Goldwyn. Only rather rarely did qualified scientists and engineers found their own businesses. Among them were the German chemist Herdwig Mond, who went to Britain in 1860, to found an industrial chemicals empire which subsequently became part of ICI; Alfred Nobel, a Swede and also a trained chemist, founder of a worldwide high explosives industry; Charles Parsons, pioneer of the steam turbine; and Rudolf Diesel, inventor of the engine which bears his name.

The dissemination of knowledge

Even if it is blurred, the boundary between science and technology is real enough and it is necessary to consider how information crosses it. The major route is through publication. The academic scientist has no desire to hide his light under a bushel: publication is the means by which his research is brought to the notice of his peers worldwide and, from the points of view of both establishing reputations and advancing careers, priority is important. Thus there is an ever-growing body of scientific knowledge on which the whole world may freely draw. Until comparatively recently most new knowledge was communicated by journals published by a multiplicity of learned societies. Not until after World War II did commercial publishers enter this special field, though they had always published scientific books. In passing, it may be noted that by mid-century the published material was so vast, and growing so rapidly, that one of the major topics of discussion among scientists was how it was to be effectively utilized: there were

not enough hours in the day to keep up with it. In the end it proved, with the application of the computer to information storage and retrieval after World War II, a largely self-solving problem. Such gratuitously available knowledge may in itself lead directly to major technological application. Thus Marconi, pioneer of radio communication at the turn of the century, was directly inspired by a casual reference in a magazine to the experiments of Heinrich Hertz. Nearly half a century later the vast Manhattan Project, which produced the first atomic bombs, stemmed directly from the published accounts of academic experiments in atomic physics, the most crucial appearing only two days before war broke out in 1939.

By the end of the 19th century, however, industry was beginning to find this source of information insufficient for its needs. As a complement, and to ensure that the research done was directly related to their own needs – and often, indirectly, that of their customers – major companies began to establish

their own laboratories. Among the first was the United Alkali Company in Britain whose Widnes laboratory was opened in 1892. Significantly, its then small staff consisted – with only one exception – of men who had trained abroad in chemistry: in Giessen, Heidelberg and Zurich.

National and international research
The research conducted in such laboratories was directly related to the activities of the parent companies. For business reasons, much of it was unpublished or made available only after protection by patenting. There remained, however, areas of research which were of general national importance but not directly related to any particular company. Such, for example, was research on corrosion and metrology. To fill this gap only government-sponsored laboratories were appropriate. Among the first great institutions of this kind were the National Physical Laboratory in the UK (1900); the National Bureau of

Standards (now National Institute of Standards and Technology) in the USA (1901); and the Kaiser Wilhelm (later Max Planck) Institute in Germany.

Thus in the opening years of the 20th century science was being promoted on three fronts: in the universities, with a strong bias toward pure research in the older ones and toward applied research in the newer ones; increasingly in the laboratories of large industrial companies with a technical basis; and in large government institutions. Though these divisions were quite real, and well understood, there was a good deal of cross-fertilization, though its nature and extent varied from country to country. Thus senior academic scientists were often consulted – frequently on the basis of a regular retainer – by their industrial counterparts. They might not only gain expert advice but also conduct a limited amount of research with the aid of their students. There was liaison, too, between academic scientists and the national laboratories.

Though the balance was to change over the years this was essentially the pattern until mid-century. Then a fourth front was opened, as the cost and complexity of research in some fields began to exceed the resources of all but the most powerful nations. To this era belong such institutions as CERN (now the European Center for Particle Research) in Geneva, for nuclear research; and the European Molecular Biology Laboratory in Heidelberg; the European Space Agency; and JET (Joint European Torus) for fusion research.

The public perception of science and technology
So much, then, for the broad interrelationships within science itself. Finally, we must consider the no less important relationship between science and the world at large, the man or woman in the street. Then, as now, the impact was largely that

▲ The world's first sustained atomic reaction, Chicago, 2 December 1942.

13

of science rather than scientists. A few names, but not necessarily of the most outstanding scientists, became widely known: Röntgen, Marconi, Edison, Einstein and Zeppelin are among the few who come readily to mind. But in the main it was the social impact of science and technology that the public was most conscious of. In the first decade of this century this would have included the automobile, radio communication, electric lighting, the cinema, the gramophone, frozen food, the vacuum cleaner and dry cleaning. On the reverse side of the coin, they would have noticed the disappearance of things which had been a normal feature of life since time immemorial. The advent of petrol-fueled road vehicles, for example, ensured the virtual demise of the horse for general transport.

It is a measure of the speed and extent of change that comparatively few people can recollect a world in which radio and television broadcasting were not powerful media for the dissemination of information. Until after World War I neither of these even existed and lectures to live audiences and the partial words of newspapers and magazines were still the prime means of communication. So far as science was concerned the popular press devoted little space to science and technology as such, a situation not destined to change substantially for more than half a century. The situation with regard to lectures was, however, different. In the 19th century most large cities and many smaller ones in Europe and America had local societies to promote a wide range of intellectual interests, such as music, literature and art; science, too, figured prominently among them. The Royal Institution in London attracted fashionable audiences, often including royalty, for its Friday evening discourses: the Manchester Literary and Philosophical Society was matched by the Literary and Philosophical Society of New York. Such societies, which still flourish in small communities, were numbered in hundreds. Though individually small, their collective membership was numbered in tens of thousands and they thus represented a significant means by which laymen might be informed about, and discuss, scientific developments of the day.

Over and above these local bodies, however, there existed a number of important national bodies. The prototype of these was the British Association for the Advancement of Science, founded in 1831 to provide a forum enabling scientists not only to talk among themselves, but to inform the public, a task which many felt the Royal Society was badly neglecting. It was a peripatetic body, meeting annually in different cities in Britain and, very occasionally, in the British Commonwealth. It attracted audiences numbered in thousands and – rather an exception to their general indifference to matters scientific – the proceedings were widely reported in the press. It still flourishes, though in recent years it has turned its attention increasingly not to science as such but to its social implications. Similar national bodies were established elsewhere: the American Association for the Advancement of Science, founded in 1848, still attracts very large annual gatherings. A French Association was founded in 1872, an Indian Association in 1876, and one in Australia and New Zealand in 1888.

Growing specialization
The earliest scientists took as their province the whole of the natural world – from the celestial bodies revealed in the early 17th century by Galileo's telescopes to the world of microorganisms explored in the late 17th century by Antony van Leeuwenhoek and his microscopes. By the 19th century, however, and into the 20th, the age of the polymath was over: the sheer expansion of knowledge had made a growing degree of

specialization inevitable. By 1900 four main strands had emerged: chemistry, physics, biology and geology. The first two were designated physical sciences, dealing with the non-living world: a distinguishing mark was that they were essentially quantitative, results being interpreted in mathematical terms. This relationship is succinctly expressed in the dictum that mathematics is the handmaiden of science. It was much less true of biology – embracing the whole of the plant and animal kingdoms – which was largely observational and descriptive. Geology was slightly anomalous; concerned largely with inanimate materials, the rocks and minerals of the earth's crust, much research depended on the observation and classification of the fossil remains of once-living organisms.

As the 20th century progressed this divergence into specialized fields became more pronounced – so much so that specialists within a major field could no longer understand each other clearly. Chemistry, for example, became divided into inorganic, and physical chemistry. Yet, paradoxically, at the same time a complex pattern of anastomosis or interconnection emerged, as one specialty found it had something in common with another. Biologists and chemists, for example, found common ground in the 1920s in biochemistry, the study of the life processes. After World War II, this had a powerful offshoot in molecular biology, explaining the nature of living organisms at the molecular level.

Thus 20th-century science inherited and expanded its own internal social pattern with groups acknowledging each other's existence but finding difficulty in understanding their esoteric approaches. Superimposed on this, however, was still a different kind of caste system. Academic scientists, with their "pure" approach, still often affected a superiority to those who applied science for useful ends. Yet increasingly it was the latter who were making the discoveries that were transforming the economies of the Western world and, in the last analysis providing the funds by which academic research could be sustained. Examples of major inventions made in industry are Bakelite plastic; the sulfonamide drugs, polyamide (nylon) and polyester (terylene) fibers; polythene; the laser; and the transistor. Recognition of the growing role of the technologist is implicit in the term technocracy and (the later (1932) term technocrat) indicating a society in which industrial resources are developed for the common good by technical experts. This term first appeared in the United States as early as 1919, though it was not generally current in Europe until after World War II.

The changing face of science and technology
The main body of this book will deal, over six selected periods, with the progress of technology, the physical sciences and the biological sciences in the 20th century and with the changing relationship between science and society. As a preliminary to this review it is instructive to seek some distinctive features of each of the periods chosen.

The first of these (1900–14) can fairly be described as the age of individual genius. For the most part scientists followed their own bent, though often within small groups or "schools" devoted to a fairly sharply defined field. Such, for example, was the school of atomic physics under Ernest Rutherford at Manchester (and later Cambridge) and the school of cryogenics under Heike Kamerlingh Onnes at the University of Leiden. Throughout the following four years scientific work generally was much disrupted by World War I: academic pursuits had to be subordinated to severely practical research directed to

▶ **Students watch an operation in progress, London, 1947.**

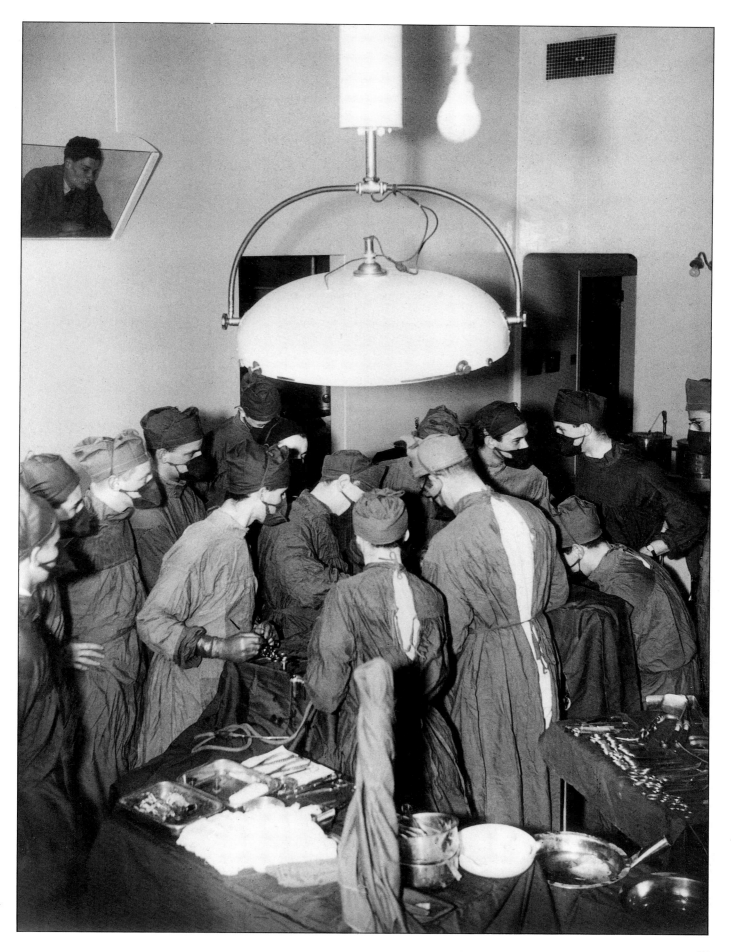

urgent wartime needs. In Britain the Department of Scientific and Industrial Research was set up in 1916 to coordinate all such work on a large scale. Its chosen title was significant, as indicating a clear alliance between science and industry. This trend continued in the 1920s with the growing influence of larger corporate enterprises based on science and increasingly employing their own research staff in sophisticated laboratories. Some – such as Du Pont, General Electric, Standard Oil and General Motors in the United States – represented a major expansion of existing small companies. Others were new creations, though often embodying smaller predecessors. Such were, for example, the de Havilland Aircraft Company (1920) and ICI (1926) in Britain and I.G. Farben (1925) in Germany.

The third period 1929–45, sees science emerging – ultimately decisively – as an instrument of power. It began with the Great Depression of 1929–34, with massive unemployment; 14 million in the United States, 6 million in Germany, 3 million in Britain. It was a grim reminder that though industrial technology had been progressively improved, social problems still remained intractable. In the USA it ended with Roosevelt's

New Deal, a massive program of public works and public loans. In Germany its end marked the rise of the Nazi Party, with all that that was to imply. By the mid-1930s war seemed more probable than possible, and once again much scientific effort was diverted into military fields. Superiority in the air was clearly going to be a major factor, and all the major powers began to strengthen their air forces, partly along conventional lines and partly on novel ones. In the latter respect two major possibilities were explored: the detection of enemy aircraft at a distance by reflected radio waves (radar) and the radical improvement of aircraft performance by the use of jet propulsion in place of the complex reciprocating action of the piston engine which was clearly near the limit of its development. In the event the British radar system was the most successful, and its contribution to the Battle of Britain in 1940 was crucial to the ultimate outcome of World War II in Europe. This was partly due to technical superiority (use of shorter wavelengths than those employed by Germany) but also to the recognition that a successful detecting device was not in itself sufficient. This had to be incorporated into a comprehensive early warning system

remarkable feature of this war was that the huge and costly Manhattan Project was based on experiments involving about half a milligram of plutonium – a scaling-up in one stage of ten thousand million times, unprecedented in the chemical industry. Not until the first successful test on 16 July 1945 could anyone be certain that the theoretical background was sound. The ethics and political wisdom of dropping the bombs on Hiroshima and Nagasaki are now the subject of high controversy, but in 1945 it seemed a miraculous deliverance. After six years of war, over 20 million military and civilian casualties, with much of Europe in ruins and crowded with refugees, the prospect of a further bitter campaign in the Pacific against a fanatical enemy had seemed unrelievedly dismaying.

Changing attitudes to science

In the first 15 postwar years the mastery of atomic fission created a general euphoria. Turned to peaceful purposes it held out the promise of cheap and limitless energy, not only for ordinary domestic and industrial purposes but for such extravagant schemes as creating fertile new land by melting the polar ice-caps. By the 1980s the economics of atomic power were in question and the consequences of the "greenhouse effect" on the melting of the polar ice was becoming a matter for general concern. The dream has become a nightmare. In those heady days, however, the promise of science was regarded as enormous. If scientists could, quite literally, stop a world war in its tracks, what could they not do in other fields? Science seemed to promise a universal panacea; the scientific bandwagon began to roll and there was a scramble to get aboard. Sensing the way the wind was blowing politicians of all parties vied with each other in their promises of support, for both scientific research and education. To the young, encouraged by both parents and teachers, science seemed to be the highroad to a successful career. However, because of its dependence on experimental work in the laboratory it was much more difficult to provide additional places for students of science and technology than for arts subjects. In many countries university applications for places far exceeded those available. Not until the 1960s, with the foundation of new universities and the expansion of old ones, was something near equilibrium restored. The number of places increased but pressure diminished as students, many politically and socially actuated, drifted away from the physical and biological sciences to the rising social sciences which in their turn were seen as potential healers of the ills of society.

While the enthusiasm for a technocracy waned there was mounting support for a meritocracy, a society governed by people competitively selected according to their innate ability. This would not, of course, exclude scientists and technologists but by broadening the field would statistically diminish their chances. However, as they have never – as a social group – shown any particular desire to participate actively in government this left them free to follow their own pursuits, though necessarily within the financial constraints set by those in power. This they did with conspicuous success: our penultimate epoch (1960–73) can fairly be described as revolutionary.

This giant leap can be ascribed to a multiplicity of factors, not least that the Western world had largely recovered from the ravages of war and was able to give increasing – but often uncritical – support to research, both pure and applied. Additionally, there was a synergistic effect within science: advances in one field made possible advances in others. For example, the technique of X-ray crystallography greatly simplified the

which could be operated by relatively unskilled operators under service conditions. While both Germany and the UK had prototype jet engines shortly before the outbreak of war, neither succeeded in developing fighter aircraft in time to affect the outcome.

Radar and jet propulsion were direct responses to perceived needs, but a third major wartime development was purely adventitious. Penicillin was the result not of a systematic search for a new chemotherapeutic agent to cure bacterial infections but of a purely academic project to investigate why certain microorganisms are mutually antagonistic. By the spring of 1940, however, as the Nazi forces surged across Europe, it was apparent that penicillin was a uniquely powerful drug with particular potential for treating infected wounds. This led to a crash program for its large-scale manufacture in the United States after Pearl Harbor. For the duration of the war it was reserved for military casualties, but afterward it quickly became available for general civilian use.

But, of course, the outstanding technological triumph of the war years was the creation of the atomic bomb. Not the least elucidation of the structure of large organic molecules. But undoubtedly two factors stood out above all others. One was the transistor (1947), dethroning the thermionic tube or valve after nearly half a century as the most important of all electronic components. The other was the microprocessor (1971) a computer central processing unit contained on one or a few minute silicon chips. This combination made possible a degree of miniaturization, and a complexity of command systems inconceivable in the previous decade, and opened up vast new fields of automatic programming. These ranged from the assembly of automobiles to speaking multilingual dictionaries that were small enough for the pocket; from automatic washing machines to economic fuel control for gasoline engines; from autofocusing cameras to cordless telephones. No less significant was the contribution to information storage and retrieval.

Beyond question the most dramatic manifestation of the new powers of science and technology was the conquest of space, culminating in the landing of the first men on the Moon on 20 July 1969. Here, too, the computer played a vital role but many other factors were no less critical. These included fuel technology, heat insulation, the overcoming of problems of weightlessness, space suits to make life possible in a vacuum, and – by no means least – remarkable management skill. The landing of men on the Moon was a uniquely American achievement – highly prestigious and very satisfying to the collective American ego after an alarming series of Soviet successes. It must be said, however, that – looked at dispassionately – the United States probably achieved little more with their manned landings than the Soviet Union did with unmanned landers and space probes.

The last epoch can fairly be described as one of doubt, as the Western world began to count the cost of progress. It began dramatically with the oil crisis of 1973. In retaliation for support for Israel the OPEC countries severely reduced the supply of oil to the United States, Japan and Western Europe, and at the same time inflated the price.

The economic consequences were enormous. The worst hit sector was, of course, transport, where fuel oil dominated on land, at sea and in the air. But as consumption of energy per capita is a widely recognized index of industrial activity, the ripples spread far. Additionally petroleum had become a main feedstock for the chemical industry, and thus the restriction suddenly affected the production and cost of a wide range of products from plastics to fertilizers, from man-made fibers to pharmaceuticals.

Science and technology could offer no short-term answers: the harsh reality was that the Western world had allowed itself to become highly dependent on oil as a source of energy without sufficient regard to continuity of supply. However, as the British author Samuel Johnson once remarked, "when a man knows he is to be hanged in a fortnight it concentrates his mind wonderfully!"

For years energy dominated political thinking. Making a virtue of necessity it was argued that the world had a duty to conserve irreplaceable fossil fuels for the benefit of future generations. This directed attention to the development of renewable energy resources such as solar energy and wind and wave power. In the event, however, rather little came of this so far as substantial overall contribution to world energy needs were concerned. A major problem is that such sources are not available on command and there is as yet no means of storing energy in large quantities to keep the supply in balance. It is, therefore, necessary to have substantial reserve capacity of conventional kinds – such as oil- or coal-burning power stations – which rather defeats the object of the exercise.

Of all the alternatives only nuclear power avoids this particular defect. There is, however, a powerful antinuclear lobby which has restricted this development and – in some countries, such as Sweden – stopped it altogether. Only France (70 percent nuclear) has consistently pursued a policy of increasing reliance on nuclear power. However, this may change over the next few years with growing concern about the greenhouse effect on world climate. A major factor here is accumulation of carbon dioxide in the upper atmosphere and to this the burning of fossil fuels in power stations makes a substantial contribution. It also contributes to the acid rain allegedly causing massive destruction of forests and affecting plant and animal life in rivers and lakes. Although of a different kind, the environmental objections are comparable to those of nuclear power. In the long term the solution may lie with atomic fusion, rather than atomic fission, but despite massive research programs which produce very little radioactive waste, there is no expectation that the technical problems will be solved before the first decade of the 21st century, at the earliest.

The green movement

Opposition to nuclear power is only one aspect of one of the most remarkable developments of the 1980s, the growth of the green movement. It concerns itself, in the broadest way, with preservation of the environment worldwide against what are seen as the consequences of modern technology: destruction of rain forests in the interests of slash-and-burn agriculture; weakening of the ozone layer through the use of chloro-fluorocarbon (CFC) aerosols and refrigerators; destruction of harmless wildlife through use of agricultural insecticides; the killing and processing of whales; and the disposal of waste, especially radioactive waste. It is a political movement in the sense that it is represented in various legislatures, such as the European Parliament. At the same time, it has no political pro-gram of a general nature. On many issues its campaigns rest more on emotion than on fact. Nevertheless, it clearly represents a widespread belief, over a broad social and political spectrum, that science and technology are mixed blessings. Because it is apolitical, and commands increasing support in

many societies among all classes, politicians of all persuasions – conscious of votes at stake – have hastened to identify themselves actively with the general concern. It remains to be seen, however, how far such political power, exercised without responsibility for the consequences, will prove beneficial.

A difficulty is that members of the public like to dine *à la carte*, to chose the benefits they fancy and suppress unpopular features associated with them. There is great enthusiasm for generating energy by massive tidal barrages, but at the same time there must be no disturbance of local wildlife. Sufficient cheap food is needed for the tens of millions who live on, and sometimes below, the borderline of starvation but there is con-demnation of the science-based agricultural practices which make it possible. And so the catalog can go on. The problem lies in the rigid and highly integrated nature of science itself. Good effects and bad effects are inseparably linked and if the latter are totally unacceptable the benefits must be forfeited.

▲ The limits of technology: the explosion of the Shuttle *Challenger*, 1986.

19

1900 · 1914

THE
INDIVIDUAL
GENIUS

Time Chart

	1900	1901	1902	1903	1904	1905	1906	1907
Nobel Prizes		• *Chem*: J.H. van't Hoff (Neth) • *Phys*: W.C. Röntgen (Ger) • *Med*: E.A. von Behring (Ger)	• *Chem*: E. Fischer (Ger) • *Phys*: H.A. Lorentz, P. Zeeman (Neth) • *Med*: R. Ross (UK)	• *Chem*: S.A. Arrhenius (Swe) • *Phys*: P. Curie, M. Curie, A. Becquerel (Fr) • *Med*: N. Finsen (Den)	• *Chem*: W. Ramsay (UK) • *Phys*: J.W. Strutt (Rayleigh) (UK) • *Med*: I.P. Pavlov (Russ)	• *Chem*: A. von Baeyer (Ger) • *Phys*: P. Lenard (Ger) • *Med*: R. Koch (Ger)	• *Chem*: H. Moissan (Fr) • *Phys*: J.J. Thomson (UK) • *Med*: C. Golgi (It), S. Ramón y Cajal (Sp)	• *Chem*: E. Buchner (Ger) • *Phys*: A.A. Michelson (USA) • *Med*: C.L.A. Laveran (Fr)
Technology	• R.A. Fessenden first transmits speech by wireless (USA) • 2 Jul: First trial flight of the airship *Star*, designed by F. Graf Zeppelin (Ger) • Marconi sets up his wireless telegraph company (UK)	• First radio transmission across Atlantic Ocean by Marconi (UK) • P.C. Hewitt invents the mercury vapor lamp arc lamp • H.C. Booth invents the vacuum cleaner (UK)	• Completion of the first Aswan Dam by B. Baker (Egy) • 14 Aug: G. Weisskopf attempts the first powered flight (Ger) • Disk brakes first fitted to an automobile by F. Lancester (UK)	• 17 Dec: First controlled powered flight by the Wright Brothers (USA) • First three-color photography process developed by the brothers Lumière (Fr) • W. Siemens develops an electric locomotive (Ger)	• Ultraviolet lamp invented • Diode valve patented by J. Fleming (UK) • C. Hülsmeier discovers the principle of radar (Ger) • J.P.T.L. Elster develops the photoelectric cell	• J.A. Fleming's wavemeter is now used in all radio equipment • The Wright Brothers' *Flyer III* is the first practical aircraft (USA) • G. Marconi invents the directional radio antenna (UK)	• L. de Forest patents the triode valve (USA) • Beginning of the Zuider Zee drainage scheme (Neth) • Amplitude Modulation (AM) radio transmission demonstrated by R.A. Fessenden (USA)	• First regular radio broadcasts made by de Forest Radio Telephone Co (USA) • R. Anschütz-Kaempfe and M. Schuler perfect the gyrocompass • L. Lumière improves color photography by the invention of the three-color plate (Fr)
Medicine	• W. Einthoven first uses the electrocardiograph clinically (Neth) • W. Reed identifies the yellow fever virus and proves its link with mosquitoes • Publication of Freud's *The Interpretation of Dreams* (Aut)	• Publication of Freud's *The Psychopathology of Everyday Life* (Aut) • E. Metchnikoff demonstrates how white blood cells fight disease (Russ) • M.R. Hutchinson patents the first electric hearing aid (USA)	• W. Bayliss and E. Starling discover the hormone secretin (UK) • First use of the electric hearing aid by Queen Alexandra at her coronation (UK) • C. Richet, with P.J. Portier, discovers cases of anaphylaxis, abnormal sensitivity to antidiphtheria serum (Fr)	• "Typhoid Mary" is discovered to be the carrier of the disease during an epidemic in New York (USA) • W. Einthoven devises the string galvanometer (Neth) • R.A. Zsigmondy invents the ultramicroscope (Neth)	• A. Glenny discovers diphtheria immunization (UK) • A. Einhorn discovers the anesthetic procaine, first used in 1905 • Ronald Ross publishes his studies of the connection between the malaria parasite and the *Anopheles* mosquito (UK)	• J.B. Murphy pioneers arthroplasty – the creation of artificial joints (USA) • Publication of Freud's *Three Treatises on the Theory of Sex* (Aut)	• C. von Pirquet introduces the term "allergy" • J. Bordet discovers the bacterium causing whooping cough (Fr) • A. von Wasserman develops his test for syphilis (Ger)	• C.L.A. Laveran discovers the role of protozoans in causing diseases such a malaria, leishmanism and sleeping sickness (Fr) • Introduction of tissue culture by R.G. Harrison (USA) • P. Ehrlich introduces chemotherapy (Ger)
Biology	• F.G. Hopkins discovers the first essential amino acid, tryptophan (UK) • Independent rediscovery by H. de Vries (Neth), K.F.J. Correns (Ger) and E.T. von Seysenegg (Aut) of Mendel's work on heredity	• Discovery in Africa of the okapi, the last large land mammal to become known to science • K. Landsteiner discovers that there are four blood types, which he calls A, B, AB and O (Aut)	• I.P. Pavlov begins his study of conditioned reflexes (Russ) • H.W. Cushing begins a study of the pituitary body (USA)	• W.S. Sutton and others argue that hereditary factors (genes) are located on the chromosomes (USA)	• R. Chittenden isolates glycogen (USA) • S. Ramón y Cajal establishes the theory that the nervous system is composed only of nerve cells and their processes (Sp)	• C. McClung discovers that female mammals have two X-shaped sex chromosomes whereas males have an X paired with a Y	• T.H. Morgan uses the *Drosophila* fruitly to study heredity (USA) • F.G. Hopkins argues that enzymes later called vitamins are essential components of diet (UK)	
Physics	• P.U. Villard first observes gamma rays (Fr) • 14 Dec: M. Planck states that substances can emit light only at certain energies, thus making the first step toward the quantum theory (Ger)	• E. Rutherford and F. Soddy discover that thorium left to itself changes into another form, later discovered to be radium (UK) • Publication of M. Planck's *Laws of Radiation* (Ger)	• E. Rutherford and F. Soddy announce their disintegration theory of radioactivity (UK) • P. Lenard conducts his experiments into the properties of cathode rays (Ger)	• E. Rutherford demonstrates that alpha particles are positively-charged (UK) • Publication of J.J. Thomson's *The Conduction of Electricity Through Gases* (UK)	• E. Rutherford and F. Soddy state general theory of radioactivity (Can) • J.J. Thomson puts forward the "plum pudding" model of the atom (UK)	• Einstein states his special theory of relativity, establishes law of mass–energy equivalence, creates a theory to explain Brownian motion and formulates the photon theory of light (Ger)	• W.H. Nernst postulates third law of thermodynamics (Ger) • J.J. Thomson begins his study of canal rays, demonstrating a new way of separating atoms (UK)	• Potassium and rubidium are discovered to be radioactive • P. Weiss develops the domain theory of ferro-magnetism (Fr)
Chemistry	• F.E. Dorn discovers radon (Ger) • M. Gomberg develops a carbon compound that has one valence location open, the first known radical (USA)	• Adrenalin extracted from animals by J. Takamine (Jap) is manufactured by Parke, Davis and Company (USA)	• W. Pope develops optically active compounds based on sulfur, selenium and tin • Herman Frasch develops a technique for extracting sulfur from deep deposits (Ger)	• W.H. Stearn and F. Topham develop the viscose method for producing artificial silk	• F. Giesel discovers "actinium X", an isotope of radium • S.A. Arrhenius proposes a theory of immunochemistry (Swe) • F.S. Kipping discovers silicones (UK)	• R. Willstätter begins his studies of chlorophyll (Ger) • Artificial silk made commercially through viscose process by Courtaulds (UK)		• E. Fischer gives the name peptide to the simplest amino acid, and publishes his *Researches on the Chemistry of Proteins* (Ger)
Other	• J.E. Keeler discovers that certain nebulae have a spiral structure (USA)	• P.V. de Camp attempts to prove that nearby stars have planetary systems (Neth)	• O. Heaviside (UK) and A.E. Kennedy (USA) discover the ionized atmospheric layer conducive to radio transmission		• J.F. Hartmann first discovers interstellar matter: stationary calcium lines in the spectrum of the binary Delta Orionis (Ger)	• P. Lowell predicts the existence of a ninth planet beyond Neptune	• R.D. Oldham deduces the existence of the Earth's core (UK)	• B.B. Boltwood discovers the use of uranium's radioactive decay to calculate the age of rocks

1908	1909	1910	1911	1912	1913	1914
• *Chem*: E. Rutherford (UK) • *Phys*: G. Lippman (Fr) • *Med*: P. Ehrlich (Ger), Élie Metchnikoff (Russ)	• *Chem*: W. Ostwald (Ger) • *Phys*: G. Marconi (It), C.F. Braun (Ger) • *Med*: E.T. Kocher (Swi)	• *Chem*: O. Wallach (Ger) • *Phys*: J.D. van de Waals (Neth) • *Med*: A. Kossel (Ger)	• *Chem*: M.S. Curie (Fr) • *Phys*: W. Wien (Ger) • *Med*: A. Gullstrand (Swe)	• *Chem*: V. Grignard, P. Sabatier (Fr) • *Phys*: N.G. Dalén (Swe) • *Med*: A. Carrel (Fr)	• *Chem*: A. Werner (Swi) • *Phys*: H.K. Onnes (Neth) • *Med*: C.R. Richet (Fr)	• *Chem*: T.W. Richards (USA) • *Phys*: M. von Laue (Ger) • *Med*: R. Barany (Aut)
• C. Swinton outlines a method of electronic scanning that forms the basis for the iconoscope, the prototype cathode-ray (TV) tube • C.F. Cross invents cellophane • Invention of the tungsten filament by Coolidge (USA)	• First commercially produced color films produced, after a process (Kinemacolor) developed by G.A. Smith the previous year (UK) • E. Forlanini develops the hydrofoil ship • 25 Jul: L. Blériot (Fr) crosses the English Channel in airplane	• C. Parson's speed-reducing gear extends use of geared turbines • Rotary intaglio printing process developed by E. Mertens (Ger) • G. Claude invents the fluorescent tube • Electrification of part of the Magdeburg–Halle railway (Ger)	• Building of the *Selandia*, the first ship with heavy oil engines (Den) • Lewis machine-gun patented by I.N. Lewis • C.F. Kettering develops the first practical electric self-starter for automobiles (USA) • First bombs dropped from German airplanes	• R.A. Fessenden develops the heterodyne radio system (Can) • An SOS in Morse code is adopted as the universal distress signal • Introduction of the regenerative or "feedback" radio receiver (USA)	• R. Lorin states basic principle of jet propulsion • Diesel-electric railway opens in Sweden • W.D. Coolidge invents a hot-cathode X-ray tube (USA) • H. Ford introduces the first true assembly line (USA)	• First traffic lights installed in Cleveland, Ohio (USA) • Cargo ship built with turboelectric engine (Swe) • Radio transmitter triode modulation is introduced • E. Kleinschmidt invents the teletypewriter • R.H. Goddard begins his rocketry experiments (USA)
• The barium meal technique shows up a gastric ulcer for the first time through X-rays (Ger) • Publication of A.E. Garrod's *Inborn Errors of Metabolism* (UK) • A. Calmette and C. Guérin develop the first tuberculosis vaccine; after 15 years further development, used as BCG in 1920s (Fr) • First international meeting of psychiatrists in Salzburg (Aut)	• H.W. Cushing discovers that acromegaly – the enlargement of the jaws, extremities and some organs – is due to overgrowth of the pituitary gland (USA) • The Meltzer–Auer tube is devised to assist breathing during mouth, throat or chest operations (USA)	• F. Woodbury discovers use of iodine as a disinfectant and antiseptic (USA) • L. Leuchs prepares the first antitoxic sera for botulism (Ger) • Discovery of sickle-cell anemia by J. Herrick (USA)	• Salvarsan developed by P. Ehrlich to combat syphilis (Ger) • W. Hill develops the gastroscope for investigating abdominal illness (USA)	• Secession of C.G. Jung (Swi) and A. Adler (Aut) from the psychoanalytic movement to form respectively the Analytical Psychology and Individual Psychology movements • W. Macewen delivers his monograph *The Growth of Bone*, in which he elaborates on the notion of bone grafts (UK)	• E. McCollum and M. Davis identify vitamins A and B in cows milk • B. Schick develops the Schick test to determine immunity to diphtheria (Hung/USA) • Diphtheria vaccine produced by E.A. von Behring (Ger)	• E.C. Kendall prepares pure thyroxin for treatment of thyroid deficiencies (USA) • R. Lewisohn discovers the anticoagulant properties of sodium citrate, and uses it to prevent blood from coagulating • A. Carrel performs first successful heart surgery on a dog (Fr)
• G.H. Hardy and W. Weinberg demonstrate the laws governing the frequency of occurence of dominant hereditary traits • I. Metchnikoff discovers the white blood cells (phagocytes) which devour bacteria (Russ)	• W. Johannsen coins the terms "genes", "genotypes" and "phenotypes" (Den) • Publication of W. Bateson's *Mendel's Principles of Heredity: A Defense*	• T.H. Morgan discovers that certain inherited characteristics are sex-linked (USA) • R. Biffen breeds Little Jos, a wheat suitable for British climate and resistant to the yellow rust fungus disease (UK)	• T.H. Morgan begins to plot the first chromosome map, of the positions of genes on chromosomes of fruit flies (USA) • A.H. Sturtevant produces the first chromosome map (USA)	• J. Loeb attempts to explain the origins of life in terms of physics and chemistry • K. Funk coins term "vitamin" (Pol)	• J.B. Watson puts forward behaviorist approach to psychology (USA) • Composition of chlorophyll discovered by R. Willstätter (Ger)	• F.A. Lipmann demonstrates the role of ATP in the release of energy in a cell (Russ) • H.H. Dale proposes that acetylcholine is a compound involved in the transmission of nerve impulses (UK)
• H. Minkowski shows that the relativity theory is more complete if time is considered as the fourth dimension of the universe (Russ) • The international ampère is adopted by the International Conference on Electrical Units and Standards	• E. Marsden determines that some alpha particles bounce back from a thin gold foil leading to Rutherford's theory of atomic structure (UK)	• R.A. Millikan determines the charge on an electron (USA) • J.J. Thomson, while measuring atomic masses of substances, provides the first confirmation that isotopes are possible (UK) • M. Curie's *Treatise on Radiography* published (Fr)	• H.K. Onnes discovers superconductivity (Neth) • C.T.R. Wilson devises his cloud chamber (UK) • E. Rutherford deduces the nature of the atomic nucleus as a very small, dense core (UK)	• Einstein formulates the law of photochemical equivalence (USA) • C.T.R. Wilson's cloud chamber photographs lead to the detection of protons and electrons (UK) • P.J.W. Debye propounds his theory of the specific heat of solids (Neth)	• Niels Bohr formulates his theory of atomic structure, with electrons in orbit around the nucleus (Den) • R.A. Millikan calculates the charge of an electron (USA) • H. Geiger invents device for counting individual alpha rays (Ger)	• J. Franck and G. Hertz confirms Bohr's atomic model by bombarding mercury vapor with electrons and measuring the frequencies of the emitted radiation (Ger) • E. Rutherford discovers the proton (UK)
• F. Haber synthesizes ammonia (Ger) • Bakelite invented by L.H. Baekeland (USA) • H.K. Onnes liquifies helium (Neth)	• S. Sörensen invents pH scale of acidity (Den) • F. Haber's process for nitrogen synthesis made a commercial operation by K. Bosch (Ger) • K. Hofmann produces synthetic rubber from butadiene		• C. Weizmann obtains acetone from bacteria involved in fermenting grain (UK) • Thermal cracking for petroleum derivatives developed by W. Burton (USA) • E. Rutherford and F. Soddy devise a scheme for the transmutation of elements (UK)		• Brevium, the first isotope of protactinium, is discovered by K. Fajans and O.H. Göring • W. and L. Bragg begin their research into X-ray crystallography (UK) • F. Bergius converts coal dust into oil (Ger)	• J.J. Abel isolates amino acids from blood (USA) • P. Duden and J. Hess synthesize acetic acid • J. Dewar elucidates the composition of air
• Giant and dwarf stellar systems described by E. Hertzsprung • A. Wilm invents the alloy Duralmin (Ger)	• A. Mohorovicic discovers discontinuity between Earth's mantle and crust (Yug)	• Halley's comet observed • Prince Albert I of Monaco founds Institute for Oceanography	• P. Monnartz discovers the anti-corrosive property of stainless steel (Ger) • Publication of F. Boas's *The Mind of Primitive Man* (USA)	• V.F. Hess discovers cosmic radiation • A. Wegener proposes an improved theory of continental drift (Ger)	• Harvard Classification of stars adopted by International Solar Union • H.N. Russell publishes his H-R (stellar magnitude) diagram (USA)	• A. Eddington shows that spiral nebulae are galaxies (UK) • B. Gutenberg discovers the discontinuity between Earth's mantle and core

23

Datafile

Although the industrial revolution began in the USA and many European countries in the 19th century, it was the dawning of the 20th century that fired an explosive expansion in almost every area of what we would now call modern technology. Personal transport for all, international flight, intercontinental communication, and the mass production of food were all unknown at the turn of the century. The future was undreamt of except in the minds of a few men: Henry Ford, Orville and Wilbur Wright, Louis Blériot, Guglielmo Marconi; men whose work made the world of 1914 an incomparably more 20th-century world than the essentially agricultural world of 1900. In time new industries exploiting new manufacturing techniques and new technologies enabled the construction of thousands of machines to travel effortlessly over land, to fly over neighboring countries, to sail the oceans while staying in contact with the shore; in short to make a world ready for the first 20th-century war.

Model T-Ford price

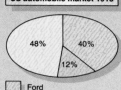
US automobile market 1913
- Ford
- General Motors
- Others

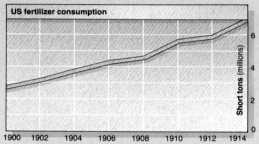
US fertilizer consumption

▲ The early 20th century saw a steady increase in the use of fertilizer on US farmland with volume doubling between 1900 and 1910. In due course there would be a ready market for the more efficient synthetic nitrate fertilizer. The introduction of large numbers of motorized vehicles onto farms greatly increased land yields.

▲ Henry Ford introduced motoring to the American people. The Model T automobile was introduced in 1908 into a select marketplace, but as manufacturing techniques improved the price fell rapidly. By 1915 sales had increased from 1,200 to 342,000 p.a. In 1921 Ford sold 56% of all American cars.

▶ Following the discovery that aluminum alloys could be made much stronger than pure aluminum but maintain lightness, demand from the expanding aircraft industry pushed production upward at a massive rate. Aluminum is electrically refined from bauxite ore dissolved in cryolite. World production has now reached 15 million tons.

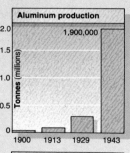
Aluminum production

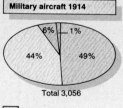
Military aircraft 1914
Total 3,056
- France
- Germany
- UK
- USA

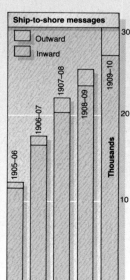
Ship-to-shore messages
- Outward
- Inward

▶ Although the first flight in a heavier-than-air machine occurred in the United States, aircraft production in the prewar years was concentrated in Europe. The Blériot monoplane (1909) was the culmination of long-standing French interest in manned flight which began with the balloon flights of the Montgolfier brothers in the 18th century.

◀ The discovery by Hertz in 1887 of long wavelength electromagnetic radiation led to the use of these "radio" waves to transmit messages across large distances through the air rather than through a cable. Communication could be made with remote targets, in particular ships at sea which were no longer out of touch with land.

In any account of the history of science and technology it is necessary to distinguish between the date of an invention or a new discovery and the time at which it begins to make a significant social impact. Sometimes the interval may be quite short, but equally it may be measured in years or even decades. Again, the impact may come more quickly in some places than in others: in very remote places it may never come at all. The opening years of the 20th century, up to the outbreak of World War I, provide examples of all possibilities.

An example of a quick reaction is provided by the history of X-rays after their discovery by Wilhelm Röntgen in 1895. They were put to practical use in medicine within a year and Röntgen was among the very first Nobel laureates in 1901. The cinema exemplifies a slower rate of growth. The first public showing of a moving picture was in Paris in 1895. Pittsburgh's famous Nickelodeon opened just ten years later, and when war broke out there were over 3,500 cinemas in Britain alone. Within two decades a wholly new form of mass entertainment had firmly established itself.

The revolution in transport – perhaps the most socially significant development of the century – proceeded unevenly. In 1900 most road transport was horse-drawn. Yet by 1908 Henry Ford was making the first of 15 million Model T automobiles which were to make motoring a popular pastime. But that was in America: in Europe, private motoring made little headway until the 1920s. In the air the story was different again. Although the first aeroplane flew in 1903, by 1914 there were probably no more than 5,000 in the world. Some 200,000 military aircraft were built during the war but civil aviation scarcely made an impact before the 1930s.

Henry Ford and the automobile revolution

Perhaps the greatest revolution produced by technology in the early 20th century occurred in transport. Interestingly, this revolution stemmed less from technological advance than from highly efficient processes of manufacture. In 1900 the automobile was already fully fledged, in that the basic features of all 20th-century automobile technology had been established, but worldwide there were still only a few thousand automobiles. The horse was still dominant. A dramatic change came in 1908 when the American automobile manufacturer Henry Ford launched his famous Model T. Initially sales were modest, only 1,700 in the first 15 months, but by 1916 they had soared to an annual rate of sale of 250,000 and the original price of 850 dollars had been substantially cut. By 1927, when production ceased, no less than 15 million of these remarkable cars had been sold: within a decade motoring had been brought

INVENTORS OF THE FUTURE

to the masses, though more so in the USA than in the rest of the world.

Henry Ford was born in Dearborn, Michigan, USA, in 1863 and as a young man was concerned with the repair and installation of agricultural machinery. At one time he was interested in manufacturing a cheap farm tractor. However, he abandoned this in 1890 in favor of automobiles and worked for a time as chief engineer of the Detroit Automobile Company, for which his 80hp, four-cylinder "999" model was a conspicuous success. In 1903 he set up his own company, with the idea of making not a cheap tractor but a cheap automobile – simple, strong and reliable. Although his adoption of tough vanadium steel as a constructional material was an important technical factor, the main reason for his success was the introduction of the conveyor-belt system of assembly and the standardization of parts. His famous quip that "customers can have any color they like provided it's black" did not, in fact, reflect a take-it-or-leave-it attitude but was a consequence of this method of manufacture. Up to 1914 a choice of colors was offered, but by then the only paint that would dry quickly enough to keep the production lines moving was black.

The system of using standardized interchangeable parts was by no means novel to American industry: Samuel Colt, for example, had used it since the 1850s to manufacture revolver pistols. Neither was the assembly line principle, for Sears, Roebuck – the great mail order pioneer – was already using it at its vast warehouse in Chicago. Henry Ford's genius lay

▶ ▲ **In contrast to the later glossy and superficial advertisements, the early automobile manufacturers sought custom on the basis of detailed specifications. This advertisement for the Model T Ford lists a couple of dozen technical features.**

▼ **Early automobile manufacturers favored distinctive styles. This 1906 Renault has the characteristic "coal-scuttle" hood (bonnet) retained into the 1920s.**

Pictorial Enumeration of the Advantages of the UNIT POWER PLANT of the Model T Automobile

in combining the best and most appropriate of available techniques and, perhaps even more, in recognizing that the automobile was not merely for the wealthy but had a great future for the general public if it could be produced cheaply enough. Yet not even he could have foreseen that within 30 years the USA alone would be manufacturing some 4 million vehicles a year – more than the whole of Europe – and that 20 percent of these would come from the Ford company and its associates.

New metals and alloys

To a considerable extent the development of industry has been closely linked to the development of metals technology. The world's first industrial revolution, in England, began when Abraham Darby used coke to produce high-quality iron in large quantities (from 1709). In the 1850s the introduction of the Bessemer process (named for the British inventor Sir Henry Bessemer) made possible mass production of steel (a harder and tougher form of iron). Within a short time, in the USA and UK, steel largely replaced iron in such diverse fields as shipbuilding, railroad tracks and civil engineering. (The Eiffel Tower in Paris, built in 1889 to commemorate the centenary of the French Revolution, was the last major building in the world to be constructed in iron.) However, the dawn of the steel age did not itself enlarge the engineer's metallurgical repertoire: it merely made an old metal more effective. The late 19th and 20th centuries were to see the introduction of new metals and an explosion in the number of alloys available.

The first major new metal was aluminum. It was first introduced into industry in the 1880s, but at first its lightness was compromised by its softness. A turning point in its fortunes was the discovery in Germany in 1909 that the strength of aluminum alloyed with small quantities of copper and magnesium could be much increased by age-hardening (leaving to harden spontaneously), to produce an alloy called Duralumin. This was

► The Osram Azo lamp signifies two major technical advances. The first word indicates the use of high-melting metals, such as osmium, in place of the old carbon filament. The word Azo indicates replacement of the vacuum by inert nitrogen gas (Fr. *azote*): later, argon was used. This lamp with its Edison screw fitting was destined for the Continental or North American market: Britain used a "bayonet" fitting. Even 90 years later there was no international standardization.

◄ For many years, skyscrapers were a peculiarly American form of building, their construction prompted by soaring land values. Increasingly steel displaced cast iron. The first was the 10-storey Home Insurance Building in Chicago (1884–85) and within ten years 30 similar buildings had been erected there. The introduction of the Otis electric elevator – with built-in safety brake – in 1889 encouraged even taller buildings. This picture shows the 90 meter Flatiron building erected in New York in 1902. By 1913 the Woolworth Building towered nearly three times as high, to 230 meters.

I was perfectly aware of what had happened. This was the great Louis Sullivan moment. The skyscraper as a new thing beneath the sun ... with virtue, individuality, beauty all its own, as the tall building was born ...

FRANK LLOYD WRIGHT

quickly adopted by the German Count Ferdinand von Zeppelin for his airships, in place of aluminum itself, and was destined to be the main construction material for the aircraft industry. Similar Y-alloys, containing nickel, were developed in the UK. Other metals were required also in substantial quantities for nonconstructional purposes. Thus world production of zinc in 1900 was around half a million tonnes, much of it used to galvanize iron to protect it from corrosion when used, for example, as corrugated sheeting for roofs or wire for fencing.

Other metals, such as nickel and chromium, were used to improve the appearance of less attractive metals by electroplating. Certain metals were used in their own right, though in very small quantities, for special purposes. Early electric lamps, for example, used incandescent carbon filaments but these had the double disadvantage of being brittle and discoloring the glass by evaporation. Around the turn of the century attempts were made to use instead the rare metals osmium, tantalum and tungsten, but because of their very high melting points (tungsten's, for example, is 3,380°C) – which is what made them interesting to lampmakers – these were extremely difficult to fabricate. In 1908, however, W.D. Coolidge in the USA devised a powder-metallurgy technique for manufacturing tungsten rod, which could then be drawn into hair-fine wires. Another rare and expensive metal, platinum, found a new use, as the basis of catalysts for a wide range of industrial chemical processes. The demand for higher performance steels for increasingly sophisticated machine tools was first met around 1861 when the British steelmaker Robert Mushet introduced tough steels containing the unfamiliar metals vanadium, tungsten and molybdenum: these were a great improvement on the plain carbon steels previously used. But perhaps the greatest single advance

was that made at the turn of the century, when the Americans F.W. Taylor and M. White introduced steels containing vanadium, tungsten and chromium, which had twice the cutting speeds of Mushet's: these made a great impression at the Paris Exhibition of 1900.

Until the latter part of the 19th century the development of new alloys was still largely empirical. Mixtures were varied, new ingredients added, and the effects noted. The conditions of manufacture were also well recognized as being important. By 1900, however, pioneers such as the British geologist H.C. Sorby and the British metallurgist W.C. Roberts-Austen had laid the foundations of a new science of metallurgy, by which the properties of metals were related to both their composition and their physical structure, especially at the micro level, and to physical treatment such as hammering and heat shock. As the 20th century advanced the growing importance of alloys had political repercussions. Although certain constituent metals might be required in only quite small quantities they were nonetheless essential. It therefore became necessary to ensure their availability at all times, either from militarily secure sources or by stockpiling.

National and International Measurement Standards

From the beginnings of the Scientific Revolution in the mid-16th century science was international in character, though for long essentially European. Major discoveries were made in Italy, France, England, Holland and in many other countries. Scientists exchanged information through informal correspondence and through the great national academies such as the Royal Society in London (founded in 1660) and the Academy of Sciences in Paris (founded in 1666). Increasingly, especially in the physical sciences, the emphasis was on the quantitative rather than qualitative observations. To avoid confusion it was important that observers at widely different centers should be able to express their results in the same terms. This need for precision was further made necessary by advances in technology, especially in mechanical engineering.

The basic physical units are length, mass and time and to obtain absolute, reproducible values recourse was made to unvarying natural features. At the beginning of the 20th century the standard length was the meter, defined as one ten-millionth of the distance between the equator and the north pole. This was inscribed on a

platinum bar in Paris, maintained at 0°C. From this as many substandards as were required were derived. The unit of mass was the gram defined as the mass of a cubic centimeter of water at 4°C (its maximum density). For time, the second was defined as 1/86,400th part of the mean solar day. On these basic units of centimeter, gram and second was based the C.G.S. system, from which many other units – of force, acceleration, power, etc. – were derived. For measuring temperature two fixed points were defined: the freezing-point and the boiling-point of water at standard atmospheric pressure.

The need to ensure that the wealth of instruments in general use all conformed to the same standards was one of the factors that led to the foundation of the National Physical Laboratory in England (1900), the National Bureau of Standards in America (1901) and the Kaiser Wilhelm Institutes in Germany (1910). With the need for ever greater precision new standards have been introduced. Thus in 1960 the meter was redefined in terms of the velocity of light and in 1983 in terms of the wavelength of emission from krypton gas.

The beginnings of powered flight

By 1900 the conquest of the air was already more than a century old. It had started with the first manned flights of the Montgolfier brothers in 1785. But balloons found little practical application – save for special purposes such as military observation – because they were at the mercy of the wind. Not until the second half of the 19th century was a motor available with a power : weight ratio that might make it possible to propel and steer an airship. In 1884 C. Renard and A.C. Krebs, in France, successfully completed an 8km (5mi) circuit in an airship with a propeller powered by a 9hp electric motor. In 1903 the French semirigid airship *Lebaudy*, powered by a 40hp Daimler petrol engine, made a 65km (40mi) flight from Moisson to Paris. But the real pioneer of airships was Count Ferdinand von Zeppelin, who between 1900 and 1914 built 160 rigid craft in which the gas (hydrogen – a flammable gas) was, for reasons of safety, contained in a large number of separate cells within the hull. Airships remained continuously in use throughout the 20th century – mainly for surveillance purposes by police, naval fleets, coastguards and so on – but they proved not to be on the main line of development in powered flight. This lay with heavier-than-air machines, of which the first successful example was that built by the Wright brothers which made its first historic flight on 17 December 1903.

But the idea of flying machines was much older than this: among early designers were the Italian painter and engineer Leonardo da Vinci (1452–1519) and the Swedish mining engineer and theologian Emanuel Swedenborg (1688–1772). In retrospect it is clear, however, that their designs were not practicable. This cannot be said however, of the series of gliders designed and built between 1808 and 1857 by the English gentleman Sir George Cayley (1773–1857). He was not only well read in mathematics but a skilled mechanic – with ample means to indulge his interest – and he addressed the problems of flight systematically. He recognized the lift that could be given by a cambered wing and distinguished this from drag. He had a clear concept of rear elevators and airscrew for propulsion. The same grasp of fundamentals made him aware, too, that no power unit available in his day would be light enough for propulsion, and so he limited himself to gliders, one of which carried his coachman on a flight of 450m (500yd) in 1853. In Germany, in the 1890s, Otto Lilienthal pioneered gliding as a popular sport – thus arousing public interest in heavier-than-air machines – and himself made more than 2,000 flights before being killed in an accident in 1896.

In the USA the Wright brothers, Orville and Wilbur, were well aware of Cayley's theoretical studies and always acknowledged their indebtedness to him. Where Cayley had failed and they succeeded was in devising a motor light enough to lift their craft off the ground: it was a 10hp petrol engine built in their own bicycle workshop.

Their first flights in 1903 passed almost unnoticed but in 1905 they built a bigger machine, which they patented, which successfully completed a 40km (25mi) circuit. This led to a US government contract for a machine of their design – a tail-less biplane with a propeller at the rear pushing the aircraft forward.

Although successful, this design was soon abandoned in favor of that of the Frenchman Louis Blériot (1872–1936) for a monoplane with a tractor engine at the front. His flight across the English Channel on 25 July 1909 attracted wide publicity and led to a seven–day airshow at Reims in August 1909 and to, among others, one at Los Angeles in 1910.

It was soon realized that the aeroplane was not something for mere sport but a potential new form of transport and one with military possibilities. World War I was to demonstrate them – some 200,000 aircraft were built in the course of it – but civil aviation had to await its end. Indeed, flying as a significant new form of public transport really began with the 1930s.

▼ In 1909 the French aviator Louis Blériot made history by flying across the English Channel – exciting the comment "Britain is no longer an island" – and in the same year a seven-day airshow was staged in the French city of Reims. Early in 1910 a ten-day Aviation Meet was organized in Los Angeles for American and foreign aviators.

Submarines, Torpedoes and Naval Warfare

Diving-bells and other underwater devices have been experimented with for at least three centuries, for research and naval warfare. In 1864, during the American Civil War, a submarine sank the Federal sloop *Housatonic* in Charleston harbor. Nevertheless, for all practical purposes the submarine is almost entirely a 20th-century development.

The main problems that needed to be solved were finding a statisfactory form of underwater propulsion and devising a maneuverable vessel. The pioneers were the French, whose electrically propelled prototype, the *Gymnote*, was built in 1888, but even by 1901 their entire fleet consisted of no more than a dozen 30-tonne vessels. In 1873 J.P Holland, an Irish inventor who had emigrated to America, began experimenting with submarines. Allegedly, he believed that such vessels might serve the cause of Irish independence: his first submarine, the *Fenian Ram* (1881) was financed by the American Fenian Society. Although this embodied most of the features of modern designs, it was not a great success; nor were several others he built for the US navy. In 1898 he designed and built his own vessel, the 120-ton *Holland*. This was far more satisfactory – it included an internal combustion engine for surface propulsion – so much so that the US navy immediately commissioned five.

The submarine literally introduced a new dimension into naval warfare. Whereas all other ships moved in two dimensions on the surface, the submarine could move in three, and was thus more elusive. Underwater currents and changes in the buoyancy of the sea, however, demanded elaborate controls to keep it on course at a fixed depth. Until sophisticated detection devices began to appear, its great advantage was that it could approach its victim unseen. (The torpedo had already been developed as an underwater weapon – its effectiveness was demonstrated by the Japanese attack on the Russian fleet off Port Arthur in 1904.) Electric motors, powered by storage batteries, provided a source of power that did not require precious oxygen from the submerged hull. They could be recharged by the internal combustion engines used for cruising on the surface with the hatches open, but during the recharging process the submarine was highly vulnerable to attack.

The earliest years of the 20th century saw the submarine develop from an effective prototype into a vessel that almost dictated the outcome of World War I. But the submarine was not invincible: because a damaged hull could be so rapidly fatal it was even more vulnerable than surface vessels.

▲ J.P. Holland in the tower of one of his early submarines.

▼ Early Holland submarine – with sail and steam behind.

Radio and international communications

Heinrich Hertz's experiments with wireless waves were designed to test the validity of James Clerk Maxwell's theory of electromagnetism. Similar experiments were carried out in England by Oliver Lodge, who demonstrated them to the British Association for the Advancement of Science in 1894, and in Russia by A.S. Popov in 1896. Although Popov apparently achieved transmissions up to 3.2km (2mi) by 1898, the initiative for making the new discovery the basis of a novel telecommunications systems came not from the world of science but from a young Italian aristocrat who had little formal technical training.

At the turn of the century the time was ripe for such an innovation. The development of an international telegraph network, and more recently the advent of the telephone (patented in 1876), had enjoyed much popular support. In 1900 some 400 million telegraphs were sent in Britain alone and in the USA there were already a million telephones. But while these systems were socially revolutionary in their day they had considerable practical disadvantages, notably in requiring tens of thousands of kilometers of wire and many hand-operated exchanges. Moreover it was possible to communicate directly only with points on the network: in particular, it was not possible to contact ships at sea. Wireless telegraphy had an obvious appeal in eliminating all these problems.

In 1894 Hertz's work had come to the notice of Guglielmo Marconi, an Italian student, then only 20 years old. Very soon he had not only transmitted signals over distances of 3.2km (2mi) but had pulsed them in Morse code. Gaining no support in Italy he went to Britain and set up his own company in 1897: in 1900 this became Marconi's Wireless Telegraph Company. By 1899 he had transmitted across the English Channel and in 1901 crossed the Atlantic. The latter achievement was surprising, for if radio waves were indeed electric waves they should have traveled in straight lines out into space. The explanation for their return to Earth was not found until 20 years later, with the discovery by the British physicist Edward Appleton of an electrified layer in the upper atmosphere which reflects the waves back to earth. In 1909 Marconi was awarded a Nobel prize and the citation stated that already 300 merchant ships and liners and most of the world's navies had been equipped. Among the liners was the *Campania* on which the notorious murderer H.H. Crippen and his mistress had embarked in Antwerp en route for Canada. The captain's suspicions were aroused by a broadcast news bulletin and he was able to inform British police of Crippen's presence on his vessel. The police traveled to Canada on a faster ship and arrested Crippen and his mistress on their arrival. Crippen was eventually hanged. The case attracted enormous publicity and gave a powerful boost to Marconi's business.

Technologically, there were important developments. An essential feature of a wireless receiver is a tube (or valve) which permits electricity to flow in only one direction and thus makes it easier to manipulate the current. Originally this was of the cat's whisker type, revived half a

century later in different guise as the transistor. But in 1904 the British electrical engineer J.A. Fleming invented the two-electrode (diode) tube or valve and this was followed in 1906 by the American inventor Lee De Forest's triode. With these aids, and tuning to narrow wavelength bands, more powerful signals could be transmitted, making it possible to replace headphones with loudspeakers and to transmit sound – speech or music – rather than a simple pulsed signal.

Technically the triode – known originally as an audion – was a major technical advance, especially when included in "regenerative" circuits. Invented independently in 1912 in the USA and Germany, audions permitted large amplification of weak signals by a cascade system. Within ten years crystal radio sets had become almost totally superseded. In the same year R.A. Fessenden and E.H. Armstrong in the USA invented the heterodyne circuit. Until then the role of the receiver was to respond to an incoming signal by simply switching on or off a direct current. In the heterodyne circuit the weak received signal modulated a strong wave generated locally, giving much more powerful output. Nineteen-twenty saw the advent of the even more sophisticated superheterodyne circuit. Meanwhile matching advances had been made in transmission. In 1913 Alexander Meissner, in Germany, combined the triode with an oscillator to generate much more powerful signals.

Thus by 1914, although equipment was still cumbersome, transmissions and reception were reliable over considerable distances, but this created its own problems. The first transmitters were simply spark-gap devices, radiating over a wide range of wavelengths. A single transmitter could effectively blanket a large area. Their merit was that they were extremely easy to operate and

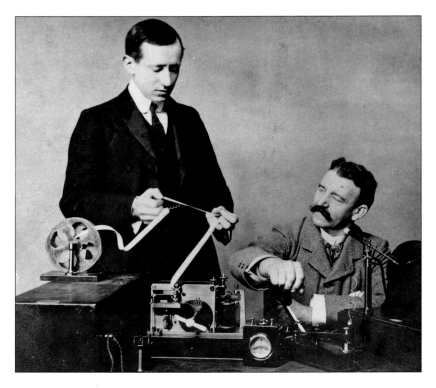

▲ In 1895 Marconi made his first practical radio transmission over a distance of 3.2km (2mi), the length of his family estate. Six years later he spanned the Atlantic and became world famous overnight. He is seen here with his assistant George Kemp in 1901, examining experimental apparatus for recording signals on tape.

It was impossible in those days to foresee the eventual results of the attempts I was then making to evolve a method of communicating across space without the use of a material conductor, but even when I had only succeeded in sending and receiving signals across a few yards of space by means of Hertzian waves I had the vision of communication by this means over unlimited distances. To have made such claims at that time would have been to invite the ridicule of scientists, as, indeed, was proved when, five years later, I had the faith to believe that by means of the system I had evolved it would be possible to send and receive signals across the Atlantic Ocean

GUGLIELMO MARCONI

they were not in fact banned internationally until 1930. Long before this, however, around the turn of the century, tuning devices had been devised whereby a particular operator could be assigned a specific wave band, leaving other bands free for others. With radio beginning to be used worldwide, especially between ships at sea, this opened the way to progress, but it was largely frustrated by the monopolistic policy of Marconi's company. This insisted that its equipment should be worked only by its own operators, who were forbidden to communicate with any stations not licensed by Marconi. Two international radio-telegraph conferences in Berlin, in 1903 and 1906, sought to break the grip of "Marconism", but with limited success. Two major disasters at sea – coupled with the Crippen affair – helped to resolve the issue. In 1909 the American ship *Republic* and the Italian *Florida* collided in thick fog off the east coast of the USA. Signals from the *Republic* brought a rescue vessel to the scene within 30 minutes and 1,700 passengers were saved. The *Titanic* disaster in 1912 was a very different story. Although the *Carpathia* picked up distress signals and arrived at the scene some twelve hours later, rescuing only 710 survivors from a total complement of over 2,000, help was

◄ Radio telegraphy was revolutionized with the invention of the two-electrode (diode) tube or valve in 1904. A further great advance was made with the advent of the triode, invented in 1906 by Lee De Forest, who is seen here examining one.

in fact close at hand. The *California*, barely 30km (18mi) away, had encountered ice and sought to warn the *Titanic* but was told by the latter's wireless operator to "shut up" as he was busy.

Another conference, held only three months after the *Titanic* disaster, easily reached agreement that radio communication should not depend on the equipment used. Marconi had, in fact, already bowed to the inevitable and abandoned his restrictive practices.

▼ By 1900 Edison's phonograph of 1877 had been translated into the gramophone, opening up a huge new market for entertainment. A recording by Enrico Caruso from Leoncavallo's *Pagliacci* in 1902 sold a million copies. Up to World War I, however, equipment was still primitive as this 1913 picture of Indian songs being recorded shows.

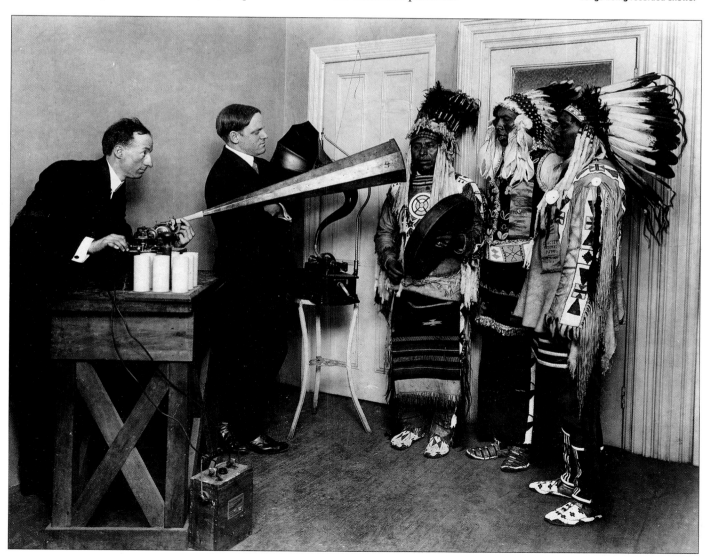

Agriculture, fertilizers and agrochemicals

Agriculture is both the oldest and the most conservative of the world's great industries. This is not altogether surprising, in that the consequences of crop failure can be so disastrous that there is little incentive to abandon well-tried methods, even if inefficient, in favor of others demanding unfamiliar techniques. Nevertheless, while the 19th century continued to see a heavy dependence on long-established empirical methods, even in the western world, science and technology were beginning to make themselves felt. New machines replaced hand-tools and steam engines and gas engines began to threaten the horse as the main source of power. But the real innovation was beginning to take place off the farm, through the medium of chemistry.

A major need for the healthy growth of plants is nitrogen (a major constituent of all living things) but this must be available in some combined form, such as nitrate. If the soil is not worked too intensively the natural content, augmented by the manure heap – the central feature of every traditional farmyard – would suffice. But the explosive growth of population from about 1800 called for more productive agriculture, some of it not involving livestock at all, and increasing amounts of caliche (sodium nitrate) were exported from vast natural deposits in Chile: by 1900 world demand had risen to 1.35 million tonnes. These deposits, not paralleled anywhere else, faced exhaustion and eventually the world would face starvation unless new sources of nitrogenous fertilizers could be found. That there should be a problem was ironic, for three-quarters of the earth's atmosphere consists

▲ The vast wheatlands of the United States and Canada lent themselves – indeed demanded – mechanization of the basic agricultural operations. Here wheat is being harvested in Canada using a battery of reapers towed in parallel by a traction engine.

of nitrogen. The technical problem was to "fix" this limitless supply of nitrogen in a form that plants could use. In Norway, where hydroelectric power was cheap, an electrochemical process was worked on a modest scale from 1904, but the real solution was found in Germany. That this was so is not surprising, for Germany was the biggest European importer of caliche and, as a major military power, also required fixed nitrogen to make high explosives. For strategic reasons, a source of synthetic nitrate was thus of particular importance to this country.

There, during the years 1907–09, the chemist Fritz Haber investigated the possibility of using the reaction between atmospheric nitrogen and hydrogen to form ammonia: the latter can then be oxidized to form nitric acid. However, the nature of the reaction is such that a useful yield of ammonia can be obtained only at far higher pressures – some 200 times atmospheric – than were then used in the chemical industry. Further, the reaction proceeded quickly only at high temperatures, but then the yield was reduced through

◄ The Haber-Bosch process for "fixing" atmospheric nitrogen was perhaps the most important industrial chemical process ever invented. Here Fritz Haber is seen in his laboratory (1913). The process was first worked at Oppau in Germany on the eve of World War I.

► In the early years of the 20th century the mechanization of agriculture seemed to hold out endless possibilities for the application of science and technology. This vision of agriculture saving the world from starvation in 1970 was conceived in 1907.

partial decomposition of the ammonia formed. This made it necessary to find a catalyst which would speed it up at lower temperatures. Haber's process was developed by Carl Bosch of BASF (Badische Anilin- & Soda-Fabrik) and first came into operation at Oppau in 1913. For his outstanding achievement Haber was awarded a Nobel prize in 1918.

After World War I, the Haber-Bosch process changed the pattern of world agriculture. Cheap nitrogen – often made cheaper still by government subsidies – meant that crop values would be raised far in excess of the extra fertilizer cost: 1.25kg of nitrogen per h (1.1lb per acre) could increase yields of rice and wheat by 15 percent, and of potatoes by an astonishing 75 percent.

But the productivity of the soil depends not only on promotion of growth but on controlling the pests and diseases that afflict both growing crops and the stored harvest. At the beginning of the 20th century the agrochemical industry was in its infancy. The best weapon against weeds, insect pests and molds was good husbandry but some chemicals were already in use. These included plant extracts (such as pyrethrum, rotenone and nicotine) but – because of their cost – these were used largely by horticulturists rather than farmers. For field use a range of inorganic chemicals – such as compounds of copper and arsenic, sodium chlorate and sulfur – were used. The copper-based Bordeaux mixture (named for its place of origin) is typical. Originally devised to control mildew on grapevines, it was also used to control blight on potatoes and tomatoes. A few cheap organic chemicals were used (such as naphthalene, a by-product of the gas industry, for soil sterilization and tar-oil to spray fruit trees) but in this field the day of the synthetic chemical was yet to come.

It was, however, not only the growing use of fertilizers and agrochemicals that was increasing agricultural productivity. Mechanization of the basic agricultural processes was making headway in two fronts: implements and sources of power. Well before 1900 machines were in use for reaping, binding and threshing – tasks done by hand since the dawn of civilization. The horse was still the main source of power though steam engines were increasingly used – usually locomotives which traveled from farm to farm with contractors. More particularly, they were used for the heavy work of plowing, working in pairs drawing specially designed plows across a field with steel cables. For light work about the farm, as for cutting chaff, stationary gas engines were used.

For any sort of traction, stationary engines were unsuitable, especially for small enclosures. A major advance came in 1908 when Holt of California began to manufacture tractors fueled by gasoline and fitted with caterpillar tracks to give a good grip on the land. Although slow, they could pull the wide heavy machines appropriate for the vast open areas of North America. Additionally, they required little manpower. In Europe, by contrast, where manpower was abundant and fields mostly small, the tractor made little headway until the 1930s. As late as 1939, Britain still had a million horses – almost all used for traction.

INVENTORS AND INVENTIONS

In the 19th century Thomas Edison had demonstrated the potential of "inventions factories", with his pioneering laboratories at Menlo Park. Large companies like General Electric and Du Pont followed suit. Invention became more and more the province of such companies, who could provide the resources needed to make fundamental breakthroughs, and put teams of scientists to work on a problem. Yet vital discoveries were still made by lone geniuses. Ample funding alone did not make inventions, as the United States War Department found when it funded S.P. Langley's attempts to build an aircraft. Langley's "Aerodrome" failed to fly in 1903, the same year as the Wright brothers achieved powered flight.

As science and technology grew ever more specialized, the chances of a lone inventor making an entirely new discovery grew less and less. Making the right invention at the right time and at the right price was no longer enough. Quality control and marketing became more and more important ingredients for success. Even though lone inventors could succeed for a time, large corporations might soon outclass them once a market had been established. This was what happened to John Logie Baird, whose mechanical television system proved technically inadequate, and was replaced by electronic systems developed by RCA and EMI.

However, if an inventor hit on the right technological fix, and held key patents to protect the invention, it was possible to lay the foundation of a new company, and repeat the successes of Marconi and Edison. In 1937 the American John Chester Carlson patented the principle of the photocopier, and out of his work grew the Xerox Corporation. Part of Carlson's success was that he made contact with a firm that was looking for a new product, and was willing to put in the resources to transform Carlson's ideas into the Xerox photocopier. Similarly, Edwin Land developed the Polaroid camera in the 1940s, and founded the Polaroid corporation.

Key patents, however, were of little use unless a market existed, or could be created for the innovation. The rise of the electronics industry created new opportunities for inventors. The key breakthrough, the transistor, was produced by John Bardeen, Walter Brattain and William Shockley at Bell Laboratories in 1947. Shockley's development led to miniaturized electronic components, and to the microchip. Availability of specialized chips made it possible to create new products based on the chips produced by the large corporations.

Such inventors succeeded in exploiting markets the large corporations had overlooked. In the 1970s one such market was the home computer. A number of small home computer companies grew up in California's "Silicon Valley". Perhaps the best known was Apple Computers, founded by Stephen Wozniak and Steven Jobs. Starting from a garage workshop, Apple dominated the home computer market and grew into a multinational company.

◀ Some inventions have become ubiquitous, yet their inventors are virtually forgotten. The zipper fastener was originally patented in 1893 by American Whitcomb L. Judson, though the modern fastener was not developed until 1913, by the Swede Gideon Sundback. Mass production began during World War I.

▼ Otto Lilienthal, of Germany, made pioneering studies of gliders in the 1890s, building a hill from which to test his devices. He made more than 2,000 flights, and died in 1896 when his glider crashed. His work resulted in much basic aeronautical knowledge.

▶▶ The British inventor Clive Sinclair made his fortune in the 1970s from cheap electronic devices sold through mail order. His attempt to branch out into a battery-powered tricycle brought disaster.

▼ The patent for Carlson's xerox dry photographic copier. The paper is wrapped around a charged drum and the image focused onto it. Where light falls, the image is conducted away. Toning powder adheres only to the charged particles.

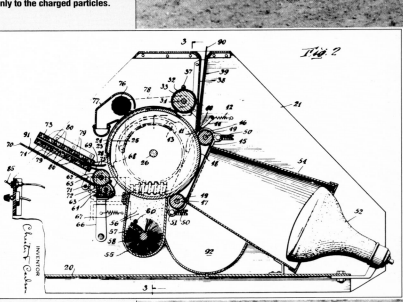

► The American inventor Thomas Edison made many pioneering discoveries of his own, and developed the concept of the research laboratory at Menlo Park, in New Jersey. Here he and his assistant (on the ladder) study the filament of a lightbulb.

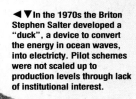

◄▼In the 1970s the Briton Stephen Salter developed a "duck", a device to convert the energy in ocean waves, into electricty. Pilot schemes were not scaled up to production levels through lack of institutional interest.

Datafile

The first decade of the 20th century saw a series of brilliant breakthroughs in sciences related to cell biology, such as biochemistry, bacteriology and immunology, heredity and neurophysiology. Discoveries included the cellular constituents of the nervous system, the cellular process of resistance to infection, the role of such cellular constituents as genes and chromosomes in heredity, and the hereditary nature of certain diseases.

▼▼ The development of bacteriology was supported by Imperialist great powers, such as France. Yellow fever exemplifies control of a disease by understanding the role of an insect carrier (or vector). A saline solution was eventually discovered as a simple cure for cholera.

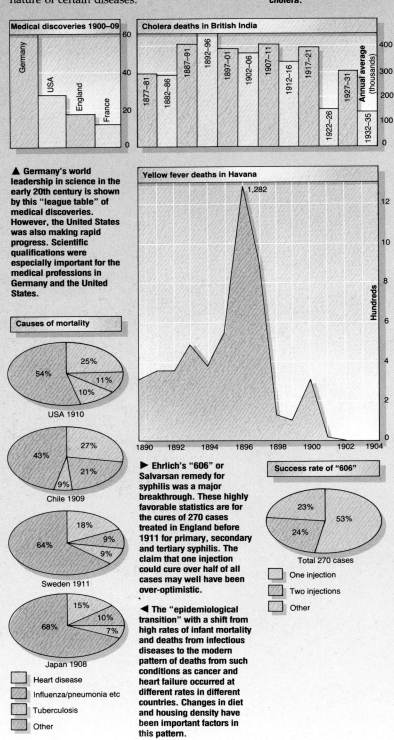

Medical discoveries 1900–09

Germany, USA, England, France

▲ Germany's world leadership in science in the early 20th century is shown by this "league table" of medical discoveries. However, the United States was also making rapid progress. Scientific qualifications were especially important for the medical professions in Germany and the United States.

Cholera deaths in British India

1877–81, 1882–86, 1887–91, 1892–96, 1897–01, 1902–06, 1907–11, 1912–16, 1917–21, 1922–26, 1927–31, 1932–35

Annual average (thousands)

Yellow fever deaths in Havana

1,282

Hundreds

1890 1892 1894 1896 1898 1900 1902 1904

Causes of mortality

USA 1910: 54%, 25%, 11%, 10%

Chile 1909: 43%, 27%, 21%, 9%

Sweden 1911: 64%, 18%, 9%, 9%

Japan 1908: 68%, 15%, 10%, 7%

Heart disease
Influenza/pneumonia etc
Tuberculosis
Other

▶ Ehrlich's "606" or Salvarsan remedy for syphilis was a major breakthrough. These highly favorable statistics are for the cures of 270 cases treated in England before 1911 for primary, secondary and tertiary syphilis. The claim that one injection could cure over half of all cases may well have been over-optimistic.

◀ The "epidemiological transition" with a shift from high rates of infant mortality and deaths from infectious diseases to the modern pattern of deaths from such conditions as cancer and heart failure occurred at different rates in different countries. Changes in diet and housing density have been important factors in this pattern.

Success rate of "606"

23%, 53%, 24%

Total 270 cases

One injection
Two injections
Other

Compared with the physical sciences, where so many new discoveries were being made in the fields of electromagnetism and atomic structure, biological science was relatively quiescent at the turn of the century. Much important but undramatic work was being done in the fields of morphology (the study of the form and structure of living organisms) and taxonomy (classification of relationships): there was an almost unlimited amount of work to be done in simply cataloging the almost infinite number of plant and animal species that populate the earth, to say nothing of those known only by their fossilized remains.

Morphology stimulated work on morphogenesis and embryology: how do organs and complete organisms, respectively, develop? As the physicists were increasingly directing their attention to the smallest particles of matter, so biologists, with the aid of the optical microscope, were investigating the smallest units of life. Cytology was emerging as a science directed at the specialized cells of which the organs of all plants and animals are formed. In Spain, Santiago Ramón y Cajal (1852–1934) had done pioneer work on the nervous system of vertebrates and had demonstrated its immense complexity: the human brain contains some ten billion cells, each linked to as many as 50 others. For this he was awarded a Nobel prize in 1906, sharing it with the Italian histologist Camillo Golgi (1844–1926) who had devised a technique of staining nerve cells black with silver.

In Britain, similar work was done by C.S. Sherrington whose classic work *The Integrative Action of the Nervous System* appeared in 1906. Sherrington, also a Nobel laureate (1932), has been called "the William Harvey of the nervous system".

No less important was the study of microorganisms, especially those which cause disease. The foundations of the new science of bacteriology (the study of bacteria) had been laid in the latter years of the 19th century, by the Frenchman Louis Pasteur (1822-95) and the German Robert Koch (1843–1910), but it was the 20th century that was to see its full flowering. Koch's research on the tubercle bacillus (the cause of tuberculosis) earned him a Nobel prize in 1905. Three years earlier Ronald Ross (1857–1932) had received the same distinction for identifying the causative organism of malaria in the salivary glands of the female *Anopheles* mosquito, thus demonstrating the mode of transmission of the disease and pointing the way to its prevention. Similarly, a trypanosome was identified as the cause of the sleeping sickness. The turn of the century saw also the discovery of even smaller infective organisms, the viruses. These cause foot-and-mouth disease in cattle and yellow fever.

NEW SPECIALTIES AND CONNECTIONS

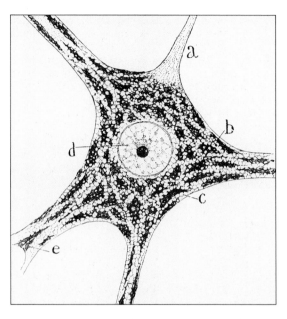

Important small-scale work of the early 20th century

The origins of biochemistry

Paul Ehrlich's breakthrough

The use of X-rays

The introduction of psychology to physiology

The rediscovery of Mendelian genetics

◀ Ramón y Cajal's own drawing of a nerve cell from the spinal cord of a rabbit.

▼ Robert Koch (center) in Africa with the German Sleeping Sickness Commission, 1906–07.

The emergence of biochemistry

Among the most important systems of the body is the blood; it is also one of the most vulnerable. At the beginning of the 20th century, blood transfusion was an extremely hazardous operation. The outcome might be successful but equally it might be fatal. The transformation of blood transfusion into a routine procedure is largely due to the pioneer work of Karl Landsteiner, who in 1900 identified three of the major human blood groups and devised techniques for identifying them. By ensuring that the donor's blood group was compatible with that of the recipient transfusion could be made safe. Since then other groups have been identified: in 1940 Landsteiner himself made another major contribution with the discovery of the Rh (Rhesus) factor (the presence or absence of the Rhesus antigen) which is particularly important in the case of blood transfusion in pregnancy.

In the late 20th century the availability of vitamins to balance diet was taken for granted; in

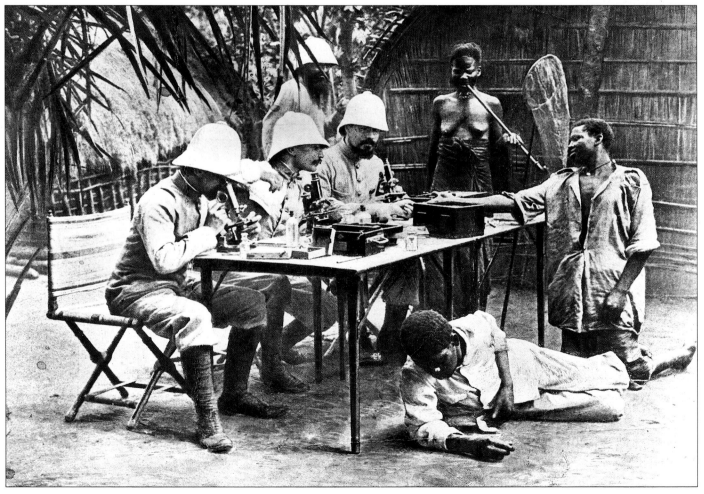

The Search for the Magic Bullet

Louis Pasteur's discovery in the 1860s of the true nature of infectious disease – an invasion of the body by destructive germs – led to the development of two alternative lines of treatment. One, devised by Pasteur himself, arose from his chance observation that infection with an attenuated strain of a germ can give rise to a mild attack of the disease and confer subsequent immunity. This was, of course, no more than a reasoned development of Edward Jenner's empirical technique of vaccinating against small pox with the relatively mild cow pox (developed in England in the late 1790s). In increasingly sophisticated forms, immunization remains an exceedingly important method of preventing many major diseases, such as poliomyelitis.

While immunization is a very effective means of containing an epidemic, the treatment of an established infection is a very different matter. As causative organisms were identified, and could be grown in pure cultures in the laboratory, another possibility came to be considered: was it possible to find substances deadly to germs but harmless to the tissues of the human body? At first the quest seemed hopeless: if substances were toxic they seemed to be equally so toward all living cells. Among the few who questioned this pessimistic view was the German physiologist Paul Ehrlich. After many years of patient research he discovered in 1906 that a synthetic compound of arsenic, Salvarsan, was a specific against syphilis. Enthusiastically he called it a "magic bullet" which could seek out and destroy its target without doing damage en route. In fact he was overoptimistic, for Salvarsan not only had serious side effects but in unskilled hands could be fatal. Neosalvarsan, introduced in

Docteur EHRLICH

1912, was an improvement but still far from satisfactory. However, Ehrlich's research did establish a vital principle: that there could exist compounds differentially toxic to bacteria and the human body. It thus paved the way to the sulfonamides in the 1930s and, later, penicillin and other antibiotics.

▲ Over the period 1909–11 Henry Wellcome, collector of medical memorabilia on a grand scale, commissioned Richard Cooper to paint eight pictures depicting the worst serious human diseases. This is *Syphilis* (1910) and depicts, half allegorically and half realistically, the reaction of an Edwardian sporting man to the knowledge that he is infected. Paul Ehrlich's newly discovered "magic bullet", the arsenical drug Salvarsan, held out the hope of cure but had dangerous side effects.

◀ Ehrlich's search for a "magic bullet" lethal to an infective organism but harmless to its host began in 1905 with atoxyl, an arsenical drug. But he had to ring the chemical changes on this 606 times before devising Salvarsan, active against the syphilis spirochaete, discovered in 1905. Hence this caricature by the German painter Max Liebermann.

► That certain diseases are directly linked with specific dietary deficiencies was becoming clear early in this century. This picture shows a group of German children suffering from rickets (vitamin D deficiency) in the immediate aftermath of World War I, when food had long been scarce.

▼ At the beginning of the 20th century advocates were still to be found of vitalism, the doctrine that living processes are not wholly explicable in terms of the ordinary laws of physics and chemistry. Increasingly, however, biological processes were being subjected to scrutiny by the techniques of the physical sciences. This photograph of the Pathological Laboratory of the Great Northern Hospital, London, taken in 1912, shows a typical range of physico-chemical apparatus: microscopes, bunsen burners, burettes, pipettes and filters.

1900, however, the term did not exist but it was already suspected that some foods contained special health-maintaining factors. In 1890 Christiaan Eijkman, a Dutch army medical officer serving in the Dutch East Indies, had observed that the incidence of beriberi – a paralyzing and often fatal disease – was linked with diet. It was prevalent among those who ate polished (white) rice but not among those who ate brown rice, with the husk on: the disease could be cured by adding rice bran to the diet. Although his observations were quite correct, Eijkman's interpretation of them was wrong: he supposed that the bran contained some sort of antidote to a poison. The right answer was found by the British biochemist F.G. Hopkins at Cambridge, whose studies on the relationship between growth and diet led him to the concept of "accessory food factors" – now known as vitamins.

This new discovery was important in itself, but possibly Hopkins's greatest contribution to science was his insistence that many biological problems can be solved by chemical methods: biochemistry became a discipline in its own right. Among the fruits of this approach was the British physician A.E. Garrod's discovery in the first decade of the century that certain diseases – such as albinism (pigment deficiency) and cystinuria (the excessive production of certain amino acids which causes stone formation in the kidneys) – are the consequence of inherited metabolic disorders. His *Inborn Errors of Metabolism* (1908) aroused much interest.

The importance of the chemical approach to

biological problems is also seen in the work of the German bacteriologist Paul Ehrlich (1854–1915). In common with others then using the microscope to study biological problems, he used dyes to differentiate particular structures in material under examination. He used them, for example, to stain Koch's newly discovered tubercle bacillus and white blood corpuscles. Reasoning that if bacteria could thus selectively absorb dyes they might also selectively absorb substances toxic to themselves but harmless to tissues they had invaded. This line of thought led him eventually, in 1909, to the arsenical compound Salvarsan, which proved to be an effective treatment for syphilis (the causative organism of which, *Treponema pallidum*, had been identified by the German zoologist Fritz Schaundinn in 1906). Ehrlich also did important work on combating bacterial infection by immunization (stimulating the body to resist infection) – the main hope of clinical physicians in the prewar years – and developed the "side-chain" theory named after him. According to this, the protoplasm (content of cells) consists of complex molecules with a stable core and less stable side-chains, and it is the latter that take part in immunological reactions. For his contributions to medical science Ehrlich was awarded a Nobel prize in 1908, sharing it with the Russian-French biologist I.I. Mechnikov, who had done important research on the role of white blood cells in attacking invading bacteria.

The impact of X-rays on medicine
X-rays were originally discovered, by W.C. Röntgen in November 1895 (see p.43). Because of their unusual property of passing virtually unchecked through some materials opaque to light, it is not surprising that it was this property that was first turned to practical account in medicine. For the first time it was possible to look inside the body without traumatic surgical intervention. At first, of course, there were limitations: only objects very opaque to X-rays, such as metal, could be clearly detected. Different kinds of soft tissue could not be differentiated. That the medical profession so quickly adapted X-rays is perhaps due to the fact that Röntgen, a physicist, announced his discovery not in a physical scientific journal

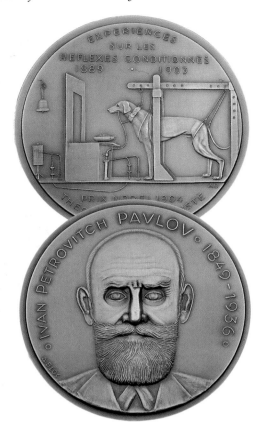

► Ivan Pavlov's researches on the physiology of digestion spanned nearly half a century (1891–1936) and earned him a Nobel prize in 1904. Today he is remembered largely for his work – commemorated on this medal – on conditioned reflexes, which he did not begin until 1902. The dog salivates at the sound of the bell even when no food is offered.

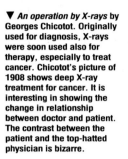

▼ *An operation by X-rays* by Georges Chicotot. Originally used for diagnosis, X-rays were soon used also for therapy, especially to treat cancer. Chicotot's picture of 1908 shows deep X-ray treatment for cancer. It is interesting in showing the change in relationship between doctor and patient. The contrast between the patient and the top-hatted physician is bizarre.

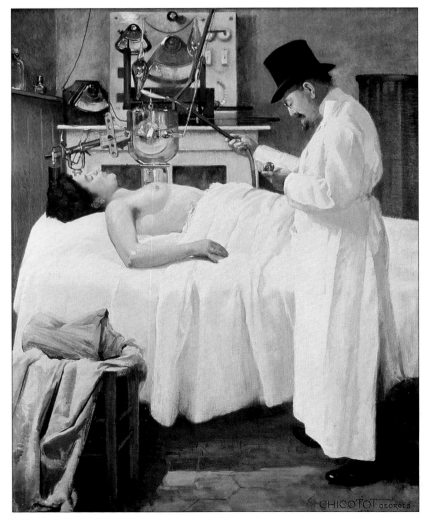

but in a report to the Würzburg Physical Medical Society. Within a year a Canadian surgeon, R.C. Kirkpatrick, had described the use of the "new photography" to locate a bullet in a patient's leg and in the same year the American physicist Elihu Thomson invented a stereo X-ray camera to allow more precise location. Within three years military surgeons were using X-ray equipment in the field in the Spanish-American war. It was quickly discovered that X-rays also had therapeutic possibilities for some skin diseases and, more particularly, certain kinds of cancer. In 1903 the German surgeon G.C. Perthes used deep X-ray therapy to treat a carcinoma and the American surgeon Nicholas Senn used it to treat leukemia, esssentially cancer of the blood.

Psychology and physiology

In 1904 the Russian biologist Ivan Pavlov was awarded a Nobel prize. The citation states that this was for his research on the physiology of digestion, which he investigated mainly in dogs. This led him to the discovery of the conditioned reflex, for which he is far better known. Briefly, he discovered that if a bell is rung before food is offered a dog will acquire the habit of salivating even if the bell is rung without food. He thus introduced, with far-reaching effects, a psychological element into physiology.

The theory of genetics

In the mid-19th century Charles Darwin's theory of evolution by natural selection (expounded in his *Origin of Species*, in 1859) had excited fierce controversy. By the early 20th century this had largely died down and biologists were concerned more with the mechanisms of evolution than with the validity of Darwin's theory.

That certain characteristics can be inherited from one generation to another must have been noted since the dawn of civilization, for familial resemblances are a matter of everyday observation. Animal breeders have long exploited this phenomenon in attempts to improve the value of their stock. But until the late 19th century the methods used were entirely empirical. The famous 18th-century English breeder, Robert Bakewell, for example, seemingly achieved his remarkable successes by outbreeding (breeding with unrelated stock) to introduce desirable traits and stabilizing these by inbreeding. Effectively genetics did not emerge as an exact science, until 1900, but in fact the foundations had been firmly laid some 40 years earlier by Gregor Mendel, of the monastery of Brno in Moravia. From 1851 to 1853 he was sent by his order to study science in Vienna: on returning to the monastery (where he became abbot in 1868) he began a series of plant breeding experiments with the edible pea (*Pisum*). He studied seven specific characteristics, including shape of the seed, color of the flowers, and height of the stem. Keeping careful records relating to more than 20,000 plants, he found that these characteristics were always inherited in a ratio close to 1:3. Crossing tall plants with dwarf ones gave either tall or dwarf seedlings – never plants of intermediate height. He interpreted this as meaning that the characteristics were

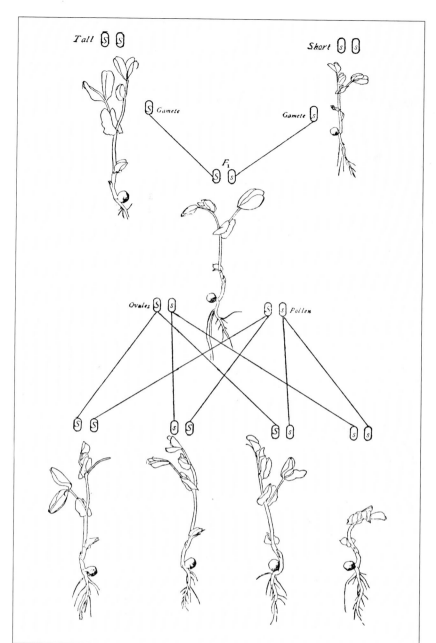

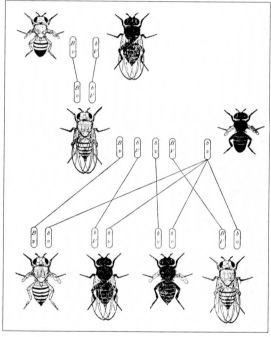

transmitted by specific hereditary factors and that these become segregated in the germ cells.

This was substantially correct, but unhappily the results passed unnoticed as they were published only in the journal of the Brno Natural History Society. Disappointed, he sent a copy of his paper to the distinguished Swiss botanist Karl von Nägeli (1817–91), who failed to recognize its importance. There was a sad irony in this, for Nägeli had himself in 1842 described in minute detail the process of pollen formation in the lily family Liliaceae, including the separation in the nucleus of what he called "transitory cytoblasts"; these were in fact the chromosomes which carry Mendel's "hereditary factors" (genes).

Thus was a great opportunity lost. In 1900, 16 years after Mendel's death, the Dutch botanist Hugo de Vries published the results of a long series of plant–breeding experiments in which he too got a 1:3 ratio. Publishing his results in 1900,

he quoted Mendel's work of 34 years earlier. Within a few weeks two other botanists, C.E. Correns in Germany and E. Tschermak von Seysenegg in Austria-Hungary published similar results. Thus the new century began with a confirmation of Mendelian ratios which laid a firm foundation for genetic theory.

In the USA the zoologist T.H. Morgan began to study evolution and heredity in 1903. At first he was skeptical of Mendel's results but his research with the fruit fly *Drosophila* – a particularly convenient experimental animal because it breeds rapidly and has very large chromosomes in its salivary glands – soon converted him. He identified genes, arranged on the chromosomes like beads in a necklace, as the units of heredity and by 1911 he and his colleagues had published the first "chromosome map", showing the location of five sex–linked genes. Ten years later more than 2,000 genes had been mapped.

▲ The basic principles of heredity were worked out by the Austrian friar Gregor Mendel in the 1860s, using peas (above left). He showed that when a tall and a dwarf plant are crossed, the next generation is of tall plants but one-quarter of the second generation is of dwarf plants: Inheritance transmitted specific characteristics rather than blended them. Mendel's results passed unnoticed until the Dutch botanist Hugo de Vries (top) rediscovered them. One of the first to enter this new field was T.H. Morgan, in the USA, who mostly used the fruit fly *Drosophila* (above).

Datafile

An essential feature of science is its international character: the results of experiment and observation are made freely available for all to use. The natural philosophers of the 17th century exchanged information through meetings of their national academies, through journals, and by correspondence. Inevitably, however, science was, for a variety of reasons, cultivated more fruitfully in some countries than others. One major reason was the availability of a leisured class able to indulge an interest with little prospect of direct return. Even then, leisure might be otherwise occupied: as a generalization, it is fair to say that experimental science appealed more to the West than the East, who preferred philosophical speculation. Thus by the 19th century the center of scientific gravity was in Europe with the USA, with its strong European affiliations, growing steadily in strength. Within this group Germany forged ahead: by 1900 it was the leading scientific nation.

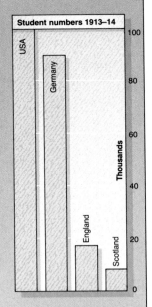

Student numbers 1913–14

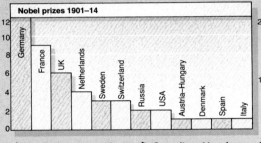

Nobel prizes 1901–14

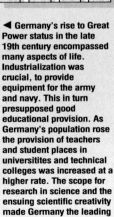

Student ratios 1913–14

▲ Nobel prizes, named for Swedish manufacturer Alfred Nobel, were first awarded in 1901. Scientific communities were at first suspicious of them, but the prizes steadily became accepted as they were given to internationally recognized outstanding scientists and as the "league table" of awards confirmed estimates of national standing.

► Commitment to science in education provides one guide to the strength of science, but different measures must be considered. In 1913–14 the USA had the largest number of students, but a strong provision of universities and colleges gave Scotland the highest proportion. Germany scored well on both counts.

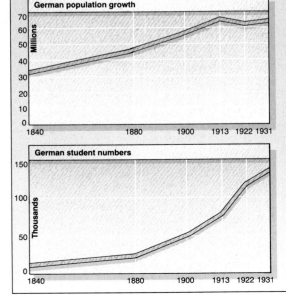

German population growth

German student numbers

◄ Germany's rise to Great Power status in the late 19th century encompassed many aspects of life. Industrialization was crucial, to provide equipment for the army and navy. This in turn presupposed good educational provision. As Germany's population rose the provision of teachers and student places in universitites and technical colleges was increased at a higher rate. The scope for research in science and the ensuing scientific creativity made Germany the leading scientific power by 1900.

Singling out individuals for praise can be invidious, but most people would agree that the two greatest physicists of all time were Sir Isaac Newton (1642–1727) and Albert Einstein (1879–1955). Both had extraordinary insights into the nature of the physical world and were gifted with remarkable powers of imagination. Einstein himself had a profound respect for Newton: "In one person he combined the experimenter, the theorist, the mechanic and, not least, the artist in exposition. He stands before one strong, certain, and alone". At first sight, therefore, it seems something of a paradox that these two men of extraordinary genius should seemingly interpret the universe in quite different ways. In fact, their views were not contradictory but complementary.

The advent of electromagnetism

In 1900 Newtonian physics was still dominant. For over two hundred years it had sufficed to explain and quantify a wide range of natural phenomena, from the movements of the heavenly bodies to the swing of the pendulum and the flow of liquids. The great achievements in mechanical and civil engineering of the 19th century assumed the validity of classical physics. Nevertheless, in the second half of the 19th century new phenomena were being encountered for which existing theory was inadequate. The main new factor was electromagnetism (magnetism produced by electric currents). This was, of course, not new in itself for electricity and magnetism had been investigated experimentally since the 17th century and in the western world were widely familiar through their applications – notably in the telegraph, the telephone and electric lighting. Yet although electricity could be generated and utilized its nature was not understood nor its true relation to magnetism.

Among the first to develop a comprehensive theory relating electricity to magnetism was James Clerk Maxwell (1831–79), a brilliant British mathematician and physicist who in 1871 became the first Cavendish Professor of Experimental Physics at Cambridge. He was responsible for designing the famous Cavendish Laboratory in which, in the 20th century, so many major discoveries in atomic physics were made. Building on the ideas of Michael Faraday (1791–1867) and William Thomson (Lord Kelvin; 1824–1907) he produced an elegant mathematical interpretation of all the electromagnetic phenomena then known. His celebrated *Treatise on Electricity and Magnetism* appeared in 1873. This was no mere intellectual tour de force, for it had very important practical implications. One of these was that light was an electromagnetic disturbance and should travel at a speed equal to the ratio of the electrodynamic to the electrostatic units of electricity.

THE NEW PHYSICS

This ratio had already been experimentally determined to be 3.1×10^{10} cm per second, very close to the best measurement of the velocity of light. Thus emerged a clear link between optics and electricity and the suggestion that there existed a broad spectrum of electromagnetic radiation of which visible light was only a part.

Not all of Maxwell's contemporaries were persuaded of the validity of his theoretical proposals. Among those who sought to test them experimentally was the German physicist Heinrich Hertz (1857–94). Using an induction coil, he charged a rectangle of copper wire in which a short gap had been left: when the coil was activated a spark jumped the gap. He then found that a spark could simultaneously be induced in a similar, uncharged circuit at a distance. The first coil was evidently generating electric waves that could be picked up by the second, tuned circuit. He showed that these waves were virtually the same as light waves. They traveled at the same speed, and they could be reflected and refracted. Hertz's experiments (1886–88) were purely academic, designed to put Maxwell's theory to practical test, and he could detect the electric waves over no more than 65ft (20m). But in the resourceful hands of the Italian entrepreneur Marconi they were to be the basis of an immensely important new means of communication.

Such experiments proved the truth of Maxwell's main conclusions but one major difficulty remained. His theory indicated that light was an electromagnetic wave which traveled with a fixed

▼ Unlike many scientific discoveries, X-rays quickly became known to the public and aroused much interest. This German postcard of an X-ray beach party is typical of its time. The medical value of X-rays was also quickly recognized. By the turn of the century they were beginning to be used to examine fractures and locate foreign bodies, such as bullets. Deep X-ray therapy for the treatment of skin diseases and cancer was introduced in 1903.

velocity independent of the motion of the source and of the receiver, whereas common sense suggested that such movement ought to add a corresponding amount to the velocity. Various ingenious theories were advanced to reconcile these two propositions but not until 1905 was a satisfactory explanation offered by Einstein with his special theory of relativity.

The foundations of atomic physics
Concurrently with these developments other lines of research were undermining one of the basic tenets of 19th-century physics: that atoms, the basic units of all matter, were indivisible and indestructible. From the middle of the century the spectroscope (an instrument for analyzing light according to wavelength) had been increasingly used for chemical investigations, and to induce gases to emit light for examination it was common practice to pass an electric discharge through them at very low pressure. It was in the course of such an experiment that, in 1895, the German physicist Wilhelm Röntgen by chance discovered X-rays. They had remarkable powers of penetration – which were very quickly put to practical use in medical diagnosis and, later, treatment – and he concluded that they must be electromagnetic waves of very short wavelength. Röntgen was awarded a Nobel prize in 1901, the first year in which they were given.

Röntgen's experiments on the discharge of electricity through gases at low pressure had been inspired by those of William Crookes in Britain

and by his own compatriots J. Plücker and J.W. Hittorf. In 1879 Crookes had described in detail the properties of what he called "molecular rays", showing that they could cast shadows, warm objects on which they fell, and be deflected by a magnetic field. Suspecting that he was not dealing with some form of electromagnetic radiation but rather with a stream of particles, Crookes referred to the examination from his discharge tubes as "radiant matter". This extended the earlier work of Hittorf and Plücker who had shown that such tubes could emit two types of radiation, according to the nature of the electric discharge – either a glow representing a continuous range of wavelengths or one containing relatively few spectral peaks.

In retrospect, the results achieved by these pioneers – which aroused much interest at the time – were inconclusive in that they could not be specifically interpreted. Nevertheless, they opened up an immensely fruitful new area of research and attracted attention to this area of physics.

The new X-rays aroused much interest, and among those who investigated them was J.J. Thomson, who in 1884 had been appointed Cavendish Professor in the University of Cambridge, and his research student from New Zealand, Ernest Rutherford. It was apparent that the X-rays were not, as it were, primary products but were generated by cathode rays, so called because they were emitted by the cathode (negative electrode) of a gas discharge tube.

By 1897 Thomson had established that the cathode rays were not a form of electromagnetic radiation – they traveled 1,600 times slower than light – but a stream of negatively charged particles. He was able to measure the ratio of the charge to the mass of individual particles: quite soon, however, he measured the charge and this gave him the mass. The result was startling. The particle – soon to be known as the electron – had a mass 1,800 times smaller than that of the lightest atom (hydrogen). Clearly, therefore, atoms were not after all the smallest particles in nature.

Meanwhile important new discoveries were being made in France. There, in 1896, the physicist Henri Becquerel discovered that the metal uranium emitted a penetrating radiation capable of blackening a photographic plate. Two scientific colleagues in Paris, Marie and Pierre Curie, wondered whether other substances might not have similar "radioactive" properties and in the course of extremely laborious experiments isolated small quantities of two exceedingly active elements, polonium (named for Marie's native country) and radium.

Rutherford at once turned his attention to radioactivity and quickly discovered that two kinds of rays were being emitted, which he called alpha and beta rays. He later found in research at McGill University in Canada that the atoms from which they were derived were spontaneously disintegrating. Returning to England (Manchester), he proved by 1908 that the alpha particles were in fact positively charged atoms of helium. In 1908 the German physicist H.W. Geiger – famous for his particle counter – and E. Marsden, working with Rutherford showed that when a

stream of alpha particles struck gold foil almost all passed straight through but a few – about one in 8,000 – suffered major deflection. Rutherford concluded that the positive charge in the gold atoms must be concentrated in a small region at the center of the atoms. On this basis he formulated his concept of the atom as containing a relatively small positively charged nucleus, corresponding to nearly all the mass, surrounded by a cloud of electrons whose combined negative charge made the atom as a whole neutral.

Why such a structure was stable, and did not collapse upon itself, was explained in 1912 by the Danish physicist Niels Bohr on the basis of Max Planck's quantum theory of 1905. The foundations of atomic physics were thus firmly laid by these outstanding scientists.

▲ The founding of the Radium Institute in Paris in January 1914 was hailed by the press worldwide as opening a new era of hope for cancer sufferers (above left). Its director was Marie Curie, seen here with her husband in their laboratory in 1904.

▼ Two of the great pioneers of atomic physics – Rutherford (right) and H.W. Geiger – who collaborated at Manchester University in Britain, 1907–12. With E. Marsden they developed in these years the nuclear theory of the atom. Geiger is still widely remembered as the inventor of the "Geiger counter" for measuring alpha particles.

The Inauguration of Nobel Prizes

When he died in 1896 the Swedish chemist and industrialist Alfred Nobel bequeathed the greater part of his vast fortune, some £2 million, to provide annual prizes for those who had "conferred the greatest benefit on mankind". They were to be in five categories: physics, chemistry, physiology or medicine, literature and peace. The first four categories were not surprising, for Nobel himself was an able scientist and a man with some literary pretensions. The fifth did, however, excite comment, for most of his money had been made in armaments. In his later years, however, he had reached the conclusion that armaments might serve better to preserve peace than "revolutions, banquets and long speeches".

Surprisingly, for as an industrialist he was well aware of the need for precision in legal documents, he drafted his will himself, without advice, and in terms so vague and controversial that it took several years to set up the Nobel Foundation in Stockholm. The first prizes were not awarded until 1901: the value of each was £8,000, a very handsome sum in those days, and since greatly increased.

The prizes quickly attracted popular interest worldwide, which grew in 1903 when the physics prize was awarded to the pioneers in radioactivity, Marie and Pierre Curie and Antoine Becquerel. Scientific communities,

▲ **Alfred Nobel in his laboratory at San Remo, Italy, c. 1895, a portrait by the Swedish painter Karl Emil Österman.**

▼ **The scene in the Royal Academy of Music, Stockholm, as the Crown Prince of Sweden awards the first Nobel prizes in 1901. The recipients were Röntgen (Physics), van't Hoff (Chemistry) and Von Behring (Medicine).**

however, were slow to value the awards. At first they feared that the selectors would be influenced by national or political interest groups. Two factors gradually commended the prizes: first, that most were given to scientists who had already been honored by their national scientific communities; second, the size of the sums given which could be used to fund research and stimulate the award of grants from other sources.

The first prize winners in science and medicine were W.C. Röntgen for his discovery of X-rays; J.H. van't Hoff, for his work on stereochemistry; and E.A. von Behring, who discovered antitoxins and pioneered their use in the treatment of diphtheria. The presentations were made by the Crown Prince of Sweden before a distinguished international audience, and the ceremony has since been repeated annually, apart from interruptions during the world wars.

By 1990 some 400 such prizes had been awarded and they had become universally recognized as the supreme scientific accolade. Only four people have been awarded two separate Nobel prizes: J. Bardeen (1956, 1972), Marie Curie (1903, 1911), Linus Pauling (1954, 1962) and F. Sanger (1958, 1980). In the early years of the century the prizes were almost always awarded to individuals but from the 1950s onward they were often shared between two or three candidates.

German Science: Depth and Creativity

Around 1900 Germany attained a position of world leadership in both scientific education and research. After the unification of Germany in 1871 rapid industrialization had been accompanied by state investment in higher education. The increasingly prosperous middle classes were keen that their sons should gain academic qualifications. University professors enjoyed high social status and were civil servants. Technical education had been fostered with the foundation of new Technische Hochschulen, university-level technical colleges. Industrialists like Alfred Krupp, the Siemens brothers, and Emil Rathenau were responsive to new science-based technologies. Specialized institutes for research and teaching were also established. The achievements of such intellectual leaders as the physiologist Emil Du Bois-Reymond and the physicist Hermann von Helmholtz show extensive interaction between different disciplines such as physics with biology. Experimental techniques were combined with philosophical generalities to develop new sciences, such as Robert Koch's bacteriology. Professors campaigned for improved funding. The education ministeries of the various federal states (such as Bavaria, Prussia and Saxony) and of Austria and Switzerland competed with each other to attract the best professors. This competitive situation stimulated intellectual creativity and diversity.

Professors and the state took much nationalistic pride in the nation's scientific achievements. Yet after 1900 there was unease that Germany might lose its prominence. There were ever-increasing numbers of students, and yet gifted young researchers faced a lack of career prospects as the old guard of professors now sought to prevent the development of new disciplines. In response to the emergence of the Carnegie and Rockefeller Foundations in the USA, German industrialists like Gustav Krupp von Bohlen und Halbach financed new institutions for research and technical applications of science. Named Kaiser Wilhelm Institutes, they marked a break with the 19th-century model of state-financed posts.

The declaration of war in 1914 was supported by the majority of German scientists. There was only a handful of pacifists such as Albert Einstein and the cardiologist Friedrich Nicolai. The chemist Fritz Haber applied his expertise to the production of poison gases, and the physiologist Max Rubner set dietary standards.

The defeat of Germany and Austria-Hungary in World War I and the ensuing collapse of the German economy resulted in international isolation of German science until the mid-1920s, and a crisis of funding. The priority of national reconstruction inspired scientists to seek additional state resources through an Emergency Committee for German Science, which established the principle of peer-group review for research projects. The intractable political conflicts which threatened the stability of the Weimar Republic also found expression in scientific controversies. Nationalist priorities resulted in innovations in genetics becoming oriented to eugenic priorities. But the freer atmosphere of Weimar culture provided a stimulus for new ideas, such as the development of quantum mechanics.

The Nazi takeover in 1933 and the union with Austria in 1938 resulted in the the dismissal, and often enforced emigration or murder, of many

▲ The Kaiser Wilhelm Institute for Experimental Therapy was one of several such institutes founded in 1913 in Berlin-Dahlem. The first director was August von Wassermann, the pioneer of a diagnostic test for syphilis. There were departments for chemistry and bacteriology. The illustration shows the room for study of metabolism.

◄ The teaching of nutrition in a Berlin state school in 1908. Until concern with the declining birth rate prompted the introduction of domestic science, women had little access to science education.

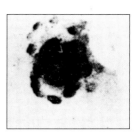

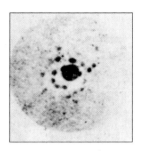

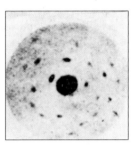

X-ray crystallography

Although so much had been learnt about the size and structure of atoms, the evidence was all circumstantial: no one had actually seen an atom. In theory this would be possible by using a sufficiently powerful microscope but in practice the resolving power of a microscope is limited by the wavelength of the light used, and small though this is, it is still vastly longer than the dimensions of atoms. An obvious alternative was to use "light" (radiation) of much shorter wavelength and a possibility was the X-rays discovered by Röntgen in 1895. Unfortunately, however, glass is opaque to them and there is thus no possibility of adapting an ordinary microscope to their use.

In 1912 the German physicist Max von Laue, showed that X-rays were scattered by a crystal of zinc sulfide. The scattered rays formed a characteristic pattern on a photographic plate, thus confirming that X-rays were a form of electromagnetic radiation. At Cambridge, England, W.H. Bragg, working with his son W.L. Bragg, perceived that this scattering provided a means of identifying the exact position of atoms within crystals of many kinds. The crystalline state is unique in that atoms and molecules are not arranged at random, as in liquids and most solids, but according to a strict geometric framework, which is reflected in the visible form of the crystal. Thus if a little table salt (sodium chloride) is examined under a lens it will be seen to consist of small cubes. This corresponds to an internal structure in which sodium and chlorine atoms are symmetrically located at the corners of cubes. The Braggs' X-ray diffraction method made it possible to determine the geometrical pattern of the crystal structure and to measure precisely the distance between atoms.

X-ray crystallography was to prove of particular value in organic chemistry, which concerns itself with the chemistry of compounds containing carbon. This element is unique in that it has the capacity to link atom to atom in long chains, sometimes – as in the use of polymers (a substance whose molecules consist of long chains of carbon atoms) – extending to many thousands of atoms. It can also form itself into rings, and can form links with other elements such as hydrogen, oxygen, nitrogen and sulfur. The overall result of this versatility is that carbon compounds are numbered literally in millions. In nature they comprise such basic substances as carbohydrates, fats, proteins, vitamins and hormones. In addition vast numbers have been synthesized – dyes, pharmaceuticals, man-made fibers, plastics and explosives – to mention only a few major classes.

In manipulating these complex molecules a first essential is to know how the constituent atoms are arranged in the molecule in relation to each other. The classical method, evolved in the 19th century by such brilliant chemists as the Germans Emil Fischer (1852–1919) and J.F.A. von Baeyer (1835–1917), was to break complex molecules down into simple identifiable ones and then confirm by synthesis the structure so deduced. X-ray crystallography eventually provided a means of short–circuiting this complex procedure.

▲ Early X-ray diffraction photographs of a copper sulfate crystal taken by W. Friedrich and P. Knipping, colleagues of Max von Laue in Munich, in 1912. The technique they developed made it possible to locate precisely the position of constituent atoms in a wide range of crystals. It opened up great new possibilities for determining the structure of organic compounds.

Jewish and socialist scientists. Although there was now much racial ideology in German science, many institutes and individuals continued to undertake innovative research.

After World War II Germans, now divided between West and East Germany, attempted to reestablish links with the international scientific community and to rebuild their scientific institutions. In West Germany the Kaiser Wilhelm Institutes were reorganized under the auspices of a newly founded Max Planck Society and renamed as Max Planck Institutes. They remain autonomous but depend on state finance. In certain fields Germans lagged behind. In the life sciences in East Germany, Soviet support for Lysenkoism inhibited the development of genetics. Yet it should also be noted that until the 1960s molecular biology was slow to develop in West Germany. Although East Germany imposed Soviet organizational models in higher education, its science-based industries, like the optical manufacturer Carl Zeiss Jena, continued to thrive. West Germany's performance as a science-based economy has been outstanding and its massive investment in science has produced a renaissance of scientific activity.

It was an unforgettable experience when late in the evening all alone in my laboratory in the Institute, I stood before the developing dish and saw the traces of the diffracted rays appear on the plate. Next day my first action was to go to Knipping and show him the plate. We hurried to Laue and to my Chief, where naturally there was the liveliest discussion of the picture.

W. FRIEDRICH, MUNICH, 1912

Toward absolute zero

That matter exists in three states – solid, liquid and gas – is a matter of everyday observation. Water is the classic example, familiar as ice, liquid water and steam. It can pass from solid to gas without liquefying in between, as when snowdrifts slowly disappear in warm sunshine even when the air temperature stays well below freezing point.

In the second half of the 19th century the conditions governing transitions from one state to another began to be systematically investigated, particularly by the Dutch school of H. Kamerlingh Onnes, who studied the properties of liquids and gases over a wide range of temperatures, but most particularly at extremely low temperatures. In 1894 he founded the famous Cryogenic Laboratory (from Greek *Kryos*, icy cold) at the University of Leiden. In 1911 he discovered a very remarkable and quite unsuspected effect at exceedingly low temperatures: metals then lose their resistance to the flow of an electric current, so that a current once started continues to flow indefinitely. This phenomenon he called superconductivity, and for its discovery he was awarded a Nobel prize in 1913. He thus pioneered a new branch of physics, known as cryogenics, which has thrown new light on the structure of matter and has had far-reaching technological applications. However, to understand the significance of this research it is necessary to say something about the temperature regions concerned, for they are far removed from those of ordinary human experience, even in the depths of a Siberian or Antarctic winter.

Heat is essentially a form of motion – the agitation of particles at the molecular level – and temperature measures the degree of agitation. As the temperature falls, this motion slows down and it follows that it will eventually cease altogether at some corresponding zero temperature. In 1851 William Thomson (Lord Kelvin) pointed out that thermodynamic constants indicated that this absolute zero corresponds to approximately −273°C. This led to the introduction of a new temperature scale – the Kelvin scale – which has since been much used in scientific work. This starts at absolute zero, designated 0°K.

At that time such a temperature was completely unattainable but nevertheless it provided a benchmark for those interested in low-temperature physics, especially those concerned with the liquefaction of gases. It was known that certain gases could be liquefied by subjecting them to high pressure, provided they were below what was called the critical temperature. Some gases, like ammonia, could be liquefied by compression at ordinary temperatures but others failed to do so however much they were cooled first. Such gases were known as permanent gases, but it was suspected that they would not liquefy only because their critical temperatures were very low. New techniques of intense refrigeration were called for if they were ever to be liquefied.

One device was to utilize the so-called Joule-Thomson effect: the gas was cooled, compressed and then allowed to escape through a small orifice, which has a further cooling effect. In everyday life this is observed, for example, when the gas in a soda-water cartridge is suddenly released: the empty cylinder is icy cold. Even at the turn of the century, however, there remained a residue of gases with very low critical temperatures yet to be liquefied. The most intractable was helium with a critical temperature of 5°K (−268°C). Onnes finally tamed this in 1908 and his pupil W.H. Keesom then succeeded in solidifying it.

At this time liquid gases of this kind were confined to the laboratory but the German engineer Carl von Linde, director of a refrigeration company in Paris, believed that they had industrial potential and in 1895 he devised a large-scale plant to manufacture liquid air. Thus was created, in the opening decade of this century, what was to become a vast new liquid gas industry which had no 19th-century predecessor. Liquid oxygen, distilled from liquid air, was widely used in the oxyacetylene burners used for welding and, on a smaller scale, in medicine as a breathing aid. Its greatest use, however, came after World War II, when oxygen began to be used in place of the air blast in the manufacture of steel.

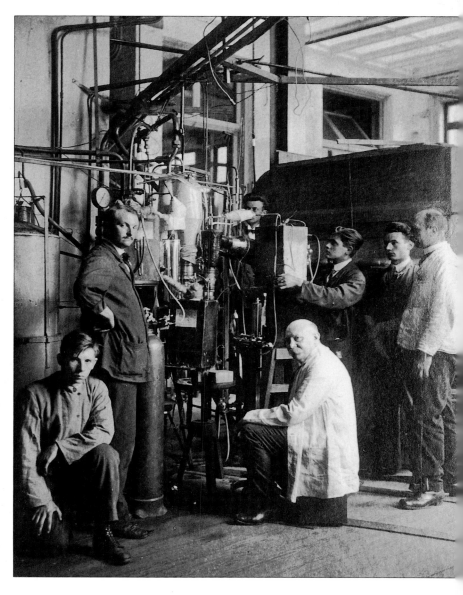

▼ H. Kamerlingh Onnes (seated) in the Cryogenic Laboratory he founded in Leiden, The Netherlands, in 1894. Here temperatures were reached that were far lower than any previously achieved and many gases, hitherto intractable, were liquefied. Helium was liquefied in 1908 and soon afterwards solidified. This research led to the discovery of a wholly new phenomenon: superconductivity. At very low temperatures the resistance of some metals falls to zero, so that an electric current once started flows indefinitely.

The Scandinavian Scientific Tradition

In 1900 there were only two independent countries in Scandinavia: Sweden and Denmark. (Norway was then united with Sweden, though it became a sovereign country in 1905; Finland belonged to the Russian Empire and only achieved independence after World War I.) Of the two, Sweden had the strongest scientific tradition and institutions. The Royal Swedish Academy of Sciences, founded in 1739, had remained vigorous in the 19th century. Under distinguished permanent secretaries it had maintained close contacts with scientists elsewhere in Europe and America. Within Sweden it was active in sponsoring research in numerous fields by maintaining facilities and promoting expeditions and surveys. Other developments had also stimulated scientific activity. Under the influence of developments in Germany, the university of Uppsala had reformed and expanded its science faculty from the 1870s onward, and in Stockholm an independent university-level science-oriented institution had been founded in 1878, the Stockholm Högskola. When the Swedish dynamite-manufacturer Alfred Nobel left money for the establishment of major prizes in science in 1896, Sweden was able to provide the institutions and scientists who could give effect to his plan and who were able to gain international preeminence for the prizes.

Sweden's scientific institutions remained largely unchanged until 1960, when Stockholm University was reorganized, the Academy's research institutions were transferred to it and the Högskola was given university status and became a state institution. The Academy continues to play a vital role by promoting contact between scientists at Sweden's universities. It is also in regular contact with scientists in other Scandinavian universities, eg Helsinki and Åbo in Finland, Oslo in Norway, and Copenhagen and Århus in Denmark.

One field in which Scandinavian scientists have shown a strong interest is earth sciences. The first man to reach the South Pole was the Norwegian explorer Roald Amundsen (14 December 1911). Much important research was done in the north polar region by his fellow countryman Fridtjof Nansen, who was professor of zoology and then oceanography at Christiana University (1896–1917). Polar expeditions have been mounted at regular intervals and permanent field stations established for studying biological conditions in the subarctic environment, such as the geophysical institute at Kiruna in northern Sweden. Such facilities have become even more important with rising concern about environmental issues. In addressing these, Scandinavian scientists have played an important role. The United Nations held its International Environmental Conference at Stockholm in 1972, which led to the establishment of the International Institute for Energy and Human Ecology in 1977.

Scandinavian countries have not only made important contributions to basic scientific research but also have a tradition of high-level technical education and, in certain fields, of notable technological development. When Sweden entered the 20th century it was well-endowed with technical schools and colleges: the military school for artillery and building techniques; the civil equivalent in the Royal Institute of Technology; and the Chalmers

PORTRÄTT AF SAMTLIGA PROFESSORER VID STOCKHOLMS HÖGSKOLA. HÖGSKOLANS NUVARANDE LOKAL, KUNGSGATAN 30, SAMT DEN PROJEKTERADE HÖGSKOLEBYGGNADEN I HÖRNET AF DROTTNING- OCH KUNGSTENSGATORNA.

Technical Institute in Gothenburg (founded in 1829). Many new institutes have since been founded such as the technical institutes at Helsinki in Finland and at Trondheim in Norway.

The exploitation of natural resources has stimulated important technological progress. In the early 20th century huge deposits of high-quality iron ore were discovered in northern Sweden. Exploitation of the deposits required the construction of a large railroad network (including the line to Narvik in Norway). It was decided to electrify the system, which in turn led to advanced hydro-electric schemes to generate electricity from the big rivers of the north. The resulting developments in dam-building and power transmission (through three-phase systems) have been exported worldwide.

Throughout the 20th century Scandinavia has produced a stream of Nobel prizewinners. Twenty-four awards have been received (Denmark 6 prizes, Finland 2, Norway 1 and Sweden 15). Early winners included the Swedish chemist Svante Arrhenius (1903) and the Danish physicist Niels Bohr (1922). In the 1980s awards were made to the Swedish chemist Kai Siegbahn and the physiologists Bengt Samuelsson and Sune Bergström.

▲ The Stockholm Högskola is a remarkable example of how private initiative made provision of high-level education in science in the late 19th century. It was an independent institution dependent on donations from wealthy individuals and foundations, open to both men and women. This celebratory picture (produced for the 25th anniversary in 1903) shows the Högskola's 10 professors and its rented and planned premises. Its students included many from other Scandinavian countries.

► The theory of continential drift was first formulated in a specific form by Alfred Wegener, here seen in his laboratory in 1912. Its basic assumption was that an original land mass (Pangaea) fragmented and over aeons of time drifted apart to form the present continents.

▼► Wegener's maps (right) show how the continents were disposed in the Carboniferous (top) Eocene and Quaternary periods (approximately 300, 45 and 2 million years ago). Evidence of the instability of the earth's crust is afforded by earthquakes. The disastrous San Francisco earthquake of 1906 (below) arose from the city's location on the San Andreas fault, as shown in one of Wegener's publications (inset).

Continental drift

From prehistoric times miners in their search for metallic ores necessarily acquired a great deal of empirical knowledge about the structure of the earth's crust: how different rocks lay in strata one above the other; how these strata were often distorted and dislocated by faults; how veins of mineral ore might strike upwards through strata; and so on. But the acknowledged founder of geology as a science was James Hutton, who worked in Scotland in the second half of the 18th century. His ideas were developed in the 19th century by pioneers such as the British geologists Charles Lyell and Archibald Geikie. Their research conflicted with long-accepted beliefs about the age of the earth and the forces that had shaped it. The prevailing opinion was that geological history could be interpreted only in terms of a series of catastrophes, like Noah's great flood. But the new geologists preached a doc-

trine known as "uniformitarianism" according to which the history of the earth's crust is to be explained simply in terms of ordinary forces acting continuously over immense periods of time.

If only for reasons of difficulty of travel, early geologists tended to restrict their studies to relatively small accessible areas, but there were some who were ready to think on a global basis. From around 1600, as maps of the world became more reliable, geographers had noted that the west coast of Africa would fit the east coast of America like two pieces of a jigsaw puzzle. This suggested, in a very general sort of way, that at some distant date the two Atlantic continents had been joined and had since drifted apart. This hypothesis was put forward more specifically by the French scientist A. Snider-Pellegrini in 1858 and just 50 years later H.B. Baker suggested that 200 million years ago all the continents had been located about Antarctica but had then broken

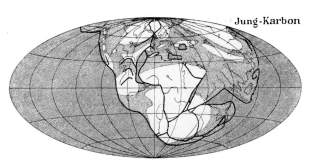

Jung-Karbon

Eozän

Alt-Quartär

away: F.B. Taylor, an American geologist with a special interest in the Great Lakes, independently put forward similar views in 1910.

Thus by the first decade of this century the idea that, over vast periods of time, even continents might be mobile was not wholly novel. The man most closely identified with the theory of continental drift – or continental displacement as he first called it – was Alfred Wegener, a German meteorologist.

When he first considered the theory he was inclined to discount it, but his interest was revised by paleontological evidence that in the distant past some kind of land bridge must have joined Africa and Brazil, much as Britain was linked to the Continent 20,000 years ago via the English Channel or Asia to North America via the Bering Strait. But these had been relatively short land bridges, and in the case of the wide Atlantic the theory of continental drift appealed to him more strongly, and from 1912 he devoted himself to developing it.

He postulated that originally there was one supercontinent, Pangaea, which began to break up in the Permian Age, over 200 million years ago. America then moved westward from the European/Asian land mass, leaving a vast gap to be filled by the Atlantic. Australia moved away northward and India drifted eastward from Africa. Later, in the Quaternary (2 million years ago), Greenland separated from Norway. Some of the major island areas such as Japan and the Philippines were identified as fragments left over from these colossal separations. It all added up to a very plausible explanation of why the world's land masses are distributed as they are, but some sort of underlying mechanism had to be found. For this he supposed that the land masses floated on some sort of plastic magma, such as that spread out from great depths in volcanic eruptions: the constant rotation of the earth would impose a westward drift.

Wegener pursued two further lines of argument. As a meteorologist, he was interested in the long-term history of climate, and he found evidence that climatic changes were in accordance with his ideas. His second line was less satisfactory. Arguing that if continental drift had indeed occurred there was no reason to suppose that it had now ceased. Accordingly, he tried to find evidence for it through precise determinations, over intervals of time, of the distances between points on different continents, using both precise astronomical methods and the duration of radio transmissions. His results were negative but this could be taken as meaning no more than that the rate of drift was too slow to be detected by the relatively crude methods there were then available to him.

Indeed, this is not surprising if the drift between America and Africa is thought of as proceeding regularly since Permian times: the average speed would then be no more than 1m (3.3ft) in 30 years. In the late 20th century, however, the use of laser beams and satellites made it possible to measure the rate of drift with remarkable precision, which supported Wegener's theory.

Mohorovičić and the structure of the earth

On 8 October 1909 a severe earthquake occurred 40km (37mi) south of Zagreb in Croatia (then part of Austria-Hungary). An earlier earthquake in Zagreb in 1880 had led to the installation of a seismograph in the Meteorological Observatory, directed by Andrija Mohorovičić. In this capacity he received from stations all over Europe records of the 1909 earthquake. When he came to examine these in detail he made an interesting discovery. The record showed, as he expected, two types of wave: compressional (P) waves in which the particles oscillate along the line of propagation and distortional (S) waves in which the motion is at right angles to the line of propagation. He then perceived that there were in fact two P waves. At a short distance from the epicenter the first wave to arrive travels at a speed within the range 5.5–6.5km (3.4–4mi) per second. At a distance of around 170km (105mi) from the epicenter this is overtaken by a second wave travelling at 8.1km/s (5mi/s). Beyond this, up to 800km (500mi), both waves can be detected but then the slower wave fades away. Mohorovičić interpreted this as meaning that the slower wave traveled direct to the seismograph and that the faster one had been refracted at a depth of around 50km (31mi). In his honor this refractive layer was called the Mohorovičić discontinuity, or Moho. Later research showed that the depth of the Moho – the boundary between the earth's crust and the upper mantle – varies from 30 to 50km (19–31mi).

THE THEORY OF RELATIVITY

The special theory of relativity, published by Albert Einstein in 1905, was based on two assumptions, or postulates: the relativity principle, and the constancy of the speed of light. According to the long-established relativity principle, experiments are completely unaffected by a laboratory's motion, provided it continues to move in a straight line at a constant speed; therefore, no experiment carried out inside a closed laboratory can reveal the speed at which the laboratory is moving. All motions are relative, none are "absolute".

The second postulate, that the measured speed of light is always the same (300,000km per second) regardless of the relative velocity of source and observer, seems to be absurd, but was confirmed by a crucial experiment conducted by the Americans Albert Michelson and Edward Morley in 1887. If two vehicles, each traveling at 100km per hour, have a head-on collision, the speed of the impact will be 200km per hour; but if an observer is approaching a light source at 150,000km per second, he will measure the speed of that light to be 300,000km per second, not 450,000km per second as common sense would suggest.

Einstein showed that among the startling consequences of accepting these postulates are length contraction, mass increase and time dilation. As the speed of a body approaches the speed of light, its length (along the direction of motion) decreases, its mass increases, and time measured on its clock passes more slowly than time measured on a "stationary" observer's clock. All of these predictions have since been confirmed to high precision, notably by experiments carried out in particle accelerators which can accelerate subatomic particles to large fractions of the speed of light.

Einstein showed that the change in mass of a moving body was a measure of the energy imparted to it and so argued that all mass was equivalent to energy. Energy (E) is equivalent to mass (m) multiplied by the speed of light (c) squared: $E = mc^2$. The principle that a tiny amount of mass can be converted into a huge amount of energy is the basis of nuclear power, nuclear weapons and the nuclear reactions which power the Sun and stars.

In 1915 Einstein extended the theory to encompass accelerated observers and laboratories. This "general" theory of relativity treats gravity not as a force which acts directly across empty space between bodies (the "classical" or Newtonian view), but as an apparent force which arises because space is distorted, or curved, in the presence of massive bodies. Therefore, bodies, particles and even rays of light follow curved paths near massive bodies. The theory also predicts that clocks run slow where gravity is strong and that light becomes redshifted (stretched in wavelength) when moving up through a gravitational field. These effects have been confirmed by observation and experiment.

▲ The children think they are at rest while the train rushes by, but observers on the train could equally well regard themselves as being at rest and the children as moving. The first postulate of relativity requires that the laws of physics be exactly the same inside the train as in the field. No experiments carried out inside the train can reveal its speed.

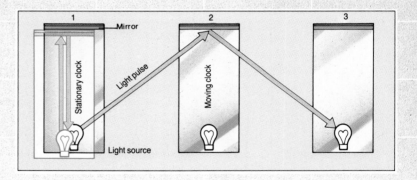

▲ The logic of relativity gives rise to phenomena contrary to common sense, such as time dilation. For a stationary clock consisting of a light source and a mirror, the period of the clock is the time taken for a light pulse to travel from the source to the mirror (1) and back. From the viewpoint of a stationary observer a light pulse in a moving clock has to travel further to reach the mirror (2) and return to its source (3). Since the speed of light is constant, the stationary observer will conclude that the pulse period of the moving clock is longer and that the moving clock is running slow.

1.0
0.8
0.6
0.4
0.2

0   0.1   0.2   0.3   0.4   0.5

Proportion of speed of light

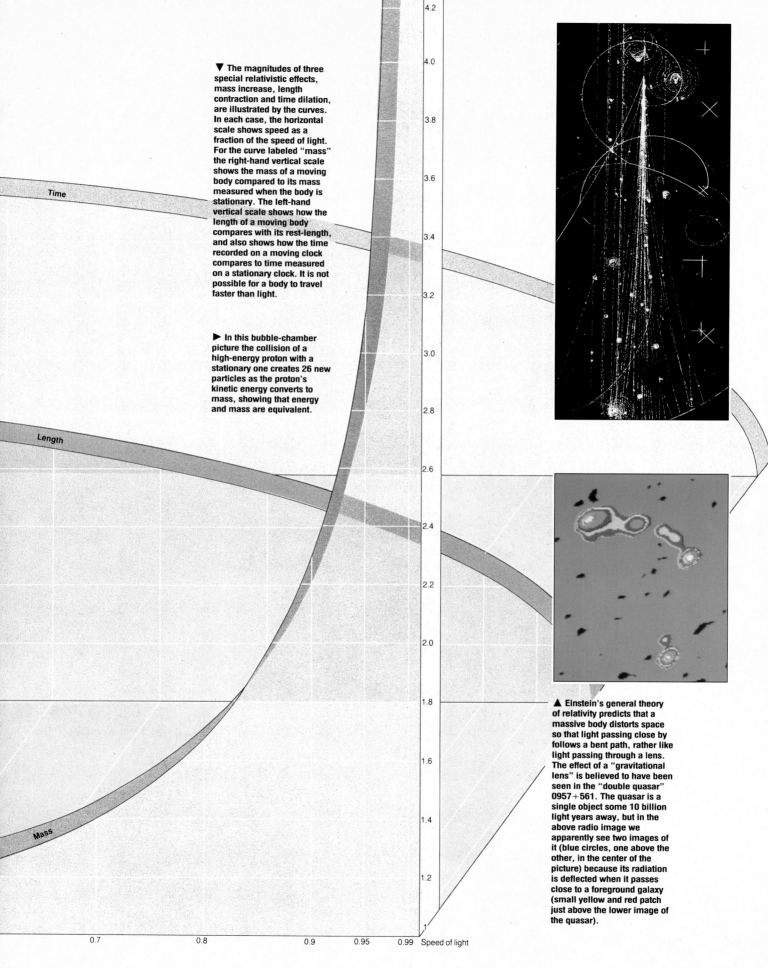

▼ The magnitudes of three special relativistic effects, mass increase, length contraction and time dilation, are illustrated by the curves. In each case, the horizontal scale shows speed as a fraction of the speed of light. For the curve labeled "mass" the right-hand vertical scale shows the mass of a moving body compared to its mass measured when the body is stationary. The left-hand vertical scale shows how the length of a moving body compares with its rest-length, and also shows how the time recorded on a moving clock compares to time measured on a stationary clock. It is not possible for a body to travel faster than light.

▶ In this bubble-chamber picture the collision of a high-energy proton with a stationary one creates 26 new particles as the proton's kinetic energy converts to mass, showing that energy and mass are equivalent.

Time

Length

Mass

4.2
4.0
3.8
3.6
3.4
3.2
3.0
2.8
2.6
2.4
2.2
2.0
1.8
1.6
1.4
1.2
1

0.7   0.8   0.9   0.95   0.99   Speed of light

▲ Einstein's general theory of relativity predicts that a massive body distorts space so that light passing close by follows a bent path, rather like light passing through a lens. The effect of a "gravitational lens" is believed to have been seen in the "double quasar" 0957+561. The quasar is a single object some 10 billion light years away, but in the above radio image we apparently see two images of it (blue circles, one above the other, in the center of the picture) because its radiation is deflected when it passes close to a foreground galaxy (small yellow and red patch just above the lower image of the quasar).

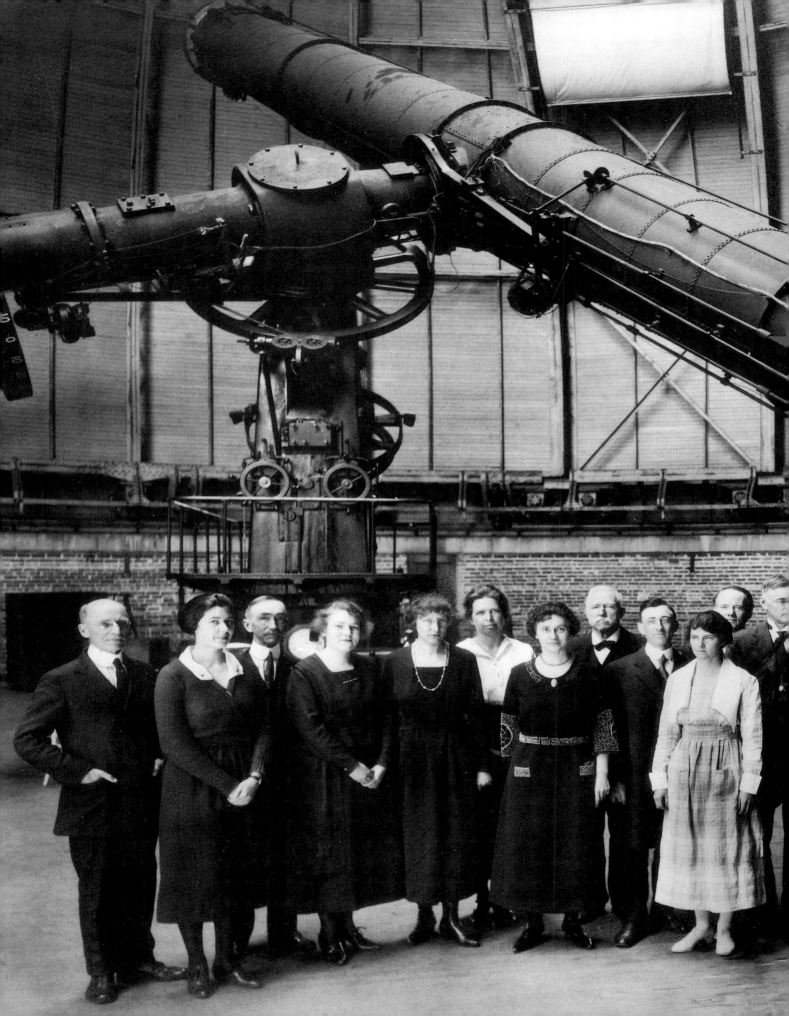

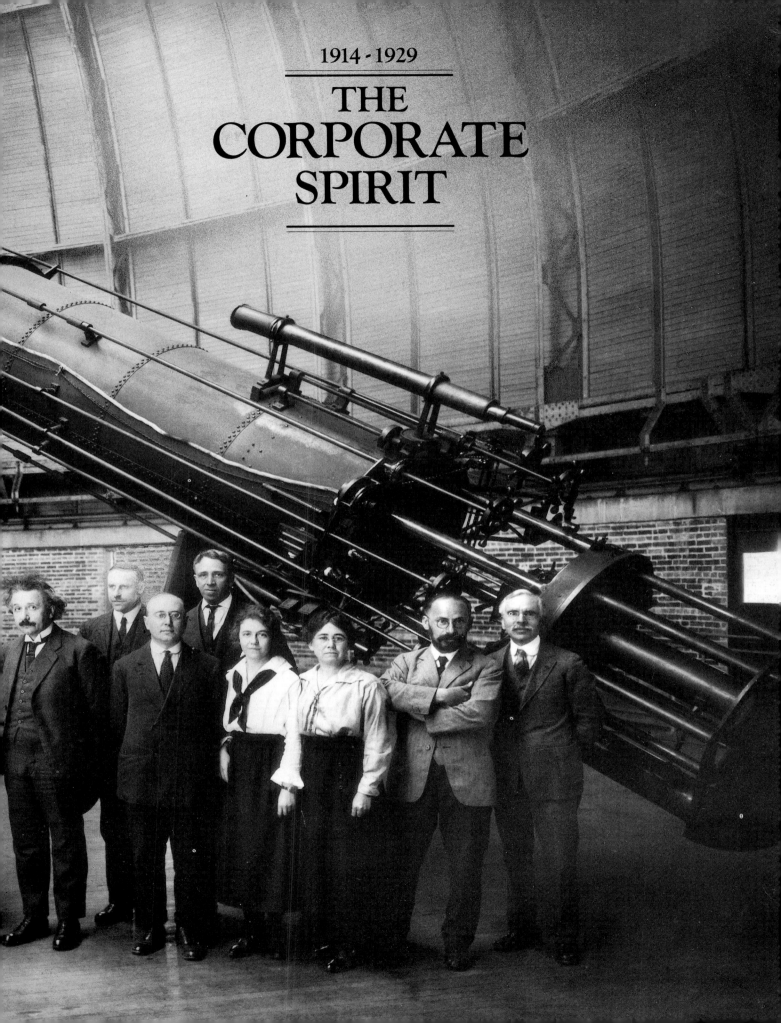

1914 · 1929
THE CORPORATE SPIRIT

Time Chart

	1915	1916	1917	1918	1919	1920	1921	1922
Nobel Prizes	• *Chem*: R. Willstätter (Ger) • *Phys*: W.H. Bragg, W.L. Bragg (UK) • *Med*: (no award)	(no awards)	• *Chem*: (no award) • *Phys*: C.G. Barkla (UK) • *Med*: (no award)	• *Chem*: F. Haber (Ger) • *Phys*: M. Planck (Ger) • *Med*: (no award)	• *Chem*: (no award) • *Phys*: J. Stark (Ger) • *Med*: J. Bordet (Bel)	• *Chem*: W. Nernst (Ger) • *Phys*: C.E. Guillaume (Swi) • *Med*: S.A.S. Krogh (Den)	• *Chem*: F. Soddy (UK) • *Phys*: A. Einstein (Ger/Swi) • *Med*: (no award)	• *Chem*: F.W. Aston (UK) • *Phys*: N.H.D. Bohr (Den) • *Med*: A.V. Hill (UK), O. Meyerhof (Ger)
Technology	• First transcontinental telephone call made between New York and San Francisco (USA) • Invention of pyrex (USA) • H. Junkers makes the first all-metal airplane (Ger)	• First tanks used in battle (UK) • P. Langevin constructs an underwater ultrasonic source for submarine detection (Fr) • Automobile windshield wipers introduced (USA) • Dodge Company produces the first all-steel bodywork for automobiles (USA)	• 100-inch reflecting telescope installed at Mount Wilson (USA) • C. Birdseye develops freezing as a method for preserving food (USA)	• First three-color traffic lights installed in New York • Radio crystal oscillator is introduced • First drill to have diamond cutting edges introduced to Shell Oil Company by van der Gracht	• First experiments with shortwave radio • The super-heterodyne radio, developed by E.H. Armstrong, the first radio allowing uniform reception of a wide range of stations, goes into mass production (USA)	• Handley Page Transport airplanes are fitted with radio direction finders (UK) • Regular public radio broadcasts begin in the UK and USA • J.T. Thompson invents the sub-machine gun (USA)	• First medium-wave wireless broadcast (USA) • Opening of the AVUS autobahn in Berlin (Ger) • W.G. Cady discovers the stabilizing quality of the quartz crystal in radio signal reception (USA)	• Development by R.A. Fessenden of the echo sounder for measuring submarine depths
Medicine	• E.C. Kendall isolates the dysentery bacillus • J. Goldberger establishes that vitamin deficiency causes pellagra (USA) • First carcinogen identified by K. Yamagiwa and K. Ichikawa (Jap)	• Treatment of war casualties leads to the development of plastic surgery • Sympathectomy for the relief of angina pectoris performed for the first time by Ionescu (Rom)	• J. Wagner von Jauregg treats syphilitic paralysis by injecting malaria (Aut) • Natural anti-coagulant heparin is discovered	• Influenza pandemic kills 15 million people in Europe (until 1919) • British Army uses mixed antitoxin to combat tetanus and the gas gangrene bacilli		• Soviet Russia is the first country to legalize abortion • K. Spiro and A. Stoll extract ergotamine for migraine treatment	• Publication of Jung's *Psychological Types* (Swi) • Ear specialists use a microscope in ear operations • A. Calmette and C. Guérin develop the B-C-G tuberculosis vaccine (Fr)	• Insulin isolated by F. Banting and C. Best (Can) • Insulin is first administered to diabetic patients • H. Chick shows how rickets is curable by cod liver oil or sunlight (Vitamin D) (Aut)
Biology	• F. Twort discovers bacteriophages (UK) • Publication of A. Thorburn's *British Birds* (UK) • M. and W. Lewis use time-lapse cinephotography to study cell growth		• Rearrangement of chromosomes during meiosis known as "crossing over" is demonstrated • G. Lusk and R.J. Anderson discover the importance of calorie consumption in producing energy	• Publication of J.M. Coulter's *Plant Genetics*	• K. von Frisch discovers that bees communicate through bodily actions • Publication of T.H. Morgan's *The Physical Basis of Heredity* (USA)		• Chromosome theory of heredity postulated by T.H. Morgan (USA) • O. Loewi proposes a chemical mechanism for the transmission of nerve impulses (USA) • E.M. East and G.M. Shull perfect a hybrid corn (maize) strain that will greatly improve crop yields	• T.H. Morgan experiments with the heredity mechanisms of fruit flies (USA) • First use in children of the tuberculosis vaccine (Fr)
Physics	• Publication of W. and L. Bragg's *X-rays and Crystal Structure*, describing their use of X-rays to determine the structure of crystals (UK) • Einstein completes his general theory of relativity	• A. Sommerfeld amends Bohr's model of the atom, proposing elliptical orbit for the electron (Ger) • R.A. Millikan confirms Planck's constant using the photoelectric effect (USA)		• E. Noether demonstrates that every symmetry in physics implies a conservation law and vice-versa • Publication of A. Eddington's *Gravitation and the Principle of Relativity* (UK)	• E. Rutherford transmutes a nitrogen atom into oxygen by bombarding it with alpha particles, and demonstrates that protons exist in the nucleus (UK) • A. Eddington describes bending of light by the Sun, after observations of a total eclipse on 29 May, as predicted by Einstein's general relativity theory (UK)	• F.W. Aston discovers that all atomic masses are integral multiples of the same number (UK) • Publication of A. Eddington's *Space, Time and Gravitation* (UK)	• J.N. Brönsted and G. von Hevesy successfully separate isotopes • E. Rutherford and J. Chadwick disintegrate most of the elements as a preliminary to splitting the atom (UK)	• Publication of N. Bohr's theory that electrons travel in concentric orbits around the atomic nucleus (Den)
Chemistry		• G.N. Lewis explains chemical bonding and valence of chemical elements by his theory of shared electrons, and shows that the number of electrons in compounds is nearly always even (USA)			• F.W. Aston builds mass-spectrograph and confirms the phenomena of isotopy (UK)	• H. Staudinger demonstrates that small molecules polymerize to form plastics • Raschig-process uses hydrogen chloride to chlorinate benzene (Ger)	• T. Midgeley synthesizes tetraethyl lead, adding it to gasoline to eliminate engine "knocking" (USA)	• P.M.S. Blackett conducts experiments in transmutation • J. Heydrowsky introduces electro-chemical analysis (polarography) (Czech)
Other	• W.S. Adams discovers the first white dwarf star, Sirius B • Publication of A. Wegener's theory of continental drift (Ger)	• E.E. Barnard discovers what is later called Barnard's runaway star, with the largest proper motion of any known star	• K. Schwarzschild develops the equations that predict the existence of black holes from Einstein's relativity equations (Ger)	• H. Shapley discovers the dimensions of the Milky Way (USA)	• J. Bjerknes discovers that cyclones originate as waves in the sloping frontal surfaces separating different air masses (Nor)	• A.A. Michelson first measures the diameter of a distant star, Betelgeuse (USA) • M. Wolf shows the structure of the Milky Way (Ger)	• E. Dacqué initiates phylogenetically-oriented paleontology	• W.W. Coblentz obtains accurate measurements of the relative thermal intensities of star images

1923	1924	1925	1926	1927	1928	1929
• *Chem*: F. Pregl (Aut) • *Phys*: R.A. Millikan (USA) • *Med*: F.G. Banting, J.J.R. Macleod (Can)	• *Chem*: (no award) • *Phys*: K.M.G. Siegbahn (Swe) • *Med*: W. Einthoven (Neth)	• *Chem*: R. Zsigmondy (Ger) • *Phys*: J. Franck, G. Hertz (Ger) • *Med*: (no award)	• *Chem*: T. Svedberg (Swe) • *Phys*: J.B. Perrin (Fr) • *Med*: J. Fibiger (Den)	• *Chem*: H. Wieland (Ger) • *Phys*: A.H. Compton (USA), C.T.R. Wilson (UK) • *Med*: J. Wagner von Jauregg (Aut)	• *Chem*: A. Windaus (Ger) • *Phys*: O.W. Richardson (UK) • *Med*: C.J.H. Nicolle (Fr)	• *Chem*: A. Harden (UK), H. von Euler-Chelpin (Swe) • *Phys*: L.V. de Broglie (Fr) • *Med*: C. Eijkman (Neth), F.G. Hopkins (UK)
• Bulldozer introduced by LaPlante-Choate Company (USA) • J. de la Cierva develops the basic principle of the autogiro (Sp) • V. Zworykin invents the iconoscope, an early form of the cathode-ray (TV) tube (USA)	• Publication of H. Oberth's *The Rocket into Interplanetary Space*, the first book to contain the notion of escape velocity • S.G. Brown Limited bring out their crystavox loudspeakers, improving sound quality in crystal radio sets	• First Leica camera built by O. Barnack (Ger) • V. Bush develops the first analog computer to solve differential equations • First public demonstration of the photoelectric cell (USA)	• 26 Jan: J.L. Baird first demonstrates television (UK) • "Electrola", a new recording technique, is developed (USA) • Liquid-fuel rocket launched by R.H. Goddard at Auburn, Massuchusetts (USA)	• S. Junghans invents process for continuous casting of non-ferrous metal (Ger) • Introduction of the pentode vacuum tube • H.S. Block introduces negative feedback in audio amplifiers to reduce distortion • May: C. Lindbergh makes the first nonstop transatlantic flight from New York to Paris	• H. Geiger and W. Müller construct the "Geiger counter" (Ger) • Quartz clock invented by J.W. Horton and W.A. Morrison (USA) • J.L. Baird gives first transatlantic television transmission and demonstrates color television (UK) • First working robot built by Rickards and A.H. Refell (UK)	• Construction begins on Empire State Building (until 1931) (USA) • Kodak develop a 16mm color film • R. Goddard launches the first instrumental rocket, containing a barometer, a thermometer and a small camera (USA) • *Graf Zeppelin* flies around the world (Ger)
• H. Souttar pioneers cardiac surgery by attempting to widen a constricted mitral valve (UK) • First birth control clinic opens in New York • G. Ramon develops a new tetanus vaccine (Fr)	• G. and G. Dick isolate the scarlet fever streptococcus (USA) • Acetylene used as an anesthetic • German scientists synthesize plasmochin, a quinine substitute, for the treatment of malaria	• J. Collip obtains extract of the parathyroid gland for treating tetanus • L. Lazzarini experiments in bone transplants using rabbits (It)	• Liver extract first used by W.P. Murphy and G. Minot for treating pernicious anemia • H. Gardiner-Hill and J.F. Smith use pituitary extract to treat lack of growth, lack of menstruation and sudden weight gain (UK)	• G. Ramon and C. Zoellar are first to immunize human beings against tetanus (Fr)	• A. Fleming discovers penicillin (UK) • Iron lung invented by P. Drinker (USA) • G. Papanicolau develops a technique for recognizing malignancy in cells taken from the vaginal wall, hence diagnosing uterine cancer	• H. Berger publishes his paper on human electro-encephalography based on experiments conducted since 1924 using primitive amplifiers (Ger) • S. Levine is first to make the connection between high blood pressure and fatal heart disease (USA)
• O.H. Warburg develops a method for studying respiration in thin slices of tissue (Ger)	• D. Keilin discovers cytochrome, an important cell respiratory enzyme (UK) • H. Steenbock discovers that ultraviolet light increases the vitamin D content of food	• Theory of gene centers postulated by N.I. Vavilov (USSR) • First successful experiments in hydroponics – the soilless cultivation of plants fed only by nutrient-rich water (USA)	• Publication of T.H. Morgan's *The Theory of the Gene* (USA) • H.J. Muller discovers that X-rays induce genetic mutations (USA) • J.B. Sumner observes urease, an enzyme essential to the nitrogen cycle	• Publication of I.P. Pavlov's *Conditioned Reflexes* • Publication of T.H. Morgan's *Experimental Embryology* (USA) • K. Landsteiner identifies blood groups M and N (Aut)	• F. Griffith discovers that genetic information is transmitted chemically • A. Szent-Györgyi discovers vitamin C (Hung/USA) • F. Boas's *Anthropology and Modern Life* refutes the Fascist theory of the "master race"	• Adrian and Matthews, using an ultrasensitive galvanometer, are able to follow a single impulse in a single nerve fiber • E. Doisy (USA) and A. Butenandt (Ger) independently isolate the sex hormone estrone
• A.H. Compton discovers that X-rays change in wavelength when scattered by matter (USA) • Publication of A. Eddington's *Mathematical Theory of Relativity* (UK)	• W. Pauli introduces his exclusion principle, which helps to explain atomic structure statistically (Aut) • Publication of L.V. de Broglie's study on the wave theory of matter (Fr) • E.V. Appleton demonstrates that radio waves of a sufficiently short wavelength will penetrate the Heaviside layer (UK)	• W.K. Heisenberg and N. Bohr develop quantum mechanics (Ger) • P.M.S. Blackett develops technique for photographing nuclear reactions (UK) • E.V. Appleton measures the height of the ionosphere (UK)	• E.P. Wigner introduces group theory into quantum mechanics (Hung) • J.D. Bernal develops the "Bernal chart", for deducing the structure of crystals from photographs of X-ray diffraction patterns • E. Schrödinger elucidates his Schrödinger wave equation (Aut)	• W.K. Heisenberg propounds his "uncertainty principle" in quantum physics • N. Bohr states the notion of complimentarity (Ger)	• C.V. Raman discovers the "Raman effect": a change in wavelength of light that is scattered by molecules (Ind) • E.H. Land develops a polarizing filter (USA) • P.A.M. Dirac develops the relativistically invariant equation of the electron (UK)	• W. Heisenberg and W. Pauli give their formulation of quantum field theory (Ger) • I. and P. Joliot-Curie observe phenomena later identified by J. Chadwick as caused by the neutron (Fr) • G. Gamov proposes the "liquid-drop" model of the nucleus (UK) • Publication of Einstein's *Unitary Field Theory* (Ger)
• P.J.W. Debye extends the Arrhenius theory of ionization of salt in solution to the crystalline solid state (Neth) • Theory of acids and bases postulated by J.N. Brönsted		• C. Bosch invents process for preparing hydrogen on a manufacturing scale (Ger) • The Fischer–Tropsch synthesis leads to industrial development of synthetic oil (Ger)	• H. Staudinger begins work on his polymer theory of plastics (USA) • ICI begins production of compound fertilizers, containing nitrogen, phosphorous and potassium (UK)	• W. Heitler and F. London provide bases for explaining the covalent bond using quantum mechanics (Ger)	• F.A. Pareth founds radio chemistry (Ind) • P.H. Diels and K. Adler develop a technique for combining atoms into molecules, useful in forming compounds such as synthetic rubber and plastics (Ger)	• P. Levene discovers acids later known as DNA • Foam rubber developed at Dunlop research laboratories by E. Murphy and W. Chapman (UK)
• F. Lindemann investigates the size of meteors and the temperature of the upper atmosphere • L.A. Bauer analyzes the Earth's magnetic field	• A. Eddington discovers that the luminosity of a star is approximately a function of its mass (UK)	• Discovery of the Mid-Atlantic Ridge (Ger) • R.A. Millikan discovers the presence of cosmic rays in the upper atmosphere (USA)	• A.A. Michelson measures the speed of light (USA)	• Lemaître initiates the theory of the expanding universe later developed as the "Big Bang" theory (Bel)	• H.N. Russell determines the abundance of elements in the solar atmosphere by studying the solar spectrum	• E.P. Hubble measures large red shifts in the spectra of extragalactic nebulae and shows that the speed of a galaxy's movement away is proportional to its distance from Earth (UK)

Datafile

In response to the military requirements of World War I, rapid progress was made in many technological areas, building on earlier inventions and developments. Wireless telegraphy was essential in a war where the armies were spread out over a continent. Fighting machines were developed for land, sea and air, which made World War I (1914–18) the first mechanized war. The tank, the submarine and the aeroplane were produced on a huge scale. The British and French produced 100,000 planes and 6,500 tanks during the war years and would have produced 10,000 Mark V* tanks had the war continued into 1919. The German navy built 335 submarines in the period 1915–18; a rate of one every four days.

After the war great effort was put into the technology of entertainment. By the end of the 1920s talking pictures had swept the cinemas of the United States and television had been successfully demonstrated on both sides of the Atlantic.

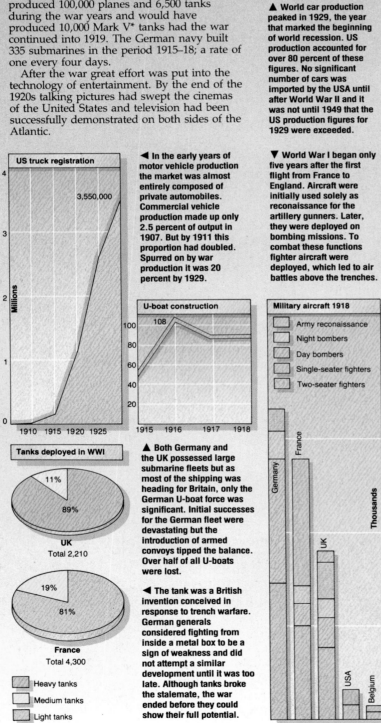

World automobile production

▲ World car production peaked in 1929, the year that marked the beginning of world recession. US production accounted for over 80 percent of these figures. No significant number of cars was imported by the USA until after World War II and it was not until 1949 that the US production figures for 1929 were exceeded.

◄ In the early years of motor vehicle production the market was almost entirely composed of private automobiles. Commercial vehicle production made up only 2.5 percent of output in 1907. But by 1911 this proportion had doubled. Spurred on by war production it was 20 percent by 1929.

▼ World War I began only five years after the first flight from France to England. Aircraft were initially used solely as reconaissance for the artillery gunners. Later, they were deployed on bombing missions. To combat these functions fighter aircraft were deployed, which led to air battles above the trenches.

US truck registration

U-boat construction

Military aircraft 1918
- Army reconaissance
- Night bombers
- Day bombers
- Single-seater fighters
- Two-seater fighters

Tanks deployed in WWI

UK
Total 2,210

France
Total 4,300

- Heavy tanks
- Medium tanks
- Light tanks

▲ Both Germany and the UK possessed large submarine fleets but as most of the shipping was heading for Britain, only the German U-boat force was significant. Initial successes for the German fleet were devastating but the introduction of armed convoys tipped the balance. Over half of all U-boats were lost.

◄ The tank was a British invention conceived in response to trench warfare. German generals considered fighting from inside a metal box to be a sign of weakness and did not attempt a similar development until it was too late. Although tanks broke the stalemate, the war ended before they could show their full potential.

In the history of science and technology World War I (1914–18) proved to be very much a turning point, for it very soon changed the course of events in these fields. With the failure of either side to obtain a quick victory in 1914, the major powers involved prepared themselves for a long struggle. The potential importance of science and technology to the war effort was recognized, but the emphasis was placed on those sectors which either serviced the armed forces directly or assisted in maintaining the domestic economy. The general effect was that progress was made, but only on a relatively narrow front: research regarded as non-essential withered both through lack of funds and because so many scientists and technologists joined the services. Later, many specialists were recalled when it was discovered that they were needed for more essential work at home. The shadow fell also over peripheral countries, such as Switzerland and Sweden, who were not directly involved. Across the Atlantic, the USA stood aloof, having formally declared neutrality, which it was to preserve until 1917.

Science and technology in the 1920s

If we are to look for the greatest impact of science and technology on society in general we shall surely find it in the world of entertainment. The 1920s saw the burgeoning of public broadcasting on both sides of the Atlantic and the rapid commercialization of radio receivers: the amateur no longer had to build his own set but could buy one ready made, increasingly sophisticated, complete with loudspeaker in place of headphones. Cinematography, too, made immense strides, with the advent of both talkies and color.

Until after World War II radio and the cinema dominated the popular entertainment world, but a rival which was to outrival both was being conceived in the 1920s. In 1926 the British inventor John Logie Baird, amidst much publicity, gave demonstrations of television in London, using a photomechanical system. In September 1929 he broadcast the first BBC television program daily. But Baird's system, like other photomechanical ones concurrently developed in the USA, was crude and not in the main line of evolution. The future lay with an all-electronic system, and the seeds of this were sown with the invention of Vladimir Zworykin's iconoscope camera in 1923.

The automobile, too, may be regarded as in part a form of entertainment, for as well as being an increasingly popular mode of transport, much pleasure motoring was done. In Britain Herbert Austin launched his tiny but immensely popular Austin 7, the antithesis of the big automobiles of the trans-Atlantic world. The measure of progress is that in 1927 the USA alone produced 3.5 million automobiles.

WAR, PICTURES AND PLASTICS

- Science in war and peace
- Aeroplanes, submarines and tanks
- From silent films to talkies
- Rival systems of television
- International cooperation in electricity generation
- Leo Baekeland and plastics

The rapid growth of the automobile industry had far-reaching repercussions. The steel industry developed continuous strip-mills to produce the vast amount of sheets necessary for bodywork and the petroleum industry expanded to keep pace with the demand for fuel. The search for petroleum became wider and, literally, deeper: in 1927 a well in the USA reached a then record depth of 2,438m (8,000ft). In 1921 tetraethyl lead, a valuable antiknock agent, was hailed as a valuable means of economizing on gasoline. Half a century later, though, it was to be condemned on the grounds of atmospheric pollution.

By contrast, civil aviation made rather slow progress, mainly due to the lack of suitable aircraft; surplus military aeroplanes did not adapt well and it took time to design and build more suitable ones. But some events gave a glimpse of things to come. In 1919 John Alcock and Arthur Whitten Brown made the first trans-Atlantic flight

▼ An American 14in gun in use during the drive on the Argonne sector, France, October 1918. Mounted on a railroad wagon, it could be moved forward to keep pace with the advance.

and in the same year the brothers Ross and Keith Smith flew from London to Australia – though the journey took 135 hours. In 1924 a US Air Force team circled the globe: a year earlier repeated mid-air refueling had kept an aeroplane aloft for 37.5 hours. In 1923, in Spain, Juan de la Cierva was developing the principles of the autogyro, short-lived but the forerunner of the helicopter.

In the main, in the 1920s the great powers were licking their wounds and trying to restore their shattered economies. But even then there were rumbles of possible future conflict. In 1928 the French began to construct the Maginot line, a supposedly impregnable line of fortifications along the Franco-German border, which the Germans were to turn so easily in 1940. In Czechoslovakia, the weapon of the infantry man was being redesigned, with the development of the Z726 light machine gun, forerunner of the famous Bren gun used in World War II and later.

Science and technology in World War I

There is some truth in the contention that new wars are initially fought with the weapons of the previous one. While peace may see steady developments in weapons and strategy, the life-or-death situation of actual warfare is inevitably a powerful incentive to innovation. So it was with World War I. On land the infantryman with his rifle, backed by artillery, was the backbone of the world's armies. At sea the heavily armed and armored battleship was the central ship used by the navies.

The weapon that more than any other determined the characteristics of World War I was the machine gun, a transportable infantry weapon that fired sustained rounds of bullets. It had been invented as long ago as the mid-19th century and had been perfected in the 1880s by the London-based American inventor Hiram Maxim. He devised a machine gun that used gas produced by the recoil after firing to prepare the gun automatically for firing the next bullet. By 1914 all the major European armies possessed machine guns, but only the German army was equipped with them in large numbers. Nobody anticipated the potential of the machine gun.

After the initial German invasion of France had been halted in the Battle of the Marne (5–10 September 1914), the German army did not withdraw but dug trenches behind the River Aisne and defended them with machine guns. A deadlock had been created and the next four years were to be spent trying to break it.

During the conflict three major new weapons came to the fore. Two, the submarine and the aeroplane, were already in existence in 1914, but in numbers so small as to have no immediate impact. The third, the tank, did not appear until 1915 and not until 1917 was it first deployed effectively in battle.

By 1914 the possibilities of the aeroplane as a new mode of transport were becoming recognized and a new industry was beginning to develop. Thus in Germany AEG, primarily an electrical manufacturer, had extended its scope to aircraft;

in the UK wholly new companies, such as the Bristol Aeroplane Company, had begun to appear. In 1911 the French had held a Military Air Show in an attempt to identify aircraft with military potential. Military leaders, however, were still far from persuaded that the aeroplane had much to offer and at the outbreak of war none of the major powers had air forces of any size: indeed there were then only a few thousand aircraft of any description, and some of very curious design, in the whole world. By the end of World War I some 200,000 aircraft had been built.

During the war four main roles had been identified for the aeroplane: spotting for artillery, photographic reconnaissance, observation and bombing. In the event, it was the first two of these that proved the most important, mainly as a result of the largely static pattern of the warfare that quickly developed on the ground. This role proved so valuable that each side sought to destroy the other's reconnaissance aircraft, leading to armed aerial combat. Here a major development, by Fokkers, was an automatic device linking machine gun and engine so that fire could be directed through the propeller blades without risk of striking them. Larger machines suitable for bombing raids appeared only later in the war (eg the German Gotha bomber of 1917): they did relatively little damage but created much alarm.

At sea the role of aircraft was limited because they had to operate from land bases and had relatively short ranges, though a few cruisers had provision for launching seaplanes by catapult. The first aircraft carrier, the British ship HMS *Argus*, a converted liner, was commissioned in 1918 but never saw service. The first custom-built aircraft carrier was HMS *Hermes* (1923).

In naval warfare the major technological development was the advent of the submarine. As with aircraft, the numbers in commission with the major powers in 1914 were quite small but they increased very rapidly. The chief exploiter of the new weapon was Germany. In the course of the war it lost over 200 submarines but inflicted such heavy losses on Allied Shipping –

Unlike a great many designers, I actually fly my planes, use them as other men use automobiles and yachts. This experience I have utilized. There is a definite reason why every part of the plane, its size, height, location, is as it is. Nothing should be left to guesswork. A good designer should be able to tell why every part was made in just that way, for every good airplane is the result of infinite compromises with aeronautic theory.

ANTHONY FOKKER

in excess of 12 million tons – that at one time (April 1917) it seemed that the new weapon might be decisive. The development that made it so influential was the introduction of diesel oil in place of gasoline as fuel. Diesel was not only safer, but the addition of larger fuel tanks transformed the submarine from a small vessel with limited range into a long-range strategic weapon.

In contrast to the aeroplane and the submarine, the tank was essentially a British innovation though one essential factor – the advent of the internal combustion engine – was common to all three. First proposed by the British army journalist E.D. Swinton as early as October 1914, the first prototype was completed one year later. The caterpillar tracks, which had been developed in the 19th century for agricultural vehicles, were designed to cross difficult ground and smash through the barbed wire entanglements that marked the enemy's front line, making a gap through which the infantry could pour. Tanks were first used in September 1916 at the Battle of the Somme – largely without effect, because their numbers were too small. But when used by well-trained troops, in mass numbers, on good terrain at Cambrai in November 1917 tanks achieved the first major breakthrough of the war on the Western Front. The French meanwhile had developed a lighter and faster tank. A massed attack by

nearly 600 Allied tanks on 8 August 1918 brought about what the German general Erich Ludendorff called "the blackest day of the German army". Had Germany not collapsed economically three months later the tank might have been decisive. Britain intended to deploy 10,000 tanks of a much improved model in 1919 and the USA, which had entered the war in 1917, had plans to make them at the rate of 100 a day. As it was, however, the tank was not to demonstrate its real potential until World War II.

The origins and growth of photography

By 1900 the basic principles of still photography had been established, and the appearance of the first Kodak mass-produced camera in 1888 had made photography a popular amateur pastime. In 1900 one person in ten in the United States and in Britain owned a camera, though the hobby grew more slowly on the European mainland. But if the principles were established, major changes in both camera and filmstock lay ahead. Even up to World War II professionals used large cameras made with wood and brass. This was by then no longer technically necessary but a reflection of short-sightedness on the part of publishers. Up to 1940 *Life*, for example, refused to accept anything but 10x8in (25x20cm) contact prints. By then the far more convenient miniature camera, of which the 1925 Leica was the prototype, could produce equally good results on 35mm (about 1.4in) film. Most amateur photographers had by then turned almost entirely to roll-film, though in shorter lengths and larger sizes than 35mm.

A major problem in photography is to ensure that the image focused on the film corresponds with the picture desired. For this some form of viewfinder is necessary. This may be no more than two frames, arranged like the sights of a rifle, or a simple external optical series which reflects an image on to a small ground glass screen. These introduce problems of parallax – the fact that the axis of the camera's optical system is not indentical with that of the viewfinder. The twin-lens reflex was introduced by Rollei in 1929. This was essentially a double camera, the upper one of which reflected on to a viewing screen exactly the same picture as the lower focused on the film. The more compact single-lens reflex, in which the visible image was reflected by a mirror which dropped out of the way at the moment of exposure, appeared in the early 1930s.

The performance of lenses was also much increased. As early as 1902 the Tessar lens had an aperture of f4.5, but Leica had reduced this to f3.5 by 1925 and in the early 1930s apertures as large as f1.8 had been achieved. Corresponding improvements were made in shutter speeds. The popular Compur camera, made by the Deckel Company of Munich, had a fastest speed of 1/250th second when introduced in 1912 but this had been increased to 1/500th by 1935.

The advent of the talkies

The beginning of the cinema can be precisely dated: on 28 December 1895 the brothers Auguste and Louis Lumière displayed a moving picture on a screen to a paying audience at the Grand Café in Paris. At the time the event attracted little notice but it marked the start of a vast new industry which grew with remarkable speed. Initially its main centers were in France and Britain, but by 1915 the United States, and Hollywood in particular, had taken a decisive lead: at the outbreak of World War I American investment was in excess of 2.5 billion US dollars. The appeal was to a mass audience (though serious full-length dramas such as The *Battleship Potemkin* directed by the Russian Sergei Eisenstein, appeared as early as 1925) and the repertoire was broad, ranging from news features, identified particularly with those produced by the French businessman Charles Pathé, to slapstick comedies such as those of Mack Sennett's Keystone Company in Los Angeles, which made Charles Chaplin's first film.

From the beginning it was appreciated that an audience did not readily take to sitting in silence watching a flickering picture with the action explained only in short written captions on the screen. From the earliest days it was, therefore, customary for cinema managers to employ a pianist who played an impromptu accompaniment, endeavoring to match the mood of the music to the action on the screen. The ultimate goal, however, was to make the characters speak for themselves as the film was shown. An obvious solution, and one quickly tried, was to record the speech on a gramophone record and to play this in step with the film. Unfortunately, the results tended to be bizarre, as it was then technically impossible to synchronize the sound and the actions: characters said one thing and did another.

◄ Many stars found it difficult to make the transition: an exception was Greta Garbo, here seen with G. Brown in *Anna Christie* (1930)

► Two types of sound track were developed: the variable-area and the variable-density (shown here).

◄ ▲ By the early 1930s talkies had almost wholly superseded silent films. A problem with early equipment was how to muffle the sound of the camera so that it was not recorded on the sound track of the film being shot. Cameras were first placed in soundproof booths as here.

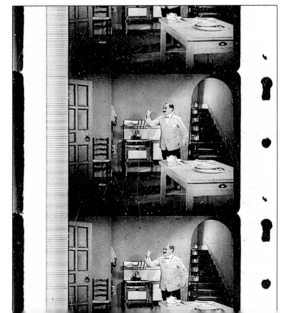

One solution was to record the sound as the film was made on a special sound track along its edge. Two systems were developed. In one the sound was recorded in terms of the varying opacity of the track; in the other a serrated track was formed, the pattern of the serration corresponding to the sound. In either case, the sound was regenerated by means of a tube- (valve-) amplifier system such as was used for radio transmissions. Not surprisingly, one of the pioneers of sound films was the American Lee De Forest, who had introduced the triode tube into wireless circuits in 1907. His phonofilm system of 1926 was unsuccessful, however, until improved as Movietone. The advent of this had dramatic success. In 1930 the number of American cinemas wired for sound increased from 8,700 to 13,500, and only 5 percent of films made were silent. The talkies eclipsed the silent film almost overnight.

▲ Although Thomas Edison's own inventions in the field of cinematography were of minor importance, he did become heavily involved in the organization of the film industry and particularly in the standardization of film and equipment. This picture shows film being shot in his studio in the Bronx, New York.

The advent of talkies had far-reaching effects on film production. On the one hand, a new kind of actor was required and many of the celebrities of the silent days found the new medium beyond their powers. There were also technical problems. Silent films could be produced in a hubbub of noise but talkies demanded a silent studio. Even the sound of the cameras was obtrusive: at first they were housed in soundproof cabins but later noiseless cameras were developed.

Although the American inventor Thomas Edison contributed little to cinema hardware, he did have a considerable influence on the organization of the rapidly growing new industry. In particular, he sensed the importance of standardizing film size and speed so that films could be shown without adaptation at cinemas anywhere in the world. At the beginning of the century he introduced a standard of 35mm (about 1.4in) width for film, with four perforations per inch (2.5cm), and 16 frames per foot (30cm). This was adopted internationally in 1909 and remained in use for 20 years. Projection speeds were not so closely monitored, however, and cinema managers anxious to give customers value for money could speed films up considerably. To cope with the new problems of sound reproduction a standard of 24 frames a second was adopted.

The World Power Conference

The worldwide concern of the late 20th century for adequate sources of power for industry and domestic uses is nothing new, but in the early years of the century the main concern was not the possible exhaustion of fuel supplies but how to keep power generation in step with demand. Anxiety about the management of electricity supplies led to the establishment of an International Electrotechnical Commission in 1906, with representatives from 19 countries.

A much more effective and influential organziation emerged after World War I as a result of the widespread desire to create effective international organizations and promote the regular exchange of information in a wide variety of fields. This was the World Power Conference, which first met in London in 1924 in connection with the British Empire Exhibition at Wembley.

It was organized by Britain and supported by all the Great Powers. The 40 countries represented included Germany and the Soviet Union. In all there were almost 2,000 delegates who considered "the power resources of each participating country and the extent to which they had been utilized" and how "to provide adequate opportunities for the cooperation of all nations in the development of power resources".
Information presented at the Conference was afterward published in five volumes of Transactions.

It was perhaps ahead of its time, in that many countries had not even established national standards for electricity generation and transmission by 1924. But it continued to meet at intervals and after World War II was reconstituted as the World Energy Conference and then as the World Energy Council.

▲ Baird's original television equipment was very primitive, based on scanning an object with a fast-rotating disk, with a series of holes punched spirally on it: the whole picture was thus scanned in one revolution. The light transmitted by each hole was then turned into an electric signal by a photocell. At the receiver the process was reversed. In the event, however, this photo-mechanical system was not on the main line of evolution: the future lay with all-electronic television.

Instant visual communications: television

In the 1920s the public accepted as a matter of course sophisticated methods of communication that had within living memory been regarded as marvels of applied science. The telegraph and telephone systems had worldwide networks based on tens of thousands of kilometers of conducting wire and the possibility of transmitting sound without any wires at all had become reality: public radio broadcasting was in its infancy but growing rapidly. The cinema was entertaining millions by recording scenes on film and reproducing them on the screen later. One gap remained to be filled, however, in this complex pattern of communication: that of seeing events as they occurred, just as radio made it possible to hear speech as it was uttered.

As with so many technical advances, it is impossible to attribute the invention of television to one individual, but there is no doubt that the person occupying the center of the stage in the 1920s was John Logie Baird. He devised a photo-mechanical system in which the picture was scanned by a fast-rotating disk containing a series of holes arranged spirally: in this way, the whole picture would be scanned in the course of one revolution. The light signal from each hole was then turned into an electric signal via a photocell and a corresponding radio pulse was generated. The receiver consisted of a similar disk system: as the varying light signals passed in succession through the holes a picture corresponding to the original was built up on a screen. There was no great originality in this, for such a rotating-disk system had been patented by Paul Nipkow in Germany as early as 1884 – though at that time, of course, with the thought in mind of a cable connecting transmitter and receiver. Nipkow did not pursue his idea, but it was taken up by a Russian, Boris Rosing, in 1906, taking advantage of the cathode ray oscillograph invented by the German physicist F. Braun ten years earlier. This made it possible to eliminate the second disk and to modulate instead the spot of light which moved at great speed along parallel lines in the tube, thus giving – through the phenomenon of persistence

... so distinct was the scene shown on the large screen that the watchers forgot the race in face of the miracle that brought it before their eyes. Many of us can remember the thrill of these first "moving pictures". They flickered and spluttered, but out of the haze we saw men move about ...

"DAILY HERALD"
REPORT ON TELEVISING OF
THE 1932 DERBY

I am afraid if this invention becomes too perfect it will cause most people to spend their evenings at home instead of visiting the theatre.
R.C. SHERRIFF, 1930

of vision – the illusion of a continuous picture. However, signal amplifiers were then insufficiently developed for clear pictures to be produced. Meanwhile, in Britain, A.C. Swinton proposed, but did not develop, yet another system, in which both transmitter and receiver were based on the cathode ray oscillograph.

Baird persevered, and on 27 January 1926 gave the first demonstration of true television before an audience in the Royal Institution, London. In September 1929 the British Broadcasting Corporation began television broadcasting with the Baird system and so, too, in the same year, did the German Post Office. Baird was a good publicist and among his more spectacular achievements were a trans-Atlantic broadcast in 1928 and the showing, in a London cinema, of the finish of the 1931 Derby horse race. Nevertheless, his success was short-lived, for his photomechanical system was not on the true line of evolution. Others had reverted to Rosing's all-electronic system. Among them was one of Rosing's students at St Petersburg (now Leningrad), Vladimir Zworykin. After the revolution in 1917 he fled to America and joined RCA, becoming director of research in 1929. In this capacity he contributed much to the development of an all-electronic transmission/reception system as visualized by Swinton.

◀ Although the best-known, Baird was by no means the only inventor working on television in the 1920s. In the United States F.E. Ives – primarily interested in color photography and photo-engraving – explored other possibilities for long-distance picture transmission. This cover of *Le Petit Inventeur* ("The little Inventor") (1928) gives an optimistic vision of a videophone system in which the picture was displayed with the aid of neon tubes. It did not prove to be a practicable proposition.

◀ The work of J.L. Baird in Britain on a photomechanical system of television was complemented in the USA in the early 1920s by that of C.F. Jenkins and P.T. Farnsworth. Both used a perforated scanning disk similar to that of Baird. The success of cinematography encouraged the belief, not to be fulfilled, that photomechanical equipment could be equally effective in television. The National Broadcasting Company became interested and made some experimental transmissions. Here Felix the Cat pirouettes in front of a television camera (1928). As the inset shows, the resulting picture was recognizable but of very poor quality. Not until the advent of all-electronic equipment in the 1930s was the future of television assured.

In November 1936 the BBC in London began a regular program of high-definition television broadcasting using the new system. In the same year RCA (Radio Corporation of America) began experimental broadcasts from the Empire State Building, New York, but the first public service there did not begin until 1939, when President F.D. Roosevelt was shown opening the New York World Fair. But television did not become a great social force until World War II was over. After a slow start as life began to return to normal, rapid progress was made. In 1947 there were only 34,000 television sets in the UK but by 1953 the number had risen to over two million.

Leo Baekeland and the birth of plastics

If we identify various periods in history in terms of the most widely used materials – as archaeologists refer to the Bronze and Iron Ages – the 20th century could fairly be called the Plastic Age. Although plastics did not become pervasive until after World War II, they had become quite familiar many years earlier.

Plastics are, in chemical terms, substances composed of giant molecules (known as polymers) consisting of long chains of simple units. Many natural products – such as rubber and cotton – are, in fact, polymers and even before the start of the 20th century new polymers had been made by chemically modifying natural ones. Thus viscose rayon, essentially a reconstituted cellulose, began to be manufactured by Courtaulds as early as 1907, and by 1919 production at Coventry amounted to some 6,000 tons a year. At that time a new form, acetate rayon, was introduced to find a use for surplus cellulose acetate which had been needed during World War I to dope aircraft fabric: after the war, light aluminum alloys were

AT breakfast, your wife pours you a cup of coffee; the handle she takes hold of on the percolator is made of Bakelite, as well as the button under the table she presses for service, and the twin-outlet plug from which are carried the wires to the toaster.

The Material of a Thousand Uses   BAKELITE

increasingly used for aircraft fuselages. Cellulose acetate was also adopted as film-base by the cinematographic industry, being far less flammable than celluloid.

In the late 20th century the main connotation of "plastics" was as solid moldings rather than fibers. As early as 1897 Galalith appeared in Germany as a semisynthetic polymer made from milk casein, but the modern plastics industry dates from 1909, when the Belgian L.H. Baekeland –

◀ Viscose rayon – introduced as a semisynthetic textile in 1907 – began to be produced in the form of thin film (Cellophane) in the 1920s. Crystal clear, it formed an ideal wrapping for food and other sensitive materials. It was the beginning of a major new industry for the packaging of goods in an attractive and hygienic fashion. This flourished particularly after World War II – when a much wider range of plastic film became available – so much so that the disposal of waste became a major problem.

◄▼► Leo Baekeland, seen here (right) in his laboratory, can fairly be described as the father of the plastic industry. His name became literally a household word through his invention of Bakelite, which he patented in 1906. Being easily molded, it lent itself to cheap mass production of endless articles for the home (left and below) and for industry. It was particularly important for the burgeoning electrical industry, with its insatiable demand for insulating components for sockets, switches, lamp-holders and other day-to-day accessories. The invention was timely also for the automobile industry, which also needed countless small items for its electrical systems.

who had already made a fortune from photographic printing papers – introduced Bakelite. This was made by chemical reaction between phenol and formaldehyde. It found a ready use in the electrical industry – including firms that produced the electrical systems of automobiles – which had an increasing demand for small, cheap, insulating compounds. For such purposes its unattractive brown color was no disadvantage, but it restricted wider use. Colorless ureaformaldehyde (Beetle) plastics were introduced in 1928: these could be produced in a variety of attractive colors by adding appropriate pigments.

The Beetle resins, and the melamine-formaldehyde resins developed shortly afterward in Germany, greatly extended the use to which plastics could be put. The physical properties of Bakelite were excellent but its drab color largely restricted its use to utilitarian purposes. How-

ever, the bright new plastics now coming on the market were attractive in their own right and found many uses in the home, from tableware to hairbrushes, crockery to hair-slides. The new materials became acceptable to leading designers and many items made in the 1930s became collectors' pieces.

The increasing use of motor transport created a corresponding increase in the use of rubber for tires. Although Germany, like other countries, still used many horses and mules for military transport during World War I, they relied also on motor vehicles. To mitigate the effects of the Allied blockade of the natural product they developed a synthetic rubber. This was rather poor, but in the 1930s – in pursuit of their strategy of self-sufficiency – they were to produce the improved Buna rubbers which were derived from acetylene.

ERROR, DECEPTION AND FRAUD

Science is a set of rules that keeps the scientists from lying to each other. Breaking these rules is unbecoming to scientists. Society wants science to be good, true and beautiful. In real life, performance does not always conform to the ideal. Over the years there have been many examples of scientific misdemeanor, ranging from inocuous error, due to faulty observation or negligence, to intentional "massaging" of data or downright fraud.

The outmoded Lamarckian view of heredity, which assumed that physical characteristics acquired during life could be inherited by subsequent generations, received support in the first three decades of this century by experiments conducted by Paul Kammerer of Vienna on salamanders and frogs. One experiment which was considered critical involved the midwife toad *Alytes obstetricans*. This toad mates on land and does not have horny pigmented pads on its thumbs (nuptial pads), like other toads which mate in water. Kammerer kept his *Alytes* toads in water for several generations, forcing them to mate there. After several generations some of the male toads developed the nuptial pads; Kammerer claimed that this new character became hereditary. There ensued a long controversy between Kammerer and the Mendelian geneticists. In 1926 the American herpetologist G.K. Noble visited Kammerer's laboratory and found that the coloration on the thumbs of Kammerer's specimen of *Alytes*, suggesting nuptial pads, was due to indian ink injected under the skin. Kammerer, though claiming innocence, was discredited, and committed suicide. Nevertheless his experiment served to bolster Lamarckian ideology in the Soviet Union during the period of Lysenko.

Trofim D. Lysenko, a soviet agronomist active during the period 1929–1960, won Communist Party support for his version of Lamarckian philosophy. His activities led to the annihilation of the science of genetics in the Soviet Union. Lysenko first gained recognition for his rediscovery of "vernalization" (when plant seedlings and seeds are kept wet and chilled during the winter, they sprout and mature earlier in summer). He denounced Mendelian genetics, and denied the existence of chromosomes as the bearers of heredity. He fanatically insisted on the correctness of his pseudoscientific ideas and completely ignored the evidence to the contrary from Western science. He succeeded in forcing collective farmers to attempt his impractical ideas on hybridization in plants, cross-fertilization in rye, transformation of one species of plants to another. Many of Lysenko's experiments were bolstered with manipulated data. By 1948 Lysenko became so powerful in Soviet agricultural science, that many of his opponents, the classical geneticists, lost their jobs and were exiled to concentration camps, only because they dared to contradict Lysenko. Lysenko was not discredited until 1965, after the fall of Stalin and Khrushchev, his main supporters. His teachings and theories then quickly became obsolete.

◄▼ In 1912 the scientific establishment was hoaxed into accepting that fossils excavated in Sussex, southern England, provided evidence for a primitive stage in human evolution. It was not until 1953 that part of the skull was shown to be an animal bone.

▼▼ The fossilized *Archaeopteryx* found in the 19th century, was hailed as evidence for the link between flying reptiles and birds. In 1985 a group of scientists claimed the fossil was a fraud. The controversy has not been finally settled.

▲ Trofim Lysenko (right) showing the results of his hybridization experiments to Soviet farmers in the 1930s.

► In 1989 two scientists claimed in a press conference to have achieved nuclear fusion at room temperature. Subsequent testing of their work revealed that the reactions they had observed were chemical ones.

Datafile

Medical advances ensured that epidemic diseases were generally kept under control during World War I. Although the influenza epidemic of 1918 showed the limitations of scientific medicine, the postwar period saw renewed activity in clinical medical research. Numerous breakthroughs were achieved in surgery, with the production of new drugs, and in preventive medicine as for tuberculosis. There was also a vogue for psychoanalysis.

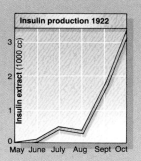

Insulin production 1922

Insulin extract (1000 cc)

May June July Aug Sept Oct

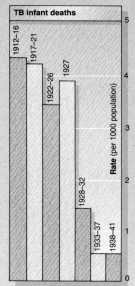

TB infant deaths

Rate (per 1000 population)

1912–16, 1917–21, 1922–26, 1927, 1928–32, 1933–37, 1938–41

▶ The discovery and production of insulin at Connaught Laboratories of the University of Toronto in 1922 (above) was quickly appreciated for its dramatic effect on diabetic patients. The Laboratories afterward remained an important center for insulin research and distributed insulin to other medical research centers (below).

◀ In 1927 the BCG anti-tuberculosis vaccine was introduced. These statistics from Gothenburg in Sweden (place of production for Sweden) indicate that it caused a rapid decline in child mortality. However, there was much opposition to the vaccine, particularly after children died in Lübeck in 1930, owing to impurities in production.

Insulin distribution

Index (1937 = 1)

1926 1934 1942 1950 1958

▼ Mortality rates for the USA in 1920, for England and Wales in 1921, for Denmark in 1921 and Portugal in 1920 show the after-effects of the postwar influenza epidemic. Cancer tumors are a new feature, although these were probably under-diagnosed before as well as in certain countries as suggested by the statistics for Portugal.

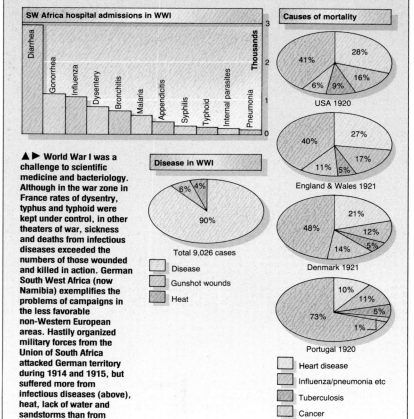

SW Africa hospital admissions in WWI

Thousands

Diarrhea, Gonorrhea, Influenza, Dysentery, Bronchitis, Malaria, Appendicitis, Syphilis, Typhoid, Internal parasites, Pneumonia

Causes of mortality

41%, 28%, 16%, 9%, 6% — USA 1920

40%, 27%, 17%, 11%, 5% — England & Wales 1921

48%, 21%, 12%, 5%, 14% — Denmark 1921

73%, 10%, 11%, 5%, 1% — Portugal 1920

Heart disease
Influenza/pneumonia etc
Tuberculosis
Cancer
Other

Disease in WWI

6%, 4%, 90% — Total 9,026 cases

Disease
Gunshot wounds
Heat

▲ ▶ World War I was a challenge to scientific medicine and bacteriology. Although in the war zone in France rates of dysentry, typhus and typhoid were kept under control, in other theaters of war, sickness and deaths from infectious diseases exceeded the numbers of those wounded and killed in action. German South West Africa (now Namibia) exemplifies the problems of campaigns in the less favorable non-Western European areas. Hastily organized military forces from the Union of South Africa attacked German territory during 1914 and 1915, but suffered more from infectious diseases (above), heat, lack of water and sandstorms than from gunshot wounds (right).

For most of the world, World War I was not a time conducive to original work in the biological sciences and medicine: many research workers were diverted to work of immediate military importance. Even in the USA, not directly involved until 1917, a slowing down was apparent. Alexis Carrel, for example, who had left France in 1906 to continue his research on organ transplantation and tissue culture at the Rockefeller Institute in New York, became involved in the treatment of deep wounds by continuous irrigation with sodium hypochlorite. In Britain, F.W. Twort was called away for army service and could not exploit his exciting discovery (1915) of bacteriophages – viruses which can infect bacteria.

However, the war years were by no means wholly unproductive. In 1915 K. Yamagiwa and K. Ichikawa reported the carcinogenic (cancer-producing) effect of coal tar, starting a new line of cancer research which remained useful for the rest of the century. In social medicine there were some curious developments. In 1916 Margaret Sanger opened the first birth-control clinic in the USA. In 1920 France made abortion illegal, while the newly founded Soviet Union legalized it. In Britain, Marie Stopes pioneered work on eugenics and birth control (her then controversial *Married Love* appeared in 1918). Three years later she founded the first birth-control clinic in Britain.

The 1920s saw a marked change of emphasis in the approach to medical problems. Before the war research was largely in the hands of professionally qualified physicians and surgeons who guarded their position jealously. Afterward, there was an increasing contribution from workers trained previously in physical and biological sciences. Hans Berger, for example, who introduced electroencephalography (EEG – electrical monitoring of brain activity) in 1929, began life as a physicist before turning to psychiatry. EEG was soon used not only to investigate mental disorders but also to diagnose physical diseases of the brain.

This trend was particularly well illustrated in the field of vitamins and hormones. Although very different in origin – vitamins are ingested with food and hormones are secreted by the endocrine glands – they have one important feature in common: though needed only in minute quantities, both are essential for maintaining the body's metabolism in proper balance. The existence of both had been recognized by the early years of the century but their identification as specific chemical substances was possible only by using a multiplicity of physicochemical techniques. Major achievements were the isolation of insulin by F. Banting (1922) and of vitamin C by A. Szent-Györgyi (1928).

TOWARD MODERN MEDICINE

The impact of World War I on medicine

Pioneering work in surgery

Calmette, Guérin and the conquest of tuberculosis

The discovery of insulin

The origins and spread of psychoanalysis

The rediscovery at the turn of the century of Mendel's classic work on inheritance, by H. de Vries and others, had initiated an increasingly intense international program of research on the mechanism of inheritance at the cellular level. In this field important landmarks were the American geneticist T.H. Morgan's *Physical Basis of Heredity* (1919) and his *Theory of the Gene* (1921). Important contributions were also made in the USA by H.J. Muller, who made meticulously careful studies of the conditions under which wholly new characteristics appear in living organisms, a phenomenon known as mutation. In 1927 he showed that by exposure to X-rays the natural mutation rate in fruit flies might be increased as much as 150 times.

Surgery and transplants

In the 19th century two major developments in surgery had much improved both the patient's comfort and his chance of survival. On the one hand, anesthesia had made it possible for surgeons to perform much longer and more com-

plex operations than were feasible when the patient was conscious. On the other, the use of aseptic techniques, and close attention to hygiene generally, much reduced the risk of postoperative infection, which was all too often fatal. In the early 20th century techniques were still primitive and great advances lay ahead, but two important principles had been established. Nevertheless, there were still areas where surgical intervention was hazardous.

The biggest problems lay with organs whose activity had to be sustained without interruption to maintain life. An important example was surgery of the chest, where opening the thoracic or chest cavity containing the lungs led to their collapse. Here an important development was one effected by Ferdinand Sauerbruch, who was appointed professor of surgery in Zurich in 1910. He designed a special operating chamber, outside of which was the patient's head and the anesthetist. His body, and the surgeon, were inside the chamber which was maintained at a reduced pressure to prevent lung collapse. Zurich

▼ On all fronts, World War I cost 6 million dead and 12 million wounded. Medical services in the field worked under primitive and difficult conditions. Here a doctor treats a casualty at a captured German ammunition dump at Oosttaverne, Belgium (1917).

▲ Many of those who survived their injuries were left badly disfigured. Plastic surgery was still primitive, so many ex-servicemen, such as this Frenchman, could only hide their injuries with masks.

▼ Sadly, the making and fitting of artificial limbs became a considerable postwar industry. Some specialist firms were founded, others diversified into limb manufacture. The big demand encouraged improvements in design, especially in articulation.

abounded in sufferers from diseases of the lung who had come to seek relief in sanatoria in the mountains, and his technique soon became widely adopted. In 1908 F. Trendelenburg attempted surgical removal of a pulmonary embolism (blockage of tissue in the lungs) but this technique was not mastered until 1924. For patients whose respiratory muscles had been severely affected – as after poliomyelitis (a form of muscular paralysis) – the "iron lung" developed by P. Drinker in 1929 was a major advance.

For the sufferer from heart disease, however, the surgeon could still do little. The medical literature contained occasional references to successful surgery after stab wounds and similar mishaps but for chronic conditions the prospects were poor. The introduction of sympathectomy (cutting off a part of the sympathetic nervous system) for the relief of angina by the Romanian surgeon Thoma Ionescu in 1916 was a small but significant development.

Although intracranial surgery had long been practised – some skulls from prehistoric sites reveal successful trepanation, perhaps for the relief of depressed fractures of the bone – even in the 19th century the mortality rate was very high. In the main, this was because the methods of general surgery were used. Progress began to be made only when more specialized techniques were developed, notably by Harvey Cushing in the USA. The secret of his success was exceptionally comprehensive diagnosis beforehand, using specialized methods, and meticulously careful surgery, often lasting many hours. He was particularly successful with brain tumors and tumors of the acoustic and optic nerves. He also made a special study of the pituitary gland located at the base of the brain. It is perhaps the most important of all the endocrine (or hormone-secreting) glands, having an influence on all the others. His reputation attracted disciples from all over the world, who returned to their own countries to set up clinics using his methods.

While neurosurgical techniques were being developed to deal with pathological disorders of the brain and nervous system, others were seek-

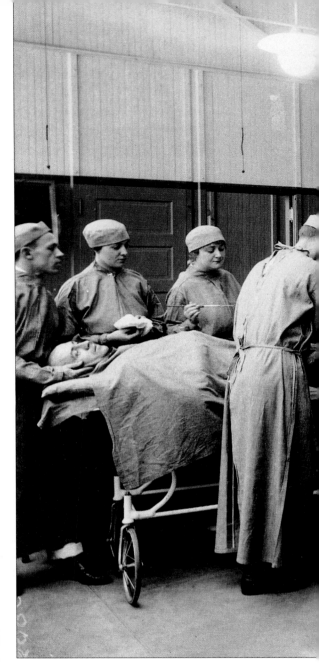

ing subtler methods to diagnose and treat disorders of the mind. In Paris J.M. Charcot (1825–93) had turned his attention from diseases of the nervous system to problems of human behavior, especially hysteria. Among his students in the latter part of the 19th century was the Austrian Sigmund Freud, pioneer of psychoanalysis. Initially derided, his concept of psychoanalysis – developed with C.G. Jung, A. Adler, and others – eventually became accepted and the International Association for Psychoanalysis was founded in 1910. Freud was elected a Foreign Member of the Royal Society in England in 1936. He continued his research and teaching for many years, until obliged to leave Vienna in 1938 as a consequence of the Nazi occupation of his country. In this field conventional experimental approaches had little validity and new ones had to be found. Among them was the famous ink-blot test for intelligence, personality and emotion devised by the Swiss psychiatrist Hermann Rorschach in 1921.

Try to recall the patient's face in every instance and a general picture of the case.

(Re patient with gastric ulcer) Hard-working, law-abiding, porridge-eating, God-fearing Scot ... Good fellow and we want to help him. Fifty percent of diagnosis in G-1 disease lies in the history so get a good one.

HARVEY CUSHING

During the 20th century organ transplant operations became a normal, albeit highly specialized, feature of medical practice. Important contributions were made in the early years of this century. Among the pioneers was Alexis Carrel, working at the Rockefeller Institute in New York. There he developed a considerable interest in organ transplantation where, among other problems, earlier workers had found difficulty in reestablishing a satisfactory blood supply to the organ transplanted: the usual outcome was thrombosis (blood clot) or stenosis (narrowing of a blood vessel). Carrel overcame this by developing new techniques of suturing (joining) blood vessels and succeeded in removing entire organs from animals and replacing them in their original position. By working with single animals he was able to avoid the rejection symptoms which remained a major problem in organ transplants in human patients.

Carrel also did original work in the field of tissue culture. He succeeded in maintaining cells alive in nutrient solution long after the animal from which they had been derived had died. Much later, in 1935, he devised a mechanical heart to maintain circulation during cardiac surgery.

In the treatment of infectious diseases the climate of medical opinion at this time favored the use of vaccines, where there had already been a long series of successes and new ones were being recorded. A major advance was the introduction of BCG vaccine in 1927 for protection against tuberculosis. By contrast, the track record for chemical agents had been disappointing. Salvarsan and neosalvarsan had proved valuable against syphilis, but side-reactions could be severe and even fatal. In 1924 German chemists produced plasmoquine, a synthetic alternative to the long-established antimalarial drug quinine.

This was an unimpressive record but with the benefit of hindsight it is possible to see that the tide was beginning to turn. In 1927 G. Domagk, director of research in experimental pathology and bacteriology with the great German chemical company I.G. Farben, was sufficiently optimistic to embark on a systematic search for chemical agents which might control some of mankind's most serious diseases such as meningitis, tuberculosis and pneumonia – the latter, in particular, dreaded as "the captain of the men of death". Progress was slow but faith and patience were rewarded in 1932 with the discovery of the first of the sulfonamide drugs, a truly revolutionary advance. In 1928 another discovery was made which passed virtually unnoticed at the time but was destined to prove even more revolutionary. In that year a British bacteriologist, Alexander Fleming discovered penicillin.

Research into microorganisms

When bacteria were first observed it seemed that they must be the smallest of all living organisms. Soon, however, just as the physicists were discovering that atoms were not after all the smallest units of matter, it became clear that there were forms of life smaller than bacteria. Around the turn of the century it was discovered that a number of major diseases – including poliomyelitis, foot-and-mouth disease and tobacco mosaic disease – were caused by infective organisms so small that they would pass through filters which would trap bacteria. Unlike bacteria, they could not be propagated in inanimate media but only in susceptible living cells, such as yolk of egg. In 1915 F.W. Twort, working in London, discovered that some viruses, which he called bacteriophages, can infect and destroy bacteria. War service made it impossible for him to continue his work at that time and it was left to Félix d'Hérelle in France to investigate the phenomenon more closely.

Until the advent of the sulfonamides and antibiotics the main weapon against infectious diseases was immunization, a system of prevention rather than cure. In the 1920s the most deadly of the endemic infections was still tuberculosis and the introduction of an effective vaccine in 1927 was a major step forward. Its inventors were two French biologists, L.C.A. Calmette and Camille Guérin and after them was named BCG (Bacille Calmette-Guérin) vaccine. It was derived from bovine tubercle bacilli whose virulence had been reduced by cultivation in ox bile.

Tuberculosis can attack many organs of the body, but especially the lungs. It has been considered as one of the great scourges of mankind since the earliest recorded history, but until the advent of vaccines and drugs such as isoniazid little could be done for sufferers. The merits of

▼ Until the development of BCG vaccine in 1927 by Albert Calmette (inset top) and Camille Guérin (below), of the Pasteur Institute in Paris, there was no effective protection against tuberculosis. The vaccine was of no value to those already infected, for whom the only recognized treatment was prolonged rest and plenty of fresh air and sunshine. The supposition was that this would stimulate the blood supply to the lungs and increase resistance to infection. A favored venue was Switzerland, but for the many who could not afford this there were many national sanatoria. Here children are treated for TB by exposure to ultraviolet lamps (London, 1930).

▲ In the conventional open-air treatment of tuberculosis, much importance was attached to plenty of sunshine. In the pursuit of this no expense was spared for those who could afford it. This sanatorium at Aix-les-Bains in southeast France had a rotating upper storey so that patients faced the sun all day. There was also — especially on the Continent — faith that spa treatment also could be helpful, and doubtless the site of this sanatorium was chosen so that patients could receive both treatments.

▶ Thanks to the development of vaccines and tuberculo-static drugs, the incidence of tuberculosis has been much reduced. Even so, in the late 20th century mortality worldwide was still around three million annually, three-quarters in developing countries. This poster, issued by the American Red Cross in 1913, is a grim reminder that TB was once one of the great killers worldwide – the white plague.

We learn from an experience of thousands of cases all over the world that BCG nearly always yields sufficient protection

ALBERT CALMETTE, 1933

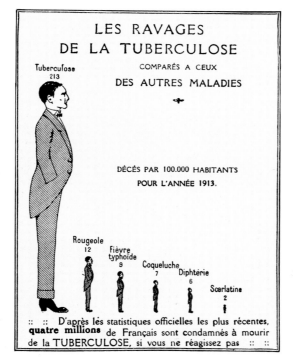

LES RAVAGES
DE LA TUBERCULOSE

COMPARÉS A CEUX
DES AUTRES MALADIES

Tuberculose
213

DÉCÉS PAR 100.000 HABITANTS
POUR L'ANNÉE 1913.

Rougeole
12
Fièvre
typhoïde
9
Coqueluche
7
Diphtérie
6
Scarlatine
2

:: :: D'après les statistiques officielles les plus récentes,
quatre millions de Français sont condamnés à mourir
de la TUBERCULOSE, si vous ne réagissez pas :: ::

fresh air and sunshine were preached, but sanatoria were palliatives rather than cures.

Vitamins and hormones
Diabetes is an age-old scourge. In many cases the symptoms may be slight, or even pass undetected, but in severe ones increased susceptibility to infection, loss of weight and impairment of bodily functions generally cause serious illness or death. The discovery of an effective treatment for this widespread disease was one of the major medical triumphs of the 1920s.

The nature of the disease was by then well understood through research in various countries. The root cause was that groups of cells in the pancreas, the Islands of Langerhans, fail to secrete a substance which regulates the metabolism of sugars. In 1920 F.G. Banting, a young

orthopedic surgeon, established himself in practice in Toronto, Canada, and also obtained a post as demonstrator in physiology in the University of Eastern Ontario, working in the laboratory of J.J.R. Macleod. There he became interested in diabetes and invited the cooperation of a young medical student, C.H. Best. Reviewing earlier work he came to the conclusion that if he tied the ducts of the pancreas the gland would atrophy except for the Islands of Langerhans, and that from this residue he might be able to extract the active substance, insulin. This approach proved effective and the first successful clinical trial was carried out with the crude extract in January 1922. Thereafter, there remained two major problems: first, to prepare insulin in sufficient quantities; second, to devise means of administering controlled doses to patients.

The question of supply was partially answered by an extraction process developed by enlisting the help of a young biochemist, J.B. Collip, who used as his source pancreas glands (sweetbread) obtained from a local abattoir. By 1926 insulin was available in pure crystalline form, which made precise dosage much easier, and the new drug became generally available for all who needed it.

In 1923 the Nobel Prize for Physiology or Medicine was awarded jointly to Banting and Macleod, an honor all the greater because it is rare for the award to be given so soon after the discovery it marks.

The discovery of insulin is important in itself because it made possible the effective control – though sadly not the cure – of a major disease. But it was important, too, as one facet of a growing understanding of human physiology and thus of the treatment of other similar diseases.

Many of the glands of the body secrete products through clearly defined ducts and have a fairly local action. But not all glands have such ducts. The so-called endocrine glands secrete physiologically active substances into the bloodstream and thus affect the body generally. Collectively, the active substances created by endocrine glands are known as hormones, a term first used in 1905. Hormones are essentially chemical messengers which serve to keep the whole complex metabolism of the body in balance. Over- or underactivity of an endocrine gland can give rise to a variety of very specific symptoms. Thus an overactive thyroid, producing thyroxine, causes exophthalmic goiter: underactivity results in myxoedema, a condition distinguished by mental slowness, sluggish metabolism and loss of hair. Such effects were beginning to be understood around the turn of the century and led to the development of a new branch of medicine, which the Italian physician Nicole Pende named endocrinology in 1909.

Development followed two major lines. First, there was increasing knowledge of the role of individual endocrine glands and the hormones they secrete. Second, and no less important, it was recognition that these glands do not act individually but in concert: unraveling their complex interaction was, and still is, a task of great difficulty.

One of the pioneers was the Argentinean physiologist Bernardo Houssay, who followed up Banting's research and discovered that the hormone of the pituitary gland, a tiny organ at the base of the skull, is closely linked with that of insulin. The sex hormones, produced in the testes and ovaries, profoundly affect sexual activity and fertility and determine secondary sexual characteristics such as facial hair.

Hormones are specific chemical substances but have widely differing structures. they are all remarkable for their extremely high physiological potency: very minute amounts produce profound effects. Thus insulin is a complex protein. Thyroxine, however, is relatively simple and is remarkable for its high content of iodine; the small thyroid gland may hold as much as one-third of the iodine in the whole body. It was synthesized in 1927.

Yet another hormone-related disease began to yield to treatment at the end of the 1920s. This was Addison's disease, an age-old scourge but one not clearly identified until 1849 by the British physician Thomas Addison. He could do no more than describe the syndrome, but it is in fact due to atrophy of the adrenocortical glands. The most obvious symptoms are weakness, loss of weight and brown pigmentation of the skin. In the absence of treatment the outcome is usually fatal. In 1929 W.W. Swingte and J.J. Pfiffuer, in the USA, prepared active extracts of the gland and a year later they found it effective in the treatment of Addison's disease. In 1934 E.C. Kendall isolated the hormone itself. Now the condition can be successfully controlled by regular treatment with corticosteroids.

Hormones are created within the body but another class of essential natural products, the vitamins, are ingested as part of the diet. Like hormones, vitamins are remarkable for their extremely high physiological activity. Although a very little goes a very long way, deficiency can have very serious, even fatal, consequences. As with hormones, the effects of vitamin imbalance were familiar long before their causes were understood. Scurvy, for example, had been well known as a scourge of seamen, more particularly on long voyages without fresh provisions. In the 18th century James Lind (1716–94) recommended the use of lemon juice for its prevention and cure. The results were dramatic, and when his recommendations were accepted – albeit very tardily – scurvy virtually disappeared from the British Royal Navy. A similar connection between diet and disease was demonstrated by Christiaan Eijkman in 1890. In 1909 the German biochemist W.U. Stepp demonstrated that chemically pure fats lack an essential food factor, which was later identified as vitamin A (1913).

By the beginning of World War I it was clear that there was a relationship between diet and certain kinds of disease – some of them very prevalent and serious – but this was very far from identifying the basis of the relationship. For health, a proper balance of the main nutritional elements – protein, carbohydrate and fat – was necessary, but it was evident that other factors were essential, through only in very small

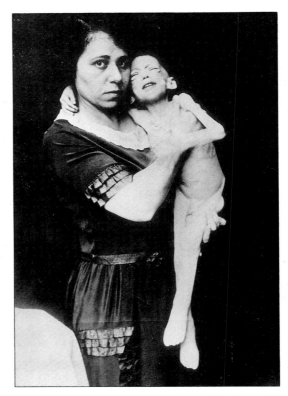

◄ Although diabetes was clearly recognized as early as the 2nd century AD, it was not effectively brought under control until the discovery of insulin by Banting and Best in 1922. This tragic photograph shows a child of three – weighing only 7kg (15lb) – afflicted with the disease in that same year.

▼ Diabetes is caused by failure of the pancreas to produce sufficient insulin. The basis of treatment is to administer the hormone regularly to balance the deficiency. The original source was animal pancreases obtained from abattoirs. This picture shows the first stage of the extraction process (grinding the animal material) at the Indianapolis plant of the American pharmaceuticals company Eli Lilly in 1923.

The Origins and Spread of Psychoanalysis

Psychoanalysis is a method of treatment of the neuroses (minor mental disorders) which developed into a general psychology. Its originator was Sigmund Freud (1856–1939). Freud began his professional career as a research worker in the physiological institute of Ernst von Brücke in Vienna, but financial pressures forced him to embark upon private medical practice (from 1886). Dissatisfaction with current methods of treatment of the neuroses impelled Freud to abandon hypnosis and other methods of suggestion in favor of "free association". By encouraging patients to reveal whatever thoughts were passing through their minds, Freud hoped to uncover the origins of their neurotic illnesses which, he was convinced, were caused by traumatic events in early childhood. The first psychoanalytic book, *Studies on Hysteria*, which Freud wrote jointly with Josef Breuer, was published in 1895.

As Freud's ideas developed, a small group of interested physicians began to meet at his apartment and, in 1907, the first psychoanalytic society was formed. An International Association for Psychoanalysis followed in 1910, and by the outbreak of World War I psychoanalytic societies existed in Zurich, Munich, Berlin, Budapest, England and America. Interest in psychoanalytic theories was fostered by the high incidence of various types of neurotic breakdown ("shell shock") during the war amongst serving members of the armed forces.

By the 1920s psychoanalysis had become influential in intellectual circles throughout Europe and America. Freud's insistence on the central importance of the individual's sexual development opened the door to freer discussion of sex. Freud's concept of the unconscious, and his rediscovery of the importance of dreams, encouraged painters, sculptors and writers to experiment with the fortuitous and the irrational. Movements like Dadaism and Surrealism owe a great deal to psychoanalysis. Although many Freudian theories have not stood the test of time, Freud has had an inescapably powerful influence upon man's view of his own nature.

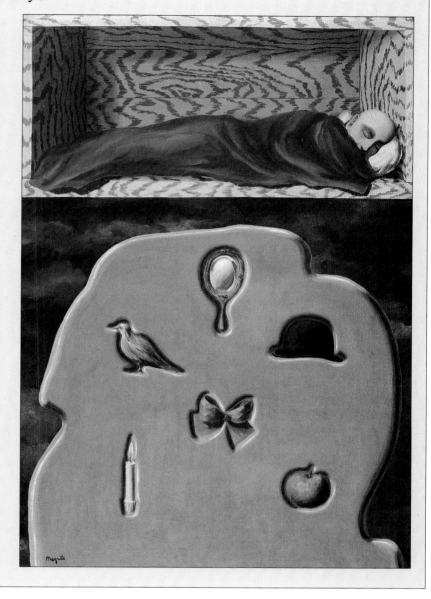

▶ Magritte's *Reckless Sleeper* (1927) with Freudian symbols.

amounts. In 1912 the Polish biochemist Casimir Funk suggested the name vitamine for such necessary food factors, in the belief that they all belonged to a class of chemicals known as amines. However, this assumption proved wrong and in 1920 the modern spelling vitamin was introduced to avoid confusion. One of the pioneers in the study of vitamins was Frederick Gowland Hopkins of Cambridge, who shared a Nobel prize with Eijkman in 1929. But although there were clear ideas abut their role, vitamins remained anonymous until 1926, when vitamin B1 – deficiency of which causes beriberi – was isolated in pure crystalline form. Two years later vitamin C – the antiscurvy vitamin – was similarly isolated and in 1933 it was made synthetically. In 1929 vitamin K was isolated, followed by vitamin D in 1931.

This emergence of the vitamins from the shadows was one of the greatest medical events of the 1920s and today a multitude of vitamins

find their place in the world's pharmacopoeias. More than this, however, it became general practice in the food industry to fortify certain foods – such as margarine and dried milk – with vitamins to ensure that the population at large had a sufficiency of them – a very important public health measure.

Vitamins differ enormously in their chemical constitution and there is thus no single technique for their isolation, which is further complicated by the fact that they occur in very low concentrations – sometimes a few parts per million – in admixture with scores of other different, and irrelevant, substances. Two considerations are paramount, however. The first is identification of a relatively rich source: for vitamin C. Szent-Györgyi used Hungarian red pepper (paprika). The second is to find a suitable test organism – usually a small laboratory animal – which can be used to follow the vitamin through successive stages of purification.

Datafile

In 1914 the periodic table (originally devised by Mendeleyev in 1869) was well established as a scheme within which the chemical elements could be ordered according to their appearance and reactivity. Atomic weight was known to increase almost linearly through the list of the elements; almost but not exactly. Studies of the X-ray emissions from metals under electron bombardment showed that it is the amount of positive charge on the nucleus which determines the place an element holds in the periodic table, not the weight of the whole atom. The discovery of isotopes explained the anomalies in atomic weights. The postwar years saw the emergence of the twin towers on which all modern physics rests: the theories of quantum mechanics and general relativity. The first describes the atomic world in terms of probability and uncertainty. The second describes mathematically the large-scale structure and evolution of the Universe. Experimental confirmation of relativity was spectacular and widely reported.

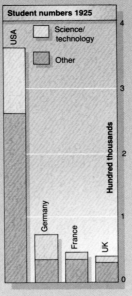

▲ Although the number of arts students was similar in the UK, France and Germany, postwar reconstruction of German industry coupled with generally high levels of unemployment encouraged many more students to study for a science or engineering qualification in the universities and technical schools.

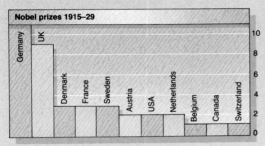

▲ The Nobel prizes awarded in science in this period were predominately given to scientists from Germany and the UK. Six scientists from Scandinavia were also honored. From Denmark Niels Bohr won the prize in physics in 1922 for his pioneering work on atomic structure. In 1920 S.A.S. Krogh had won in physiology. Working under Bohr's father Christian, Krogh had studied respiration and capillary blood vessels. Sweden won prizes in chemistry for Theodor Svedberg's work on colloids which showed visual evidence for the existence of molecules, and for studies of fermentation by Hans von Euler-Chelpin. A physics prize went to K.M.G. Siegbahn for work on X-ray spectroscopy.

▶ In the late 1920s European chemicals companies perceived that they could only compete with the giant Du Pont company in the USA if they formed larger units and sponsored large-scale research. In the UK four companies merged to form ICI in 1926. Its Dyestuffs division prospered in the 1930s.

◀ Based on highly creative work in organic chemistry, the synthetic materials industry in Germany grew rapidly to dominate world production at the outbreak of World War I. After defeat, however, the crucial German patents were siezed by the victorious powers and similar products were soon being manufactured elsewhere.

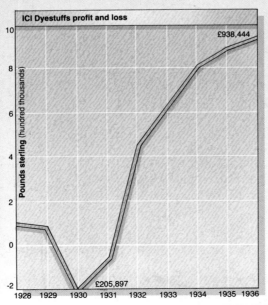

In the physical sciences the second and third decades of the 20th century had a distinctive flavor. Although research based on classical physics continued steadily, interest was increasingly being directed to the extremes of the Universe. While atomic physicists were continuing to unravel the nature of the smallest particles of matter, astronomers were looking further and further into space, making not only important discoveries by observation but developing theories about the origins of the Universe.

The atomic characteristics of elements

At Cambridge in the UK Sir Ernest Rutherford and his colleagues were still making major discoveries with what he called "string and sealing wax" apparatus. By 1914 a clear and simple concept of the atom had been established. It consisted of a small heavy nucleus, with a positive electrical charge, surrounded by negative electrons moving in orbits like planets round the Sun. The whole was electrically neutral because the number of electrons exactly balanced the nuclear charge. In order to explain why the normal attraction between positive and negative charges did not lead to the coalescence of the whole structure, Max Planck had devised his quantum theory.

Like a marksman with a rifle, Rutherford could use radioactive materials such as radium to fire alpha-particles – positively charged helium nuclei – at atoms of the lighter elements such as nitrogen. By 1919 he had shown that in the impact a proton (charged hydrogen nucleus) could be expelled from a nitrogen atom: in other words, the nitrogen atom had been split. It would, perhaps be more accurate to say that it had been chipped. A split into parts of comparable size was not achieved until 1932.

Somewhat earlier, in 1913, the British physicist H.G.J. Moseley had examined the X-ray spectra of the elements and was able to show that the property which determined their chemical individuality was the charge on the nucleus, and it was this that properly determined their position in the periodic table of the elements (formulated by D.I. Mendeleyev in 1868 on the basis of atomic weights). Largely, charge and mass went in step, but Moseley's theory satisfactorily explained certain anomalies in the Table. He was, as it were, able to call the roll of the natural elements from 1 (hydrogen) to 92 (uranium).

Sadly, Moseley was killed in World War I (at Gallipoli in 1915) and it was left to others to develop his brilliant concept. Among them was another British physicist, F.W. Aston. Working at the Cavendish Laboratory in Cambridge before the war, he had discovered that neon, with an experimentally determined atomic weight of about 20, existed in two forms. One, the

THE EXPANDING UNIVERSE

- Isotopes and neutrons
- Science in the United Kingdom
- The problem of light and development of wave mechanics
- Edwin Hubble and the big bang theory
- Atomic theory and the periodic table
- Einstein's general theory of relativity

predominant form, had an atomic weight of 20; the other, relatively scarce, one of 22. The *observed* atomic weight was the *averaged* mean of the two. Chemically, they were identical, so conventional methods of separation were not applicable: some method had to be found depending on differences in physical properties. In 1913 he achieved a partial separation based upon the difference in the speeds at which the two species diffused through the walls of a pipe-clay tube.

Aston, too, was called to the war and could not resume his research at Cambridge until 1919. There he built a "mass spectrograph" in which particles were differentially deflected by combined electric and magnetic fields. They were detected as they fell on a photographic plate. This apparatus was so called because it separated particles according to mass just as a spectrograph splits up light according to color (wavelength). He investigated 50 different elements and found that in every case – except hydrogen – the atoms had integral (whole-number) weights.

On theoretical grounds the concept that atoms could be chemically identical and yet have different atomic weights had been advanced by Frederick Soddy in 1913. To these different species he gave the name isotope because they occupied the same place (Greek: *isos topos*) in the Periodic Table. Aston's work was a striking vindication of this theory.

▼ The far-reaching consequences of the application of quantum theory to atomic physics prompted the first of a famous series of Solvay Congresses, in Brussels, attended by leading scientists in the field. The first was held in 1911. This picture shows participants in the fifth (1927). They include (from left to right): in the front row, Planck (second), Marie Curie (third), Einstein (fifth); in the second row, W.L. Bragg (third), Dirac (fifth), de Broglie (seventh) and Bohr (ninth); in the back row, Schrödinger (sixth) and Heisenberg (ninth).

Up to 1920 the atom was interpreted in terms of only two basic units: the proton and the electron. But some physicists began to suspect that there might be a third, electrically neutral, particle resulting from the combination of a positive proton and a negative electron. Such a particle would be difficult to detect in the conventional manner, since the usual techniques involving photographic emulsions or cloud chambers required ionized (charged) particles to form tracks. By contrast, an uncharged or neutral particle could move freely through matter and early attempts to demonstrate its existence all failed. In the event it was ten years before success was achieved, and then only indirectly. In France, in 1932, Irène and Frédéric Joliot-Curie reported a new phenomenon which they thought could be interpreted only in terms of "a new mode of interaction of radiation with matter". However, James Chadwick, in Cambridge, realized that no such radical assumption was necessary: all could be satisfactorily explained on the basis that the radiation concerned was a stream of neutrons, and he soon confirmed the existence of such particles.

Wave mechanics

While Rutherford, Aston, and many others were making progress on the basis of direct experiments in the laboratory, the theoreticians were also making important contributions. Thus Paul

Science in the UK: An Uneven Evolution

If one factor is to be identified which most particularly influenced the evolution of science in the UK it must surely be respect for those who pursue pure science as an intellectual exercise and disdain for those whose aim is to apply it for useful purposes. Although Britain produced the world's first industrial revolution, in the mid-19th century scientific and technical education was neglected, despite a blunt warning from a Royal Commission in the 1860s that Britain in this respect lagged dangerously behind France, Prussia, Austria, Belgium and Switzerland. The country entered the 20th century ill prepared to compete with the rest of the industrial world.

In pure science the situation was different, however, for although relatively little was spent in the universities much could be achieved with very little. From the turn of the century the Cavendish Laboratory in the University of Cambridge was the world leader in atomic physics, yet its preeminent figure, Lord Rutherford, prided himself on achieving his results with "string and sealing wax". The National Physical Laboratory was founded in 1900 but it needed World War I to establish that technological superiority, or at least parity, was necessary for survival. The Department of Scientific and Industrial Research was set up in 1916 to coordinate national resources.

In the interwar years this continued, administering a network of specialist research stations. These assisted the smaller firms who could not undertake independent research. Only a few, such as ICI, GEC and Shell could afford their own research laboratories. Though these all had good research facilities, and demonstrably performed a very valuable public service, they nevertheless found it difficult to attract the ablest scientists, except in a consultative capacity. Despite lower salaries, and often worse facilties, they preferred the socially more acceptable – and less competitive – life of academia.

The consequences of World War II were somewhat paradoxical. Whatever the recriminations that were to follow, the atom bomb had shown that scientists could change the course of history: the UK itself had made lesser, but still very considerable, contributions to victory with radar and penicillin. Jet propulsion, pioneered in Britain by Frank Whittle, was to revolutionize aviation. Yet the postwar climate did little to change attitudes. A period of socialist government (1945–51) encouraged the belief that to distribute wealth was highly creditable, but to create it through industry was somehow unseemly. The Robbins Report of 1963 advocated a substantial increase in higher education. New universities were funded and old ones expanded, but the expansion plans failed to provide any incentives for the ablest candidates to enter the faculties of science and technology where, in the national interest, they were most needed.

Though the reasons were by no means the same in every case, all the European countries found it difficult in the postwar years to compete in scientific research with the USA. With the formation of the Common Market, which Britain joined in 1973, new strength was found in the formation of European, rather than national, research laboratories and agencies. Of the laboratories, the most prestigious are CERN, in Geneva, for particle research, and the European Molecular Biology Laboratory, in Heidelberg.

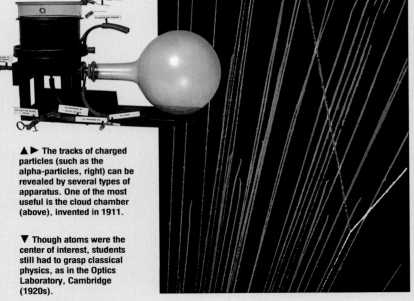

▲ ▶ The tracks of charged particles (such as the alpha-particles, right) can be revealed by several types of apparatus. One of the most useful is the cloud chamber (above), invented in 1911.

▼ Though atoms were the center of interest, students still had to grasp classical physics, as in the Optics Laboratory, Cambridge (1920s).

Dirac, the Lucasian Professor of Mathematics at Cambridge, developed in 1928 a relativistic theory of the electron, involving quantum mechanics. This theory predicted not only the existence of the intrinsic angular momentum, or "spin", of particles such as the electron, but also that of antimatter (particles identical to the familiar ones, but carrying an opposite charge). The positron (antielectron) was discovered in 1932. In 1926, in Austria, Erwin Schrödinger had begun to publish the first of a series of papers elaborating his concept of wave mechanics, describing the behavior of electrons and other particles in terms of a wave field.

As the medium through which we most directly observe the world about us, light has long been the subject of scientific speculation. That it travels in straight lines was implicit in Euclid's *Optics* in the 3rd century BC and in the 2nd century AD Ptolemy reported that the angles of incidence and reflection were equal and had formulated an approximation to the law of refraction. But it was unanimously believed that it traveled instantaneously, with infinite velocity. Interest in light then lapsed, but was restimulated with the advent of telescopes, microscopes and other optical instruments in the 17th century.

In the theoretical field controversy raged over the nature of light. Some, like Isaac Newton, believed it to be corpuscular or particulate; his contemporary, the Dutch philosopher Christiaan Huygens, ascribed to it a wavelike character. For a century, experimental evidence remained controversial. By the beginning of the 20th century, however, it seemed that the proponents of the wave theory had triumphed. Thomas Young in Britain, and Augustin Fresnel in France, had advanced wave theories which accounted for reflection and refraction and for the fact that rays of light proceeding from two closely adjacent slits "interfere": that is, there is at a distance a pattern of light which could be plausibly explained as resulting from crests and troughs of waves alternately canceling and reinforcing each other. More significantly, it at last proved possible to measure the speed of light in a medium denser than air, such as glass or water: faster according to particle theory, slower if it was a wave. In fact, it was slower. In his monumental *Treatise on Electricity and Magnetism* (1873), James Clerk Maxwell had on theoretical grounds identified light as part of a broad spectrum of electromagnetic waves, and in 1886 H.R. Hertz demonstrated experimentally the existence of radio-frequency waves which would be reflected and refracted just like light, and traveled at the same speed. Subject to clarifying a few points of detail, the evidence for the wave theory seemed overwhelming.

Then, at the turn of the century, serious doubts arose. Planck's quantum theory postulated that radiation was not continuous but was divided up into "quanta", the size of which were strictly related to their frequency. This surprising, and intellectually difficult, concept gained further credence when Einstein adopted it in 1905 in formulating an explanation for the photoelectric effect, in which light striking metals can cause the emission of electrons.

In 1923 Louis de Broglie, in France, showed that a wave could be associated with every particle and related to its momentum (the product of its mass and its velocity). Three years later, Erwin Schrödinger developed this concept further, expressing it mathematically in the form of a wave equation. In some circumstances the solution to the equation involves integers which can be identified with the quantum numbers of the particle described. In effect, Schrödinger had achieved a compromise: matter could partake of the nature of either particles or waves. The fact that electrons could behave as waves was demonstrated by Davisson, Gesmer and G.P. Thomson in 1927, by showing that they could be diffracted as they passed through narrow slits. However, the wave does not precisely identify the position of a particle but rather the probability of finding it at a certain point. In 1927 W.K. Heisenberg explained this with his uncertainty principle, which states that it is impossible precisely to define both the momentum and the position of a particle at any instant in time: the more precisely one is determined, the less precisely the other.

Although Heisenberg's principle had resolved most remaining anomalies it was so revolutionary that it was at first only reluctantly accepted. In particular, it ran directly contrary to a universally accepted dictum of the great French mathematical physicist Pierre-Simon Laplace (1749–1827). More than a century previously he had asserted that, if the position and velocity of every particle in the universe were known at any given moment in time, its past history and future development could be unequivocally determined. Heisenberg's concept challenged this, asserting that Laplace's condition could never be met. Probability had replaced certainty.

In their different ways, Einstein (in his theory of relativity) and Schrödinger both overturned and recast what had long been accepted an immutable laws of nature. Einstein showed that energy and mass were not distinct entities as had been supposed but were related by his equation $E=mc^2$; Schrödinger that there was a duality between particles and waves. Today quantum mechanics seems as unassailable as classical Newtonian physics once did. The old axiom that *Natura non facit saltus* (Nature makes no leap) is refuted: Nature works discontinuously.

◀ Heisenberg's uncertainty principle (1927) excited much public interest. Although its subtleties were not clearly understood, the thought that science was not, after all, a matter of certainty, but contained an element of chance, was intriguing. This cartoon of the 1930s shows George Gamow — one of the great exponents of the new physics — playing billiards with quantum balls. The caption reads: "The white ball went in all directions".

My original decision to devote myself to science was a direct result of the discovery which has never ceased to fill me with enthusiasm since my early youth – the comprehension of the far from obvious fact that the laws of human reasoning coincide with the laws governing the sequences of the impressions we receive from the world about us; that, therefore, pure reasoning can enable man to gain an insight into the mechanism of the latter. In this connection, it is of paramount importance that the outside world is something independent from man, something absolute, and the quest for the laws which apply to this absolute appeared to me as the most sublime scientific pursuit in life.

MAX PLANCK

The expanding universe

While the atomic physicists were thus investigating a submicroscopic world, astronomers were probing the far reaches of space, making new observations and speculating about the origins of the Universe. In 1915 the American astronomer W.S. Adams was engaged in the study of stars using a spectroscope and showed that dwarf and giant stars could be distinguished from each other by their spectra. Looking at Sirius B, he discovered that it is both exceedingly hot and very small and dense. He had, in fact, discovered the first of the so-called White Dwarfs. These are stars at the end of the evolutionary road: comparable in mass to the Sun, they have collapsed to form extremely dense objects. Because of their exceptional density they have very intense gravitational fields. In 1924 Adams found evidence of such a field in the spectral shift of the light emitted by Sirius B, giving additional confirmation to the general theory of relativity formulated by Einstein in 1915.

How the Universe came into existence has for long been the subject of philosophical speculation but in the 1920s speculation began to gain some support from experimental observation. In 1927,

▼ As this picture of the 2.54m (100in) telescope at Mount Wilson, California, shows the builders of telescopes faced formidable mechanical problems as well as optical ones. It was commissioned in 1917 and was the world's largest for 30 years.

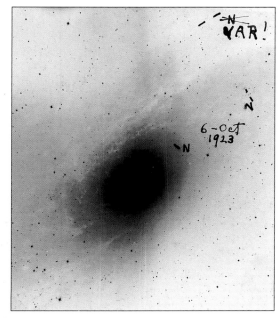

► The first known Cepheid variable star in a spiral galaxy as discovered by Edwin Hubble on 6 October 1923. Calculation of the distance of such stars proved that they belonged to galaxies outside the one to which the Sun's solar system belongs.

Hubble's work derived from that of Henrietta Leavitt (1868–1921) at Harvard Observatory and of the Danish astronomer Ejnar Hertzsprung (1873–1967). Leavitt was observing certain stars whose luminosity varies regularly, with a period of 1 to 50 days: they were named Cepheids because the first was discovered in the constellation Cepheus. Turning her attention to Cepheids in the Magellanic Clouds she discovered that the average brightness and the periodicity were related: specifically the brightness varies according to the logarithm of the periodicity. Applying her results to a few Cepheids whose distance was known, Hertzsprung was able to calculate how far away the more distant ones were, thus establishing a network of fixed points in the Universe.

From 1915 this new techinque was used by Harlow Shapley (1885–1972) at Mount Wilson Observatory, northeast of Pasadena in California, to construct the first overall picture of our Galaxy. Studying the distribution of star clusters, by means of Cepheids contained in them he found that this was not uniform but disproportionately large in the direction of Sagittarius. By 1920 his mapping had gone far enough to deduce that our own galaxy is disk-shaped, with a diameter of about 300,000 light years. This has since been shown to be something of an overestimate, but this does not invalidate his general conclusions.

The general theory of relativity
Einstein's special theory of relativity (1905) had been concerned with uniform motion, and it took him another ten years to formulate a general theory which could also accommodate accelerated motion. This brought him into the field of gravitation. Newton had interpreted planetary motion on the basis that gravitation diminished in proportion to the inverse square of the distance and Galileo had shown that bodies of different mass fall with a constant acceleration – refuting

following earlier theoretical work by A.A. Friedmann in Russia, the Belgian astronomer G.E. Lemaître deduced from Einstein's theory of relativity that the Universe should be expanding, even though Einstein himself believed it to be static. In 1929 E.P. Hubble, in the United States, was investigating the characteristics of galaxies, and showed that they were receding from the Earth with velocities proportional to their distance. Known as Hubble's Law, this was firm evidence for an expanding Universe. Knowing the rate at which the Universe is now expanding, Lemaître extrapolated backward to the instant at which the expansion began. Thus was conceived the so-called big bang theory of the origin of the Universe, now generally regarded as the most satisfactory explanation.

◄ As long as as the 18th century the English astronomer Edmond Halley identified bright cloudy patches, visible only with a telescope, which we now call nebulae. This picture shows the great nebula in Orion. The 2.54m (100in) Mount Wilson telescope provided exciting new pictures of such phenomena in the 1920s and 1930s.

Atomic Theory and the Periodic Table

Dmitry Ivanovich Mendeleyev (1834–1907) of the University of St Petersburg (now Leningrad) drew up the first periodic table in 1869. Mendeleyev took the 60 elements then known and arranged them in order of atomic weight and into eight columns that brought together those which formed oxides with the same chemical formula. Mendeleyev was confident enough to leave places for undiscovered elements and forecast their properties. The first to be discovered was gallium, found in 1875, and it was exactly as Mendeleyev had predicted.

Mendeleyev lived long enough to learn of the discovery of the electron, but not long enough to know how the electronic makeup of atoms explains the structure of the periodic table. In 1905 the great German chemist and Nobel prizewinner Alfred Werner (1886–1919) grouped the elements into the blocks that are now part of the modern table, still without realizing that these related to the electronic nature of atoms.

The final pieces of the puzzle of the periodic table were put in place in the 1920s. The discovery of the nucleus of the atom and Moseley's theory of atomic number (see p.78) had brought the realization that this was the primary property of an element, not its atomic weight. Since atomic number and weight are related it was not surprising that Mendeleyev's periodic table, based on the latter, would be essentially correct. Soddy's work on isotopes (see p.79) solved the other mystery that had puzzled Mendeleyev – why iodine with atomic weight 126.9 came after tellurium with weight 127.6.

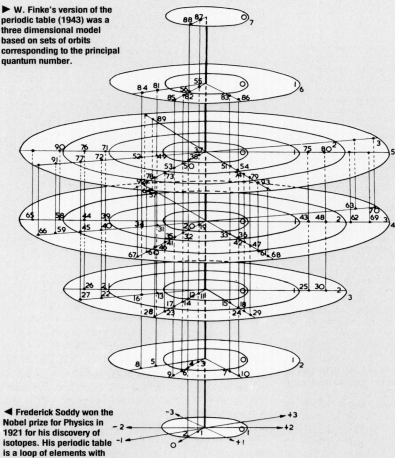

▶ **W. Finke's version of the periodic table (1943)** was a three dimensional model based on sets of orbits corresponding to the principal quantum number.

◀ **Frederick Soddy** won the Nobel prize for Physics in 1921 for his discovery of isotopes. His periodic table is a loop of elements with chemically similar ones on the same horizontal line.

▼ **Mendeleyev's version** of the periodic table continued to be popular with chemists and physicists alike, well into the 20th century. Here it is consulted by Albert Einstein and chemist John A. Miller.

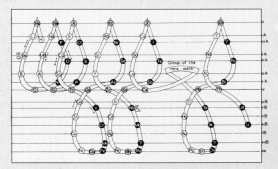

Chemists in the 1920s finally realized that their long cherished periodic table was not just a convenient way of displaying the chemical elements, but was in fact determined by fundamental properties of the atom. The eight columns of Mendeleyev's table were now seen as too restrictive and Werner's expanded form began to gain in popularity. In 1921 C.R. Bury proposed adding an extra long block to the table to hold the rare earths and radioactive metals. In 1922 Bohr supported this from his knowledge of electron shells (see p.86) and the periodic table was now virtually complete.

Bohr's theory, that electrons were in orbits around the nucleus, was the key to the periodic repetition of properties of the elements. It is the outer electrons which determine the chemistry of an element and these are in shells which surround the nucleus. As successive electrons are added to these shells the same electronic arrangements recur at regular intervals. Shells have different quantum numbers and hold only certain numbers of electrons. The s shells hold a maximum of two, the p shells hold six, the d shells ten and the f shells fourteen, and this explains why there are these numbers of columns in the s, p, d and f blocks of the so-called long form of the periodic table.

This form of the table was used by the American H.G. Deming for his textbook in 1923. Since that time the long form of the table has never been seriously challenged though many other versions were proposed. None, however, caught on.

the "commonsense" view that heavy bodies fall faster than light ones. In his new theory – developed on the basis of an abstruse form of mathematics known as the calculus of tensors – Einstein postulated that space itself is disturbed in the presence of a large mass. For more than 2,000 years the geometry of Euclid had allowed for rectilinear space, but now Einstein showed that it was curved.

This effect is of negligible importance in everyday affairs, for it is imperceptible over the short distances encountered in everyday operations on the Earth. But this is not true when distances of astronomical dimensions are involved. Thus for corroboration of Einstein's general theory scientists had to turn to the heavens, and two convincing pieces of evidence soon appeared. The first concerned the orbit of the planet Mercury, which is approximately an ellipse with the Sun at one focus. Its motion is, however, affected by the gravitational forces of other planets, the overall effect of which is to cause the axes of the ellipse to precess (rotate) very slowly. If the precession is calculated on the basis of Newtonian physics it is rather less than the value obtained by direct observation, but the discrepancy is exactly corrected if the calculation is made on the basis of general relativity.

However, this confirmation had little popular appeal: the correction amounted to only 43 seconds of arc – far less than could be perceived on an ordinary protractor – in a century. Far more interesting was the prediction validated during a total eclipse of the Sun observed in 1918 by a British expedition sent to the Gulf of Guinea. In claiming that space was curved Einstein's new theory upset another general axiom, that light travels in straight lines: according to him it should be bent in traversing a strong gravitational field. The eclipse provided a rare opportunity to put this to the test: if the beam of light from a star is bent when it passes close to the eclipsed Sun – the only time at which it can be observed in such circumstances – it will appear to be deflected from its proper position. The observation was made, the deflection was exactly as Einstein had predicted, and the result was extensively reported in the press. As a result, relativity was widely, and rightly, acclaimed as one of the great achievements of the human intellect.

New concepts in chemistry

In the late 18th and early 19th century chemistry had been set on a new course by the fundamental research of the Frenchman Antoine Lavoisier on combustion and the concept of the chemical properties of the elements being determined by the weights of their atoms which had been developed by the British scientist John Dalton. In 1868, the Russian chemist D.I. Mendeleyev marshaled the known elements systematically according to atomic weight within his "periodic table". Subsequently, with Soddy's concept of isotopes and Moseley's recognition of nuclear charge rather than weight being the determinative factor, some anomalies in the Table were ironed out and a definitive pattern emerged.

Understanding the way in which atoms react to

► Einstein's general theory of relativity upset many concepts taken as axiomatic, among them the belief that light travels in straight lines. According to Einstein, however, a ray of light should be susceptible to gravity and thus perceptibly deflected by a strong magnetic field. The total eclipse of 1918 provided a unique opportunity for testing this. Not only was the light shown to be deflected, but the magnitude of the deflection was precisely what he had predicted. This dramatic observation was widely reported in the press.

It is a natural suggestion that the greater difficulty in elucidating the transcendental laws is due to the fact that we are no longer engaged in recovering from Nature what we have ourselves put into Nature, but are at last confronted with its own intrinsic system of government.

ARTHUR EDDINGTON

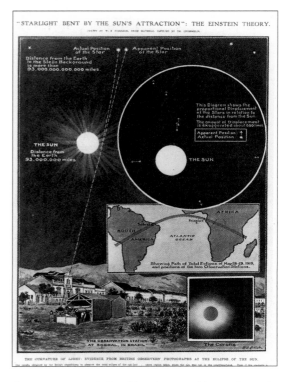

"STARLIGHT BENT BY THE SUN'S ATTRACTION": THE EINSTEIN THEORY.

form compounds is of fundamental importance to chemists, and they were quick to see the significance of the discoveries being made by the atomic physicists. One of the pioneers was Sir Ernest Rutherford's friend in Oxford, N.V. Sidgwick, who – among other theoretical advances – conceived the concept of the coordinate link, in which a chemical bond is formed by two electrons derived from one atom. His classic work *The Electronic Theory of Valency* appeared in 1927 and established his international reputation. Electronic theories of organic chemical reactions were developed in Britain by Robert Robinson and by C.K. Ingold.

Chemists also quickly realized the potential of isotopes for their research. The isotopes of a given element differ in mass but have identical chemical properties, which makes it impossible to separate them by chemical methods. Many, however, are radioactive and can thus be used to "tag" normal atoms of the same element and thus follow them through a whole series of complex reactions. It is rather akin to following the movements of a flock of starlings by fitting one with a bleeper which can be detected at a distance with a radio receiver.

The pioneer of this tracer technique was the Hungarian chemist Georg von Hevesy (1885–1966) who was led to it by his failure chemically to separate radium and lead, not at first realizing that the two were isotopes of the same element. He later developed the technique with F.A. Paneth (1887–1958), in Vienna, using labeled lead and bismuth to investigate compounds of the elements. For his contributions in this field Hevesy was awarded the Nobel Prize for Chemistry in 1943. The great advantage of the technique is that radioactivity is so easily and accurately measurable that only minute amounts of the tracer are needed.

QUANTUM PHYSICS

One of the most remarkable scientific advances of the 20th century is the development of quantum mechanics – the description of the behavior of matter on the atomic scale. It is now a powerful tool for understanding the behavior of atoms and molecules, and is vital to physicists, chemists and biochemists alike.

Its roots lie in Max Planck's discovery at the turn of the century that the radiation from a hot object can be successfully described only if it occurs with specific amounts of energy – "quanta" – rather than with a continuous range of energies. This discovery led ultimately to the description of light in terms of "particles", known as photons, the name coined in 1926 by the American Gilbert Lewis.

In 1913 the Dane Niels Bohr built on these ideas to postulate that the energy of the atomic electrons must also be "quantized". The model explained the origin of the spectra of light emitted by atoms such as hydrogen, which had long been recognized to have characteristic and separated lines of color. But the explanation of *why* the energy of the electrons should be quantized had to wait until the mid-1920s with the full development of the mathematical formulation known as quantum mechanics by the Austrian Erwin Schrödinger, the German Werner Heisenberg and the British physicist Paul Dirac.

Schrödinger's theory of quantum wave mechanics treated the electron with a wavelike description, the amplitude of the wave giving the probability of finding the electron at a given point in space and time. This wave, like the electromagnetic waves of radiation, was subject to quantization, and the energy levels (shells) in Bohr's model could be explained in terms of the allowed energies of an electron-wave, effectively caught by the electric attraction of the nucleus.

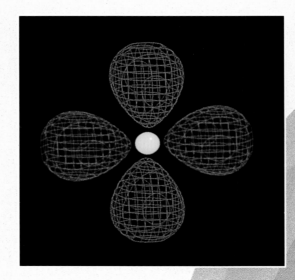

▲ In quantum mechanics the planetary orbits of Bohr's model of the atom are replaced by "orbitals" – three-dimensional regions in which the atomic electrons move, as in this example for sulfur. The orbitals come from solutions to Schrödinger's wave equation, which describes the atomic electrons as waves, trapped as if in a box by electrical attraction to the nucleus. They show the probability of locating an electron with a given energy in the space around the nucleus, and reflect the uncertainty in knowing an electron's position and velocity simultaneously.

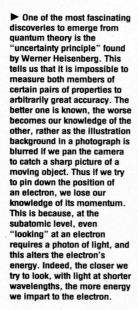

► One of the most fascinating discoveries to emerge from quantum theory is the "uncertainty principle" found by Werner Heisenberg. This tells us that it is impossible to measure both members of certain pairs of properties to arbitrarily great accuracy. The better one is known, the worse becomes our knowledge of the other, rather as the illustration background in a photograph is blurred if we pan the camera to catch a sharp picture of a moving object. Thus if we try to pin down the position of an electron, we lose our knowledge of its momentum. This is because, at the subatomic level, even "looking" at an electron requires a photon of light, and this alters the electron's energy. Indeed, the closer we try to look, with light at shorter wavelengths, the more energy we impart to the electron.

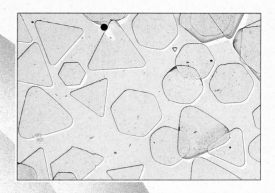

▶ Photography began in the 1830s, but a century was to pass before scientists could begin to explain how it works. One necessary step was to recognize that although light often behaves like a wave, it is sometimes more helpful to think of it in terms of "particles", called photons. This is especially true when dealing with the detailed processes that occur when atoms in a material absorb light. An unexposed photographic film is basically an "emulsion" of gelatin containing crystals of silver bromide (top). When an atom in a grain absorbs a photon, it releases an electron which migrates to the edge of the crystal and combines with a silver ion to form a silver atom. During development, the whole grain turns to silver (bottom) while those that were not hit by photons remain unchanged.

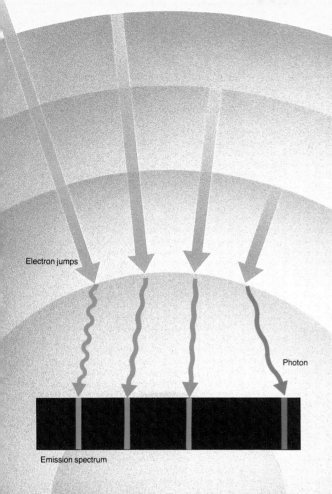

Electron jumps

Photon

Emission spectrum

Hydrogen nucleus   Energy level 1   Energy level 2   Energy level 3   Energy level 4   Energy level 5

▲ Niels Bohr postulated that the electrons inside an atom do not spiral in to the nucleus because they lose energy only in discrete amounts, or "quanta". From this he was able to explain the spectrum of light emitted by hydrogen, the simplest element with only one electron. The electron normally occupies the lowest energy level, closest to the nucleus. But when hydrogen is energized, its electrons gain energy and move to higher levels. They later jump back to lower energy levels, emitting energy (as light) as they do so. Because the wavelength of light varies with energy, being shorter for higher energies, the light is emitted at distinct wavelengths corresponding to the various gaps between energy levels. Jumps to the first energy level are at ultraviolet wavelengths, but jumps to the second level from the next four levels emit light at visible wavelengths.

▶ Just as it is often useful to describe light as particles, it can be helpful to think of a stream of subatomic particles, such as electrons, in terms of wave motion, where the associated wavelength decreases the greater the momentum of the particles. This relationship between wavelength and energy has proved particularly useful in the electron microscope. Because an electron beam can be made to have a much shorter wavelength than light, the beam can reveal smaller details. These photographs show bacteria seen through a light microscope, magnified about a thousand times (top) and through an electron microscope (bottom) where the magnification is approximately 30 times greater still.

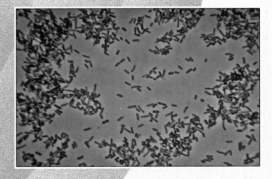

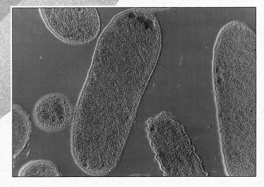

87

THE
LIBERATION
OF ENERGY

Time Chart

	1930	1931	1932	1933	1934	1935	1936	1937
Nobel Prizes	• *Chem*: H. Fischer (Ger) • *Phys*: C.V. Raman (Ind) • *Med*: K. Landsteiner (Aut)	• *Chem*: C. Bosch, F. Bergius (Ger) • *Phys*: (no award) • *Med*: O.H. Warburg (Ger)	• *Chem*: I. Langmuir (USA) • *Phys*: W. Heisenberg (Ger) • *Med*: E.D. Adrian, C. Sherrington (UK)	• *Chem*: (no award) • *Phys*: P.A.M. Dirac (UK), E. Schrödinger (Aut) • *Med*: T.H. Morgan (USA)	• *Chem*: H.C. Urey (USA) • *Phys*: (no award) • *Med*: G.H. Whipple, G.R. Minot, W.P. Murphy (USA)	• *Chem*: F. Joliot-Curie, I. Joliot-Curie (Fr) • *Phys*: J. Chadwick (UK) • *Med*: H. Spemann (Ger)	• *Chem*: P.J.W. Debye (Neth) • *Phys*: C.D. Anderson (USA), V.F. Hess (Aut) • *Med*: H.H. Dale (UK), O. Loewi (Aut)	• *Chem*: W.N. Haworth (UK) • *Phys*: C.J. Davisson (USA), G.P. Thomson (UK) • *Med*: A. von Szent-Gyorgyi (Hung)
Technology	• 7 Jan: Picture telegraphy service opened between UK and Germany • B.V. Schmidt builds the first coma-free 14-inch Schmidt mirror telescope (Eston) • Turkestan–Siberian railroad completed (USSR)	• First LP recording, of Beethoven's *Fifth Symphony*, released by RCA-Victor	• Balloon tire produced for farm tractors • RCA demonstrates a television receiver with a cathode-ray picture tube (USA) • Sydney Harbour Bridge opens, after seven years' construction (Aus)	• First all-metal wireless valve is made by Marconiphone Company • High-intensity mercury vapor lamps are introduced • Frequency modulation (FM) in radio is patented by E. Armstrong (USA)	• Introduction of a drum-scanner facsimile telegraph • A refrigeration process for meat cargoes is devised • Radar successfully demonstrated at Kiel Harbor by R. Kuhnold (Ger)	• R. Watson-Watt builds first practical aerial radar • Jan: Oil pipeline from Kirkuk (Iraq) to Haifa and Tripoli (Leb) opened • Fluorescent lighting demonstrated by General Electric Company (USA)	• First diesel-electric vessel, the *Wuppenthal*, launched (Ger) • Boulder (Hoover) Dam on Colorado River is completed (USA)	• First rocket tests performed at Peenemünde by W. von Braun and others (Ger) • W.R. Dornberger organizes the construction of the V2 rocket (Ger) • F. Whittle develops the first jet engine (UK)
Medicine	• M. Theiler develops a yellow fever vaccine (SA) • H. Zinsser develops an immunization against typhus • Uroselectan, an iodine-containing substance which can be concentrated in the kidneys, allows the kidneys to show up on X-rays (Ger)	• First clinical use of penicillin (UK) • E. Goodpasture uses live chick embryos in which to culture vaccinia virus, which can be used as protection against smallpox	• Introduction of vitallium into joint surgery (USA) • G. Domagk discovers the first sulfa drug, prontosil (Ger) • First intravenous anaesthetic, hexobarbitone sodium (Evipan), used by Weese and Scharpff (Ger)	• H.L. Marriott and A. Kekwick recommend a continuous drip technique in blood transfusion (UK) • E. Graham performs the first successful removal of a lung • Smith, Andrews and Laidlaw isolate a flu virus from the throat swabs of flu victims (UK)	• Publication of A. Fleming's *Recent Advances in Vaccine and Serum Therapy* • J.S. Lundy introduces sodium pentothal as an intravenous anaesthetic	• G. Domagk first uses prontosil for treating streptococcal infections (Ger) • A. Moniz pioneers lobotomy as a treatment for mental illness (Port)	• First use of an oxygen tent (UK) • D. Bovet discovers that sulfanilamide is as effective as prontosil in killing streptococci (Swi) • A. Carrel invents the first artificial heart	• Zinc protamine insulin is successfully used in cases of diabetes • U. Cerlutti and L. Bini develop the first form of electro-convulsive therapy (ECT) for treating schizophrenia (It) • Yellow fever vaccine developed (USA)
Biology		• A. Butenandt isolates the male sex hormone, androsterone (Ger)	• E. Adrian describes electrical activity in nerve and brain cells • H.A. Krebs discovers the urea cycle (UK)	• T. Reichstein synthesizes pure vitamin C (Swi) • Publication of L. von Bertalanffy's *Theoretical Biology* (Ger)	• A. Butenandt isolates the female sex hormone, progesterone (Ger)	• K. Lorenz describes imprinting in animal development (Aut) • R. Schoenheimer uses radioactive elements to track biochemical reactions in the body (Ger)	• A.I. Oparin argues that life developed through random chemical processes (USSR)	• H.A. Krebs discovers the Krebs cycle of respiration (UK)
Physics	• A. Eddington attempts to unify general relativity and quantum theory (UK) • W. Pauli proposes a new particle (later known as the neutrino) to account for the apparent violation of the law of conservation of energy in beta decay (Aut)	• P.W. Bridgman conducts research on materials at pressures up to 100,000 atmospheres (USA) • R.J. van de Graaff builds a high voltage electrostatic generator (USA) • J. Cockcroft develops high-voltage apparatus for atomic transmutations (UK)	• J. Chadwick discovers the neutron (UK) • J. Cockcroft and E. Walton build the first particle accelerator • C.D. Anderson discovers the positron in cosmic rays (USA) • E.O. Lawrence builds the first cyclotron (USA)	• E. Fermi proposes a theory of beta decay (USA) • E. Ruska devises a 12000 x magnification transmission electron microscope (Ger) • C.D. Anderson and R. Millikan discover positive electrons (positrons) while analyzing cosmic rays (USA)	• J.F. and I. Joliot-Curie discover induced radioactivity (Fr) • I.I. Rabi begins his work on the atomic and molecular beam magnetic resonance method for observing spectra in the radio-frequency range (USA)	• A.J. Dempster discovers the U-235 isotope of uranium (Can) • H. Yukawa proposes that a new particle causes the attraction between particles in the nucleus (Jap) • J.W. Beams separates the first isotopes by centrifuging	• E.W. Mueller develops the field-emission microscope (USA) • F. Bloch suggests a method for polarizing neutrons by passing them through magnetized iron	• C.D. Anderson discovers the mu meson (muon) in cosmic radiation (USA) • Foundation of the Nobel Institute of Physics in Stockholm (Swe) • M. Blau uses a photographic plate to examine cosmic radiation (Aut)
Chemistry	• J. H. Northrop makes pepsin and trypsin in crystallized form (USA) • Acrylic plastics are invented: Perspex in the UK, Lucite in the USA • Poly vinyl chloride (PVC) is discovered by W.L. Semon (USA)	• P. Karrer isolates Vitamin A (Swi) • ICI produce gasoline from coal (UK) • J.A. Nieuwland invents the neoprene synthetic rubber process	• H.C. Urey discovers heavy hydrogen (deuterium) (USA) • W.H. Carothers synthesizes polyamide (nylon in 1936) (USA) • R. Kuhn investigates riboflavin (Aut)	• First commercially-produced synthetic detergent is made by ICI (UK) • G.N. Lewis prepares heavy water (USA) • Vitamin B2 (riboflavin) discovered by R. Kuhn, A. von Szent-Gyorgyi and J. Wagner-Jauregg (Aut)	• A.O. Beckman develops the first pH meter • Protactinium isolated in metallic form by A.V. Grosse	• Nylon synthesized by W. Carothers (USA)	• Catalytic cracking is developed for refining petroleum	• C.A. Elvehjem discovers Vitamin A • Aneurin synthesizes Vitamin B • Basic ingredient of Milk of Magnesia is extracted from dolomite (UK)
Other	• W. Beebe and O. Barton use the bathysphere to explore the ocean floor (USA) • Discovery of Pluto by C.W. Tombaugh (USA)	• K. Gödel shows that the logical grounds for arithmetic are incomplete (Aut) • K.G. Jansky finds radio interference from the Milky Way (USA)	• K.G. Jansky establishes a foundation for the development of radio astronomy (USA)	• Scientific research in Nazi Germany is hampered by the new anti-Jewish laws	• Publication of R. Benedict's *Patterns of Culture*, a classic work in anthropology (USA)	• C.F. Richter devises scale of earthquake strength (USA)	• Lehmann proves the existence of the inner core of the Earth by observing diffracted p waves	• First intentional radio telescope built by G. Reber (USA)

1938	1939	1940	1941	1942	1943	1944	1945
• *Chem*: R. Kuhn (Ger) • *Phys*: E. Fermi (It) • *Med*: C. Heymans (Bel)	• *Chem*: A. Butenandt (Ger), L. Ružička (Swi) • *Phys*: E.O. Lawrence (USA) • *Med*: G. Domagk (Ger)	(no awards)	(no awards)	(no awards)	• *Chem*: G. von Hevesy (Hung) • *Phys*: O. Stern (USA) • *Med*: E.A. Doisy (Den), H. Dam (USA)	• *Chem*: O. Hahn (Ger) • *Phys*: I.I. Rabi (USA) • *Med*: J. Erlanger, H.S. Gasser (USA)	• *Chem*: A.I. Virtanen (Fin) • *Phys*: W. Pauli (Aut) • *Med*: A. Fleming, E.B. Chain, H.W. Florey (UK)
• Radio altimeter developed (USA) • G.H. Brown develops the vestigial sideband filter for improving TV transmitters • F. Porsche introduces prototype of the Volkswagen beetle (Ger) • L. Biró patents the ballpoint pen (Hung)	• Test flight of the first turbojet, the Heinkel He 178, designed by P. von Ohain (Ger) • I. Sikorsky constructs the first helicopter (USA) • E.H. Armstrong builds first FM station (USA)	• First color television broadcast takes place using P.C. Goldmark's system • Freeze-drying first used for food preservation (USA)	• Grand Coulee Dam, Washington, begins operation (USA) • K. Zuse's Z2 computer is the first to use electromagnetic relays and a punched tape for data entry (Ger) • Radar television developed in Allied bombers (UK)	• Magnetic tape is invented (USA) • Jul: First flight of the earliest jet-fighter, the Messerschmidt Me 262 (Ger) • First launch of the V2 rocket designed by W. von Braun (Ger) • Aqualung designed by J.-Y. Cousteau and E. Gagnan (Fr)	• Telephone repeaters improve quality of long distance calls (UK)	• First use of V1 and V2 rockets in warfare (Ger) • Unsuccessful rocket-plane, the Me 163B-1 Komet, introduced (Ger) • Greenwich Royal Observatory installs a quartz clock (UK)	• 16 Jul: Atomic bomb detonated at Alamogordo Air Base, New Mexico (USA). On 6 August an atomic bomb is dropped on Hiroshima, Japan
• H.H. Merritt and T.J. Putnam use a new anticonvulsant drug, Epantin, as a treatment for epilepsy (USA) • P. Wiles develops the first total artificial hip replacement (UK) • J. Lempert introduces fenestration as a treatment for deafness	• Gramacidin, first antibiotic in clinical use, introduced by René Dubos (Fr)	• H. Florey and E. Chain develops penicillin as an antibiotic (UK) • K. Landsteiner and A.S. Wiener discover the Rh factor in human blood (Aut)	• C.B. Huggins shows that female sex hormones can be used to control prostate cancer (Can) • S.A. Waksman coins term "antibiotic" (USA)	• S. Hertz and A. Roberts pioneer a safer treatment for hyper-thyroidism, using radioactive iodine • H.R. Griffith and E. Johnson introduce curare as a muscle relaxant during operations (Can)	• Penicillin first used to treat chronic illnesses • First kidney machine built secretly for the Dutch underground by W. Kolff • Xylocaine produced as a local anaesthetic (Swe)	• B.M. Duggar discovers the antibiotic Aureomycin (USA) • S.A. Waksman discovers streptomycin as a treatment for tuberculosis (USA) • Following work on Rh factors, H. Taussig and A. Blalock successfully operate on "blue babies"	• Introduction of water fluoridation in the USA
• C. Bridges produces a map of 1024 genes of the X chromosome of the *Drosophila* fruitfly • H. Goosen catches a living coelacanth, presumed extinct for 60 million years, in the Indian Ocean	• A.N. Belozersky begins his experimental work to show that DNA and RNA are always present in bacteria	• G.W. Beadle and E.L. Tatum begin a study of the workings of genes with the common pink bread mold, *Neurospora crassa*	• G.W. Beadle and E.L. Tatum develop the theory later known as the one-gene, one-enzyme hypothesis (USA)	• S.E. Luria obtains first good electron photomicrographs of a bacteriophage (USA)		• O. Avery, C.M. Macleod and M. McCarty isolate dioxyribonucleic acid (DNA) from bacteria (USA)	• M. Calvin begins using the carbon-14 isotope for investigating photosynthesis (USA)
• O. Hahn is the first to split the uranium atom (Ger) • F. Zernike invents the phase contrast microscope (Neth)	• Hahn and Strassman discover nuclear fission by bombarding uranium with neutrons (Ger) • N. Bohr proposes a liquid-drop model of the atomic nucleus (Ger) • J.F. Joliot-Curie demonstrates the possibility of splitting uranium-235 (Fr)	• Cavity magnetron, a new valve making radar more sensitive, invented (UK) • Cyclotron built at the University of California for producing mesotrons from atomic nuclei (USA) • Uranium-235 is isolated from the heavier isotope uranium-238 (USA)	• Dec: "Manhattan Project" of atomic research begun (USA) • G.N. Flerov discovers spontaneous fission of uranium (USSR) • J.D. Bernal investigates the physics of air raids	• H.O.G. Alfvén predicts magneto-hydrodynamic waves in plasma (Swe) • 2 Dec: E. Fermi initiates a controlled chain-reaction in the first nuclear reactor	• World's first operational nuclear reactor constructed and activated at Oak Ridge, Tennessee (USA)	• Second uranium pile built at Clinton, Tennessee, for manufacturing plutonium for an atomic bomb (USA) • New cyclotron at the Carnegie Institution, Washington, finished (USA)	• Atomic Research Centre established at Harwell (UK) • L. Jánossy investigates cosmic radiation (Hung)
• A.J. Ewins and H. Phillips synthesize sulfapyridine • Isolation of pyridoxine (Vitamin B6) • Karrer synthesizes Vitamin E • Polyamide plastic "Perlon" discovered (Ger)	• P. Müller synthesizes DDT. First use by R. Wiesman to combat an outbreak of Colorado beetle in Switzerland • ICI begins commercial production of polythene (UK) • H. Dam and E.A. Doicy isolate pure Vitamin K (USA)	• V. Du Vigneaud identifies biotin, the compound previously known as Vitamin H (USA) • M.D. Kamen discovers carbon-14 (Can) • E. McMillan and P. Abelson discover neptunium	• J.R. Whinfield and J.T. Dickson develop the synthetic polyester fiber "Terylene" (UK) ("Dacron" in the USA) • A.O. Beckman invents the spectrophotometer • E. McMillan and G.T. Seaborg discover the artificial element plutonium (USA)	• V. Du Vigneaud deduces the two-ring structure of biotin (USA) • F.H. Spedding produces two tons of pure uranium for use in developing the nuclear bomb (USA)	• Beginning of the commercial manufacture of silicones by Dow Corning Corp (USA) • A. Hofmann discovers the hallucinogenic qualities of LSD (Swi)	• Paper chromatography developed as a tool of chemical analysis by A.J.P. Martin and R.L.M. Synge (UK) • Quinine is synthesized	• Vitamin A synthesized • Patenting of the herbicide 2,4-D, closely related to 2,4,5-T (USA)
• H. Bethe proposes nuclear fusion as the reaction that causes stars to burn (Ger) • Publication of F. Boas's *General Anthropology*	• J.R. Oppenheimer discovers the properties of what is later known as a black hole (USA)	• Publication of M. Minnaert, G. Mulders and J. Houtgast's *Photometric Atlas of the Solar System*		• G. Reber makes first radio maps of the universe		• The sunflower is developed as an oil-producing crop (UK)	• First radar signals reflected from the Moon by Z. Bay (Hung)

Datafile

In the 19th century sulfuric acid was commonly regarded as "the barometer of industry", because it was used in such a wide variety of processes. In the 20th century this was no longer so, mainly because of its disproportionately high use in fertilizer manufacture. In its place power consumption, and particularly power consumption per person, came to be seen as an index of national industrial activity.

▼ Fossil fuels have been increasingly used through electricity as an intermediary. The industrialization of the Soviet Union after the 1917 Revolution relied heavily on a program of electrification, but the United States was by far the world's largest consumer.

▼ The registration of patents is a useful pointer to industrial activity. These figures for plastics patents 1931–45 reflect intense interest in Germany, stimulated by the substitutes (*Ersatz*) program. By now the inventive individual was being eclipsed by the big chemicals companies created in the 1920s.

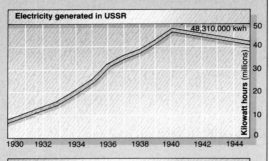

Electricity generated in USSR

48,310,000 kwh — Kilowatt hours (millions)

1930 1932 1934 1936 1938 1940 1942 1944

Plastics patents 1931–45

16%

84%

Total 5,132

☐ To firms
☐ To individuals

Electricity generated in USA

228,189,000,000 kwh — Kilowatt hours (billions)

1930 1932 1934 1936 1938 1940 1942 1944

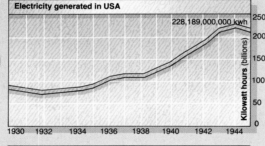

Plastics patents awarded

IG Farben
Du Pont
Eastman Kodak
Dow Chemical
ICI
American Cyanamid
CIBA
Monsanto Chemical

Hundreds

US rubber consumption

☐ Natural rubber/latex
☐ Butadiene-styrene rubber (GR-S)
☐ Other synthetic rubber

Long tons (thousands)

1941 1943 1944 1947 1949

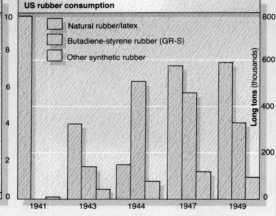

Crude oil output 1938

0.5%
6%
3.5%
13%
15%
62%

☐ North America
☐ Latin America
☐ Eastern Europe
☐ Middle East
☐ East Asia
☐ Other

US refinery products 1939

5% 3%
42%
50%

☐ Gasoline
☐ Fuel oil
☐ Illuminating oil
☐ Lubricating oil

▲ Until the Japanese largely cut off its supply of natural rubber in World War II, the USA had little interest in synthetic alternatives except for special purposes. During the war they mounted a crash program to manufacture general-purpose synthetic rubber.

◄ Before World War II the USA produced more crude petroleum and refined products than the rest of the world combined.

The 1930s and 1940s saw a significant change in the role of science in many societies. At the beginning of the period the applications of science had already profoundly influenced the pattern of life in the Western world, though with considerable regional variations. The telephone, for example, was far more rapidly adopted in Sweden than in the rest of Europe. From the 1920s most countries had national systems of radio broadcasting. Motoring was transforming the transport field. The cinema had become a popular form of mass entertainment.

The list is impressive, and could be extended, but the public gave little credit to the scientists who had played an essential role in bringing about such developments. The motor industry was identified with the industrial leaders, such as Henry Ford in the USA and William Morris in the UK, rather than with the technologists who had developed the new alloys needed for engines and chassis; with the heads of the great oil companies rather than the engineers who sank the oil wells and invented the new chemical processes essential to maintaining the flow of petrol. In radio popular broadcasters were better known than the new race of electronic engineers who invented the complex thermionic tubes and other components necessary for radio transmitters and receivers. In the film world it was the directors and stars – the Hitchcocks and Garbos – who were esteemed rather than the technologists who had developed equipment and processes.

That this should be so is not altogether surprising, for most of these scientific and technological advances had developed gradually, so the public only slowly became aware of them. Scientists were also generally ignored by the publicity services commanded by the captains of industry. Indeed, most had no desire for publicity, preferring to be judged by their scientific peers rather than by the public at large.

The war changed all this. Advances which had been kept secret, or at least not publicized, were slowly revealed. The strategic significance of radar, the degaussing of ships to protect them against magnetic mines, the proximity fuse, aerosol-dispersed insecticides to control the insect vectors of disease, even the Biro pen which pilots could use at high altitudes without leaking, were all revealed to a suitably impressed public. But the biggest single factor was undoubtedly the atomic bomb. At the time, the bombing of Hiroshima and Nagasaki in August 1945, followed almost immediately by the unconditional surrender of Japan, seemed little short of miraculous. Only later, as the full horrors began to be seen in historical perspective, and the implications for the future of the human race had to be faced, were misgivings to be voiced.

POWER FOR INDUSTRY AND WAR

Power generation and distribution

New kinds of electric light

Cracking processes in oil refining

Polymers, plastics and the impact of war

Radar

Jet engines, helicopters and rockets

The practical social consequences of research and development aimed primarily at wartime needs were considerable. Radar, for example, was immediately adaptable to the burgeoning civil aviation industry. The cavity magnetron which generated the intense centimeter-wavelength radiation for radar eventually entered the kitchen in the microwave cooker.

But the most marked and important change was in the public attitude toward scientists. Long regarded as a remote breed who carried out esoteric experiments in the remoteness of their laboratories, they emerged as real people who could change the history of the world. Young people, and those who advised them, saw science as not only the salvation of mankind but a high-road to fame and fortune. Without science, life's battle might be lost before it had properly begun. This change of attitude was strikingly reflected both in the enhanced importance attached to science in education – especially tertiary education – and in the media. For a time science rode the crest of the wave.

▼ A crucial feature of Soviet policy after the Bolshevik Revolution of 1917 was to invest heavily in science and technology. This typical poster (1932), by Alexei Kokorekin, urges the need to "Unite the power of science with the creative energy of the working class".

Electricity generation and distribution

For many purposes requiring electricity, such as telegraphy, only a low-power source was required, and various forms of battery sufficed. A more powerful sustained supply – such as that needed for lighting – was feasible only when reliable mechanical generators became available from the 1880s. Thereafter progress was steady, but haphazard. Public supplies became increasingly available in urban areas but there was no standardization of voltage or frequency – in some areas frequency was irrelevant because only direct current was available. This was discouraging for the consumer: electrical appliances could not be transferred from area to area because of their differing ratings. Moreover, with only local markets to supply the appliance industry too was fragmented.

The British experience typifies the general course of events throughout the western world. There the key event was the establishment of a government committee in 1925 to consider national electricity supply. Its recommendation

was that a gridiron of supply lines should be set up to connect all the supply areas. An eight-year program to construct the Grid was begun in 1928: by 1935, when the scheme was almost complete, nearly 5,000km (3,000mi) of primary transmission lines were in operation.

To unify supply, frequency and voltage had to be standardized. In 1926 three-quarters of the electricity supply in Britain was three-phase with a frequency of 50Hz; as this was also common in Europe it was agreed to adopt it as standard.

Nevertheless, there were considerable areas where it was different: northeast England was generally 40Hz and part of London was 25Hz. In some areas, such as part of north London, only direct current was available.

By 1935, when the Grid was largely complete, it was still not possible to transmit power freely from point to point, because of voltage differences. There were over 600 supply undertakings, providing electricity at voltages ranging from 100 to 480 volts: not until 1945 – after

▼ One of the most significant and dramatic technological developments of the 20th century was the spectacular growth of the electrical industry, an industry which had no earlier counterpart. This picture of New York in the 1930s epitomizes the prodigal use of electricity. Not until the OPEC oil crisis of 1973 was fuel conservation to be a matter for serious international concern.

generating and distribution systems and appliances had been modified – was it finally possible to standardize nationally at 240 volts. By that time generating capacity had risen to 12,000-MW, compared with 5,000MW when the National Grid was first approved. Since 1928 the number of houses wired for electricity had increased threefold (from about 2 million to 6 million).

Whenever electricity flows through a conductor, such as a metal wire, power is lost in the form of heat. Sometimes this is a desired effect – as in an incandescent filament lamp – but in the transmission of electricity it is wasted energy. This can be minimized in two ways. Firstly, by using cables that are good electrical conductors; secondly, by transmitting at high voltages.

Of the commonly available metals the best conductor is silver but this is too expensive for general use. The next best is copper and initially this was widely used. After World War I, however, the price of copper rose considerably and increasing use was made of aluminum, which although not such a good conductor was a good deal cheaper and much lighter. Two defects had to be overcome, however. The first was its poor mechanical strength, a problem when cable has to span long distances between pylons: this was overcome by wrapping it round a supporting steel core. The second was corrosion. This was solved by developing special resistant alloys.

The question of transmission voltage is complicated, but the basic fact is that losses diminish in proportion to the square of the voltage. Consequently, substantial savings can be effected by using very high voltages, but there are inherent penalties. The most important are those presented by generation and insulation. In Britain 132,000 volts was chosen for the main power lines but for the US Hoover Dam in the mid-1930s 289,000 volts was specified for its 500km (300mi) main transmission line. For local distribution these voltages must be stepped down by transformers in substations.

Sources of light
Until the 19th century artificial light was for all practical purposes synonymous with a naked flame. Not until the advent of electricity was there any fundamental change. This gave birth, well before the turn of the century, to the arc lamp and the familiar incandescent filament lamps. Of the two, the incandescent filament lamp – since made literally in billions – was by far the most familiar and convenient, but its efficiency was low.

The domestic consumer was not unduly bothered by its wasteful use of electricity, because the consumption was quite small. The case was very different for big users, however – factories and commercial premises of all kinds and, especially, local authorities obliged to provide lighting for kilometers of streets and large public places. There was, therefore, an incentive in the 1930s to develop more efficient electric lamps and interest turned to ones based on quite a different principle – the discharge of an electric current through a gas. In 1901 the Cooper-Hewitt lamp had appeared (in the UK), in which the discharge was through mercury vapor: this gives a bluish

light. In 1910 George Claude demonstrated in Paris a different form of discharge lamp, one filled with neon gas: this gave a characteristic reddish yellow light. It was developed in two forms: long tubes for advertising displays and a more compact form for street lighting. Progress was slow, however, and the now ubiquitous neon lighting was not widely adopted until the 1930s. Like the mercury vapor lamp, the light from a neon lamp is monochromatic but in a region of the spectrum to which the eye is particularly sensitive.

The neon discharge lamp has about three times the efficiency of incandescent filament lamps, but it is unsuitable for domestic use: apart from producing a monochromatic light unacceptable in this context, it is expensive and bulky.

In these circumstances interest switched back to the mercury discharge lamp. One reason for its relatively low efficiency is that part of the radiation it emits is in the ultraviolet region of the spectrum which is not only invisible to the human eye but harmful to it, so that it must be filtered off and wasted. But if this invisible light falls on certain substances, known as phosphors, they glow and emit visible light. By suitably blending phosphors the emitted light can be made to approximate to normal daylight, or to warmer colors if desired. The fluorescent lamp, which first appeared in the mid-1930s, consists essentially of a tube, coated on the inside with an appropriate mixture of phosphors, in which an electric discharge passes through mercury vapor. After World War II it was widely adopted for commercial and industrial purposes.

▲ Lenin's equation for modernizing the Soviet Union was: Electrification plus Soviet power equals Communism. The great rivers of Russia favored the exploitation of hydroelectric power. In 1922 he proclaimed that "Dnieprostrol (above) will be the great monument of the electrification of the USSR".

▼ In London a similar message – "Power: the Nerve Centre …" was spelled out on the posters of the Underground.

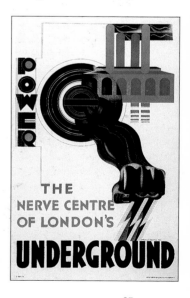

▲ In the 1920s the refining of crude oil to satisfy the growing demands of the automobile for gasoline – and of diesel engines for the heavier fractions – had become a major industry worldwide. This picture shows the Texas Company's refinery at Port Arthur, Texas, which operated the oil industry's first continuous thermal cracking process.

▼ Although it was particularly identified for many years with the United States, oil production was already an international business by the turn of the 20th century. The Nobel brothers operated oilfields in the Caspian in the 1880s. This picture shows drilling in Kuwait.

Oil refining

The petroleum industry is today so closely identified with transport that it is easy to forget that its origins had nothing to do with this. When E.L. Drake sank his first oil well in Pennsylvania in 1859, thus starting a great new industry, the first automobiles were still some 30 years in the future. His main target was lamp oil, and to obtain this viscous crude oil had to be fractionated (broken down into constituents) by distillation. This process yielded some useful by-products, including lubricants and a wax suitable for candles.

The lightest product of all, obtained in large quantities, was an embarrassment. It was highly flammable, with a low flashpoint, and was difficult to dispose of, save for limited use as a solvent. Yet its production was unavoidable, for it came off the still before the fractions actually required. However, it was these very qualities that made it invaluable when the pioneers of the automobile industry, notably Gottlieb Daimler in Germany, sought to use the internal combustion engine as a power unit for road vehicles. As a liquid it was easily handled and transported and it was so volatile that it could easily be vaporized to form an explosive mixture in the cylinder. Moreover, it was cheap and abundant.

It was not long before the tail was wagging the dog. With gas lighting given a new lease of life with the advent of the Welsbach mantle (a construction of asbestos impregnated with "rare earth" oxides which emits brilliant light when heated, invented in 1885), and increasing use of electric lighting, the demand for lamp-oil and candles slackened. At the same time the demand for petrol rocketed and, paradoxically, a situation arose which was exactly the opposite of that existing in the early days of the industry. Then, unwanted petrol had to be made to produce sufficient lamp-oil and other desired products: by the 1920s large quantities of unwanted higher fractions accumulated in order to meet the demand for petrol. Moreover, by then engine design was more sophisticated so even narrower fractions of the petrol fraction were required.

Basically, petroleum is a complex mixture of hydrocarbons, differing from each other in the number and arrangement of carbon atoms joined to each other to form individual molecules. In principle, therefore, one means of balancing supply and demand would be to "crack" some of the longer-chain molecules and so convert them into shorter ones of the required length. One way of doing this is simply to heat the higher fractions strongly – to around 500°C – in order to rupture some of the carbon-carbon bonds. Such a thermal cracking process was introduced about 1912, but was not widely practiced until the 1920s. Apart from producing more high-grade motor fuel, it also yielded as by-products hydrocarbon gases suitable as raw materials for the American chemical industry. However, these gases were not always desired in preference to a better yield of petrol and there was therefore much interest in alternative catalytic cracking processes which yielded a lower proportion of gas. The first major success here was that of the French engineer

E.J. Houdry; after some ten years of development work by a consortium of oil companies – at a cost of around 10 million US dollars – his process first became operational in 1936. In a variation introduced a little later – the so-called fluid process – the catalyst is differently handled, making it easier to control the composition of the product.

These cracking processes made it easier for the oil refineries to balance supply against demand, but by the mid-1950s the demand for high-octane fuels for new high-compression automobile engines was such that even some straight-run petrol fractions had to be subjected to catalytic reforming. But the refineries had also to provide fuel for an internal combustion engine with rather different requirements. This was the diesel engine, devised by the German engineer Rudolf Diesel around the turn of the century. In this, the fuel in the cylinder is ignited not by an electric spark but by the heat generated by the compression of the fuel mixture by the piston. It requires a rather heavier fuel, which is injected into the cylinder.

The development of polymers

The modern plastics industry effectively dates from 1907 when almost simultaneously L.H. Baekeland in the USA and James Swinburne in the UK patented the first phenol-formaldehyde plastic, subsequently known as Bakelite. The discovery was timely, for the material was a good insulator and could easily be molded into the multiplicity of small items needed for the electrical industry and for the electrical components of automobiles. For these purposes the brown color of the plastic did not matter but the later colorless formaldehyde resins based on urea and melamine found wider application as they could be attractively colored.

The 1930s saw the advent of another transparent plastic, polymethylmethacrylate, which became familiar under trade names such as Perspex and Plexiglass. Initially it was almost entirely used for the windshields and cockpit covers of military aircraft, for which it was ideally suitable, but after the war it found many other uses as a lightweight substitute for glass.

After the Nazis' seizure of power in 1933, Germany paid much attention to developing domestic substitutes for essential imports, of which one of the most important was rubber. This was required both as an insulator for electrical cables and for the tires of motor vehicles. For the first purpose PVC (polyvinylchloride) proved superior to rubber, being more durable, and it was also manufactured in the USA for this and other purposes, such as architectural sheeting and pipes. For tires, Germany developed Buna synthetic rubbers, but between the wars the USA was interested only in synthetic rubbers for a few special purposes. This situation changed dramatically when Japan conquered European and American rubber plantations in Asia (1941–42): in a crash program the USA managed to produce synthetic rubber at the astounding rate of nearly a million tonnes a year.

The Haber-Bosch nitrogen fixation process developed in Germany before World War I represented a major advance in chemical engineering. It was effective only at pressures some 200 times greater than atmospheric, far higher than had ever before been achieved in a large-scale manufacturing process. In the 1930s Imperial Chemical Industries in Britain embarked on an investigation of high-pressure chemistry generally. In special laboratory equipment they were able to attain pressures up to 2,000 times greater than atmospheric. In 1935 they succeeded in polymerizing ethylene, a simple gas consisting only of carbon and hydrogen. The new product, polyethylene (polythene), proved to be an excellent electrical insulator, repelling water. It was tough and could be easily molded. This combination of properties suggested a specialist use in submarine cables. Trials made in 1938 were so successful that ICI commissioned a commercial plant with an annual capacity of 200 tons. This came on stream on 1 September 1939, the day on which Germany invaded Poland and precipitated World War II in Europe.

This date was very significant, for the first major use of polythene was as an insulator in the development of centimeter-wave radar, for which its unique combination of properties proved virtually tailor-made. It arrived on the scene too late to be incorporated in the defensive radar chain constructed in Britain before the war but was crucially important in the shipborne and airborne radar equipment developed later.

At this stage it still appeared that polythene's main use would be in the electrical industry, as an insulator: expectations were that production would be at most a few thousand tonnes per

▶ Polymers – exemplified by bakelite and rayon – began to come into use long before there was any clear understanding of their nature. Some chemists believed them to be giant molecules, each containing thousands of atoms; others that they were aggregates of quite small molecules. The first view, powerfully advocated by Hermann Staudinger, professor of chemistry at Freiburg in Germany (1926–51), proved correct. Belatedly, he received a Nobel prize in 1953.

Man-made Fibers

The textile industry is literally as old as civilization but until the end of the 19th century there was virtually no change in the raw materials it utilized. These were primarily the natural vegetable fibers cotton and flax and those of animal origin – wool and silk. The first break with tradition came around the turn of the century, when C.F. Cross and E.J. Bevan in 1892 invented viscose rayon, made by a chemical process from cellulose. Manufacture was first taken up in Britain by Courtaulds in Coventry.

By the mid-1930s world production of rayon (largely viscose rayon) was around 750,000 tonnes a year, comparable with wool at around 1.5 million tonnes, but far short of cotton at over 6 million tonnes. Nevertheless, rayon was sufficiently big business for the chemical industry to explore other possible manmade fibers, especially wholly synthetic ones.

The first to achieve a major success was Du Pont in the USA, where in 1935 W.H. Carothers discovered the fiber-forming potential of polyamides (a kind of polymer). To transform these into a marketable product was a difficult task but Nylon 66 came on the market in 1939. It was an immediate success, especially in the stocking market: nylons and stockings soon became virtual synonyms. Later, nylon was used in many other ways – in fashion fabrics, ropes and fishing lines.

In Britain a chemically different sort of fiber – a polyester – was patented in 1941 but was not substantially developed until after World War II. It was marketed by ICI in Britain and the rest of the world, except the USA, as Terylene; in the USA as Dacron. This found many applications in the clothing industry, as fabric and knitwear but not as stockings; like rayon its relaxation time was wrong so that stockings bagged.

◄ **Model wearing artificial clothes and (inset) W.H. Carothers.**

◄ **Growing understanding of the chemistry of polymers in the 1930s made it possible to develop new ones systematically instead of empirically. One of the earliest fruits of this new approach was polymethylmethacrylate, a transparent, light-weight material marketed under various names such as Perspex and Plexiglass. Becoming available just before World War II, almost all available supplies were used to make cockpit covers for military aircraft, such as this German JU87 Stuka. Later is found many uses as a substitute for glass, for example, in roofing. Its attractive physical properties led also to other uses in opaque varieties colored with pigments.**

year. This prediction was utterly confounded by events. Although the electrical outlet remained important, polythene proved highly successful for the manufacture of a wide range of moldings, such as toys and household goods. It also made excellent film. Production worldwide soared, especially with the advent of low-temperature low-pressure processes in the 1950s: no more than 1,000 tons in 1945, it had risen to a million tons in 1960 and five million tons ten years later. What had started as a specialist chemical emerged as a basic world commodity with an importance akin to rubber or cotton.

The reason why plastics are all based on long carbon chains is simply that this element is virtually alone in being able to unite atom-to-atom in this way. The only other element which has this property – and to only a very limited extent – is silicon. During and since the war a range of silicon plastics – the silicones – have been produced for application where their anti-adhesive properties, water repellency, and heat resistance justify their high cost. They range from fluids to jellies and tough flexible solids.

Basically, plastics are of two kinds: thermosetting and thermoplastic. The former – such as urea-formaldehyde – set permanently solid when they are heated; the latter – such as polythene – can be repeatedly softened and hardened by heating and cooling. In either event they have to be shaped by molding, casting, extrusion through dies, or in other mechanical ways.

A major development in the plastic industry has been in the ability to make very large articles, such as baths or lengths of gas or water main. There is, however, an upper limit to what can be achieved by this means and if only a small number of items are to be made the considerable cost of a mold may not be justified. In such circumstances large moldings – such as the hulls of small boats – can be made by coating a former, often of wood, with a paste of plastic and reinforcing glass fiber. The plastics used are epoxy resins, generally familiar as "Araldite" and similar household adhesives. Here the setting is effected not by the application of heat but by chemical reaction between two components.

The cheapness and ease of fabrication of plastics led to some unexpected uses. In 1958 there was a worldwide craze for Hula-Hoops, most of them made of polythene. While it lasted, many millions were made: one New York store sold more than 2,000 in a single day. Later a similar boom arose with a craze for skateboards.

Radar

The phenomenon of the echo has been familiar since the earliest times. The principle is very simple: when a sound wave hits a hard boundary, such as a vertical cliff, it is reflected back much as light reflects from a mirror. Thus a handclap or any other short sharp sound comes back again after an interval of time. By timing the interval between handclap and echo the distance of the reflecting object is easily calculated. Implicit in this very simple concept are the essential principles of radar though – as is commonly the case – the transition from theory to practice was slow and laborious.

Where the story begins is ill-defined: one might go back to Heinrich Hertz's classic experiments in the 1880s when he showed that, like light, radio waves could be reflected by solid objects. More immediately relevant, however, were observations in the early 1920s that wireless waves emitted by radio beacons – increasingly used for ship and aircraft navigation – were distorted by passing aircraft or flocks of birds. By 1927 thunderstorms, a serious menace to aircraft, were being located through the severe electrical disturbances associated with them. By the early 1930s, in an uneasy peace, the military possibilities of detecting approaching aircraft were being seriously explored in the USA, Britain and Germany: the latter, indeed, had a simple radar device as early as 1933.

The most successful development was in Britain, not so much because it was technically the most advanced but because the pioneer, Robert Watson-Watt, realized that a detector was not in itself sufficient: it had to be built into a defense system that could be operated by regular troops.

Watson-Watt's original remit, in 1934, was to advise the British government whether a radio beam could be developed for use as a destructive weapon. He concluded that it could not, but his report to the UK Radio Research Board set out in detail the principles involved in the radio detection of aircraft. During 1935 detection was demonstrated at distances up to 32km (20mi). Almost immediately authorization was given for the construction of a chain of radar stations to defend

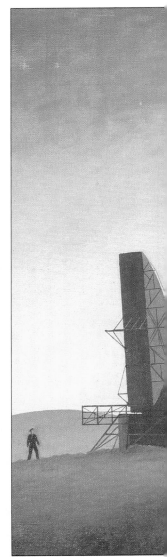

◀ In the 1930s all the major powers were experimenting, in strict secrecy, with various kinds of radio location systems. Originally only land-based detectors were feasible, but with the availability of lighter and more sophisticated equipment airborne radar became possible. This picture shows a German Messerschmitt 110 fitted with an array of radar aerials.

Britain, and this was virtually complete when war broke out in 1939.

Three major technical considerations governed the development of radar. Firstly, because of the extremely high speed of radio waves – equal to that of light – the interval between transmission of a pulse and the reception of its echo is very short: at a distance of 16km (10mi) it would be around one ten-thousandth of a second. Secondly, the pulsed radio signals have to scan a large sector of the sky and the proportion of the emitted energy reflected back from a located aircraft is exceedingly small. Thus the reflection from a

▶ The turning point of World War II in Europe was undoubtedly the Battle of Britain (1940): had the Germans won this they would have had command of the sky, essential for the invasion of Britain. The biggest single factor in determining the outcome was the chain of radar stations completed just as war broke out. This picture shows a wartime control room.

bomber at a range of 200km (120mi) would be much less than a billionth of the output energy. Consequently, extremely sensitive detectors are necessary.

Finally, the wavelength of the radio wave must be small in relation to the size of the target. The emphasis in microwave generation was on a device known as the klystron, developed at Stanford University, USA. In Britain, however, a substantial advance was made in 1939 with the invention of the cavity magnetron, capable of powerful output at centimeter wavelength. This was, in fact, a novel modification of the magnetron, developed by General Electric in the USA in 1921.

A major defect of German radar was that the wavelengths used were too long, so the resolution was poor. Nevertheless, German radar was by no means ineffective. During 1940–42 a chain similar to that in Britain – the Kammhuber line – was established across northwest Europe. They also devised for their U-boats a Metox device which gave warning of aircraft with radar equipment in the vicinity. Both sides developed a simple radar-jamming device known as "Window", consisting of strips of aluminum foil scattered from aircraft.

Initially, radar was envisaged as a land-based defensive system to give early warning of the approach of hostile aircraft, but new developments became feasible as equipment became lighter and more compact. This made it possible to install it in ships both to detect aircraft from carriers and also other warships. In the later stages of the war airborne radar became feasible both in fighters to track enemy aircraft and in bombers to scan the ground beneath them, making it possible to pinpoint targets even in adverse weather.

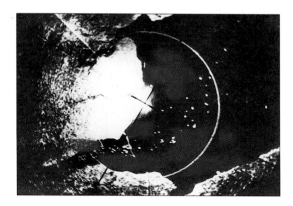

◄ As World War II developed, radar ceased to be an essentially defensive weapon used to locate the position and course of hostile aircraft and became also part of the offensive system. The use of H2S radar, and subsequent refinements, made it possible for bombers to scan the ground beneath them, even in darkness and bad weather, and to identify salient features needed for guidance. This radar screen clearly depicts a stretch of coastline and a number of ships lying offshore.

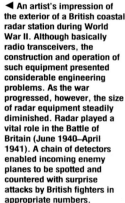

◄ An artist's impression of the exterior of a British coastal radar station during World War II. Although basically radio transceivers, the construction and operation of such equipment presented considerable engineering problems. As the war progressed, however, the size of radar equipment steadily diminished. Radar played a vital role in the Battle of Britain (June 1940–April 1941). A chain of detectors enabled incoming enemy planes to be spotted and countered with surprise attacks by British fighters in appropriate numbers.

We had the rare and almost incredible good fortune to be writing a fragment of impending world history on a clean slate...

ROBERT WATSON-WATT

The pioneers of rocketry

The Space Age may be said to have begun on 12 April 1961, when the Russian astronaut Yuri Gagarin made a single orbit of the Earth in Vostok I. This dramatic event was made possible only by combining many advanced technologies. Of these, the most important was undoubtedly rocket propulsion, for this was the only means by which a spacecraft could be put into orbit.

Although the Chinese used rockets in warfare in the 14th century, rocketry as a means of exploring the upper atmosphere and outer space is essentially a 20th-century phenomenon. The basic principles were established in 1903 by the Russian physicist K.E. Tsiolkovsky: in particular, he realized the importance of a high exhaust velocity and the need to expel exhausted fuel units as the flight proceeded. He successfully launched a multistage rocket in 1929.

Meanwhile R.H. Goddard had begun experiments at Clark University in America. Initially he used solid fuels, but he soon realized that liquid ones would generate greater power. In 1926 he successfully launched a rocket fueled with petrol and liquid oxygen. His aims were not military, but to make meteorological observations in the upper atmosphere. By 1935 his instrument-carrying rockets, with relatively sophisticated control systems, had reached a height of 2.3km (1.4mi).

In Germany events had taken a rather different course. There a leading pioneer was Hermann Oberth who, in the 1920s, designed a liquid-fueled rocket; the German Rocket Society was founded in 1927. This development attracted the

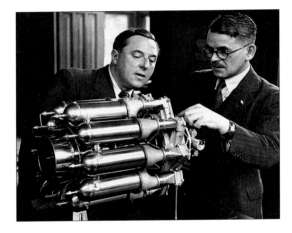

attention of the Germany army, which in 1937 established a rocket research station at Peenemünde, on the Baltic, under Wernher von Braun. It expanded rapidly, especially after the outbreak of war in 1939, and by 1945 employed 20,000 people. Here were developed a series of long-range military rockets: large-scale firing of the V2 against Allied targets began in 1944. Until the launching pads were overrun by the Allies more than a thousand of these rockets were directed against the London area and Antwerp, causing heavy casualties.

After the war, von Braun went to the USA to work on long-range missiles, and in 1960 became Director of the George C. Marshall Space Flight Center. He played a leading part in the development of the Apollo Program, which landed the first man on the Moon in July 1969.

◄ By the early 1930s it was realized that for aircraft propulsion the conventional internal combustion engine, with its multiplicity of reciprocating parts, was near the limit of its potential. In Britain and Germany attention turned to the possibility of jet propulsion. The pioneers were Frank Whittle (right in picture) and Hans von Ohain. Although Britain achieved the first successful prototype engine (1937), Germany was first in the air with the Heinkel HE178 (1939) jet-powered aircraft.

► Although Germany began to take a serious interest in rockets for military use as early as 1932, a potentially effective weapon was not developed until 1942 (the V-2, as seen here) and did not become operational until September 1944. More than a thousand were launched against London and Antwerp, causing heavy casualties. This V-2 is a captured example, being tested by US Army personnel at White Sands, USA, in 1946 or 1947.

From Autogyro to Helicopter

While the heavier-than-air machine has dominated the aircraft world since the beginning of the 20th century, it has two inherent, and related, disadvantages. To stay in the air its forward speed must be fast enough to generate sufficient aerodynamic lift; below this speed it will stall. For the same reason, a long runway is needed to achieve sufficient speed to provide lift at takeoff and room for deceleration on landing. From the early days of flying there has therefore been a considerable interest in aircraft which can take off and land vertically and hover in one position indefinitely.

The first serious attempt to make such an aircraft was made by the French aviation pioneer Louis-Charles Bréguet in 1917. He constructed a helicopter powered by four rotors, but although it hovered for fully a minute is was basically unstable and had to be steadied by assistants holding guy ropes. In 1923 the Spanish aeronaut Juan de la Cierva tried a different approach with a machine he called an autogyro. In this the lift was provided by a rotor and the forward propulsion by a conventional engine and propellor. By using hinged flowpath rotors he overcame the inherent technical difficulty that the advancing blade of the rotor travels faster than the receding one as the aircraft moves forward, giving unbalanced lift. From 1925 some hundreds of autogyros were built in Europe and the USA and continued in use well into the time of World War II.

The true helicopter, in which the rotor serves for both lift and propulsion, was not perfected until 1941, with the advent of Igor Sikorsky's VS-300 in the USA. A three-bladed horizontal rotor provided both lift and, by tilting forward, propulsion. A small antitorque rotor, turning in a vertical plane at the rear of the machine, stopped it gyrating. Although the helicopter is basically uneconomic and slow, it has unique value in many special situations. In the Vietnam War (1955–75) the United States used helicopters extensively to evacuate casualties and also developed helicopter gunships as a new tactical weapon.

▲ The development of the helicopter is particularly identified with the name of Igor Sikorsky. He built his first machine in Russia in 1909 and then left to seek his fortune in the USA. He is here seen at the controls of his successful VS-300 helicopter in May 1941.

SCIENCE AS THE FINAL FRONTIER

Although life in the 20th century has depended increasingly on the application of the products of scientific research, that activity is too technical and specialized to be appreciated fully by untrained people. Hence what the lay public perceives as science is in fact a mixture of engineering, medicine and magic. This last is not a matter of invoking supernatural agencies; any phenomenon which appears strange or wonderful, however natural its causes, conveys the effect of magic.

In this sense the exploration, or conquest, of space is not merely the last great frontier for humanity; it is an unending source of magic. This can be conveyed in unexpected ways; thus the planetaria, with their projectors resembling gargantuan ants, reproduce the night sky indoors and change its configuration at will. Magic is also reflected in imaginative literature; and here the theme of space travel has dominated "science fiction" from its beginnings at the turn of the century.

The early, primitive attempts at taking the human imagination into space were inevitably the endeavors of visionaries. As in the earlier case of flying, it was the military who provided the resources whereby these visions could become practical realities. Strange and complex bargains could be struck, as between the German rocket specialist Wernher von Braun and first the Nazis and then the Americans. The distinction between space rockets and ballistic missiles carrying hydrogen bombs was a fine one; and hence the dream of the conquest of space turned into the nightmare of Armageddon on Earth. The magic of space was blackened; the suspicion of military motives could never be lifted from this field.

But astronomy has managed to retain its innocence as a source of wonder and edification. When the astronomers and cosmologists argued about the origins of the Universe, or about the first split-second of its existence after the supposed big bang, they still attracted a large public who were not at all deterred by their lack of technical comprehension.

▲▶ The mad scientist, creating a robot capable of simulating a human being but subject to his evil will, has been a recurrent nightmare figure of the 20th century. He appeared in an early but definitive version in Fritz Lang's film *Metropolis* (1926).

▶ The threat from outer space has been domesticated in various ways. The appearance of Halley's Comet in 1910, so often the harbinger of bad tidings, was laughed away as a ghoul from a child's story. By the 1930s comics flourished, in which an intergalactic struggle between good and evil took up imagery once popular in the cowboy genre.

◄ One of the most popular television science fiction series has been *Star Trek*, created in the late 1950s, but still popular 30 years later. Here space is used as a field in which the values of modern Western man can be presented as of a genuinely universal validity – a morality play for our time. Unlike later science fiction films, the gadgetry is of little importance to the story.

▲ In the 1970s the US space agency NASA developed its own vision of life in space in the future. The fantastic space stations that were proposed differed from science fiction only in that they were officially sponsored by an organization that might genuinely aspire to build them.

◄ The planetarium, as here at the Adler Planetarium in Chicago, aims to combine a genuine educational function with a large element of show business. Once the imagination of a young audience has been captured with dreams of the stars, the future of science is assured.

105

Datafile

The hope that science could discover the root causes of diseases gave rise to new therapies and medical technologies during the 1930s. Not only did the application of physics to biological research lead ultimately to fundamental discoveries in molecular biology, but there were spin-offs for medicine. Biochemists developed new drugs such as the sulfanilamides and penicillin. Not only did these drugs control postoperative infections in surgery, but also advances in anesthesia greatly reduced pain in surgery and childbirth. Advances in embryology led to the use of tissue culture for the cultivation of viruses. Although some lines of research were rather utopian (as for artificial organs), by the 1960s kidney machines and organ transplants were becoming part of clinical medicine. World War II accelerated the process of the application of basic biological sciences to medicine. However, eugenics (the improvement of human stock) were also developed – particularly in Germany, where "inhumane" experiments sponsored by the Nazis had tragic consequences.

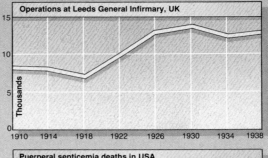

Operations at Leeds General Infirmary, UK

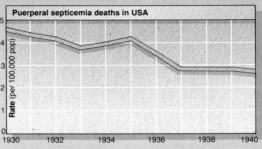

Puerperal septicemia deaths in USA

◄ Modern antiseptic surgery and the availability of painkillers have meant that hospitals have ceased to be "gateways to death". As anesthetics and the success rates of surgical operations improved there was a corresponding rise in numbers of operations, as shown in these statistics from the main hospital in the British city of Leeds.

◄ Despite the general improvement in health in the 1920s, maternal mortality remained high. The graph illustrates deaths from maternal septicemia in the USA, which only gradually began to decline during the 1930s. While sulfa drugs may have contributed, other possible factors include higher standards of maternity services.

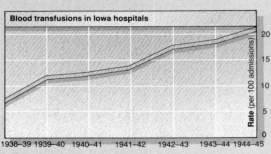

Blood transfusions in Iowa hospitals

◄ After blood groups had been discovered and blood storage perfected, regional blood banks were established. These statistics for the State University of Iowa (in the mid-West) show a steady increase in blood transfusions. Once a blood bank was established, physicians improved their understanding of how best to use transfusions.

► These statistics for aggregate national mortality show the increase in chronic degenerative diseases. Tumors now exceed tuberculosis. Heart diseases were especially high in the United States. The USA has most similarities with Australia, and France with Italy. Infant mortality continued to decline, life expectancy to increase.

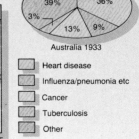

Causes of mortality

USA 1930
39% 37% 6% 9% 9%

France 1931
57% 20% 8% 7% 8%

Italy 1931
43% 28% 18% 5% 6%

Australia 1933
39% 36% 3% 13% 9%

Heart disease
Influenza/pneumonia etc
Cancer
Tuberculosis
Other

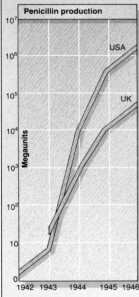

Penicillin production

◄ Although scientific understanding of penicillin was achieved in Oxford between 1939 and 1942, Britain lacked facilities for large-scale production. Clinical trials had impressive results, and US drug companies responded quickly to the possibilities. Military use preceeded civilian and commercial distribution.

► These graphs illustrate the improvement in UK medical services during World War I. Numbers of blood donations increased substantially for both military and civilian purposes (above). Diphtheria rates fell once the government introduced immunization, which had been practised in the United States and Canada since 1922.

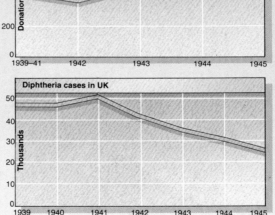

Blood donations in England & Wales

Diphtheria cases in UK

THE VICTORY OF CHEMOTHERAPY

The application of physics and chemistry to biology

A revolution in anesthetics: intravenous anesthesia

Penicillin: the medical triumph of World War II

Tissue culture and artificial organs

War and medicine

During the interwar years the three traditional scientific disciplines continued to be combined in new ways. In biology classical research in such fields as taxonomy and morphology continued, but biologists were increasingly making use of techniques derived from the physical sciences, especially physics. In Germany, for example, Hans Berger, a specialist in the relatively new field of psychiatry, found an unexpected use for his early training in physics. From 1924 he began to record and analyze the rhythmic electric currents in the brain, and used the results to diagnose mental disorders. His technique was developed in Cambridge by E.D. Adrian, and EEG (electroencephalography) soon became a widely acknowledged aid to diagnosis over a wide field, from epilepsy to brain tumors.

In chemistry a common goal is to obtain a substance in crystalline form, indicative of purity. In 1926 J.B. Sumner in the USA succeeded for the first time in crystallizing an enzyme (biochemical catalyst). In the 1930s the American biochemist J.H. Northrop crystallized a number of other enzymes, including the important digestive enzymes pepsin and trypsin. They all proved to be proteins. This was of major importance, for it allowed research to be done with pure enzymes, uncomplicated by accompanying unidentified impurities. In 1935 W.M. Stanley, also in the USA, took this work an important stage further by crystallizing a virus, obtaining tobacco mosaic virus (TMV) in the form of fine needles. It was soon shown to be a nucleoprotein (a combination of a protein with a nucleic acid). This introduced an entirely new concept: that a living organism could exist in pure crystalline form.

At about the same time the Americans G.W. Beadle and E.L. Tatum were applying biochemical techniques to the study of genetics. Producing mutants in the mold *Neurospora crassa* by means of X-rays they tested the new strains for their ability to synthesize nutrients essential for their own growth. This led them to the conclusion that every enzyme is associated with a specific gene – the so-called "one-gene one-enzyme" theory.

Similar research was done by O.T. Avery, an American bacteriologist interested in the bacteria (pneumococci) which cause pneumonia. These exist in two forms, rough and smooth. In 1928 it had been shown by F. Griffith in Britain that in mice the rough (nonvirulent) form could be transformed into the smooth (virulent) form by contact with heat-killed smooth pneumococci. Investigating this, Avery found that the transformation is effected by a substance known as deoxyribonucleic acid (DNA), which was shown in the 1950s by the American biologist A.D. Hershey to be the information carrier in the cell nucleus.

▼ Motor paralysis is a frequent sequel to poliomyelitis. When the lungs are unimpaired, but their musculature paralyzed, the iron lung – invented by Philip Drinker in 1929 – enables life to be sustained indefinitely.

The introduction of anesthesia

Although surgery has been practiced since the days of classical antiquity, there were two major limitations to the surgeon's skill. The first was the amount and duration of pain a patient could endure. The second was the high rate of mortality resulting from postoperative infection. The first obstacle was overcome with the introduction of ether anesthesia in the 1840s. The second yielded with the advent of sulfonamides, and later penicillin and other antibiotics, in the mid-1930s.

In the field of anesthesia it is generally agreed that the introduction of intravenous anesthesia was the biggest event since the adoption of inhalation anesthesia nearly a century earlier. Apart from rendering the patient unconscious almost immediately, it has obvious advantages in the case of operations involving the head and neck. Like many techniques it did not appear fully fledged, but was the culmination of earlier experiments. When barbital (veronal) was introduced

in 1902 the possibility of its intravenous use was investigated but not until 1932 did Helmuth Weese in Germany find a satisfactory drug in evipan: this was followed two years later by pentothal. By the end of the decade millions of people had successfully undergone surgery under evipan anesthesia. It proved very safe but suitable only for operations of short duration: later it was increasingly used for induction of long-term anesthesia for major surgical procedures.

During this century the role of the anesthetist has changed dramatically. The anesthetist of 1900 was little more than a technician, dropping chloroform or ether on to an absorbent pad held over the face of the patient. By the late 20th century he or she had become a highly skilled member of the operating team, monitoring the patient's heart rate and blood pressure and administering oxygen or carbon dioxide as circumstances demand. More sophisticated anesthesia made possible more advanced surgical techniques. Thus 1936 saw the first attempt, at Massachusetts General Hospital in Boston in the USA, at open-heart surgery in which the heart's function was temporarily taken over by a combined pump and oxygenator. This was the prototype of the heart-lung machine which appeared in 1953, revolutionizing surgery of the heart.

Another important new technique was that of spinal anesthesia. In 1899 cocaine was used for this purpose but proved dangerous. The advent of the synthetic product procaine in 1904 made the procedure safer and correspondingly more popular. The main hazard was then the needle – liable to break – rather than the drug. Not until mid-century was a really reliable needle devised but by then curare and other intravenously injected relaxants were becoming available.

The introduction of procaine is a reminder of the increasingly important role of the chemist in providing new and better anesthetics of all kinds. Thus cyclopropane was introduced as an inhalation anesthetic in 1934 and another important addition in this class was halothane – a volatile anesthetic containing fluorine – introduced shortly after World War II. It is widely used because induction is quick and smooth and there is a notable lack of side effects.

Overall, anesthesia is probably the most important of all advances in surgery. It made possible operations of great complexity and long duration, such as open-heart surgery and organ transplants, where the surgical team may have to work steadily for many hours on end. Not least of its blessings, of course, is the relief it can give in relieving the pains of childbirth. For this purpose chloroform is far from ideal and the advent in 1935 of trichlorethylene, chemically not dissimilar, was a major advance. This can deaden pain without causing full loss of consciousness.

Chemotherapy

Until the 1920s the medical profession had little faith in the prospect of treating with chemicals infections that had invaded the body as a whole. Quinine, of course, was a recognized specific against malaria and salvarsan and neosalvarsan were used – though not without considerable risk

– to treat syphilis. Then, in 1927, Gerhard Domagk, a director of research in pathology and bacteriology for the great German chemicals firm I.G. Farben, embarked on a systematic study of the effect of a wide range of chemicals on some pathogenic bacteria: any which showed promise were to be tested first on mice and then if of sufficient interest on human patients.

Not until 1932 did he make any notable discovery: he then found that a red dye developed by I.G. for the leather industry was remarkably successful for controlling streptococcal infections in mice. It was then established in France that the active antibacterial substance was not the dye molecule itself but a simple moiety of it known as sulfanilamide. As this substance had been discovered in 1908, though its antibacterial activity was quite unsuspected, I.G. could not patent their discovery. This discovery led to worldwide interest and clinical trials and the preparation of hundreds of compounds related to sulfanilamide. Remarkable success was recorded in the treatment of forms of streptococcal infection, such as puerperal and rheumatic fever, and of other serious infections such as meningitis, gonorrhoea and pneumonia. In the event, rather few of the new compounds matched the success of the original sulfanilamide: among the few were sulfapyridine (1938) and sulfathiazole (1940) particularly successful for treating pneumonia and sulfadiazine (1941) for cerebrospinal meningitis.

Meanwhile, in 1928, Alexander Fleming, a bacteriologist at St Mary's Hospital, London, noticed a peculiarity in a bacterial culture which had become accidentally contaminated with a mold. Around the mold colonies the bacteria had been destroyed and he deduced that during its growth the mold secreted a substance lethal to the bacteria. However, such antagonisms between microorganisms had been well known since the latter part of the 19th century (the term antibiosis had been coined in 1899 to describe the phenomenon) and there was no reason to suppose that this new example was significantly different from others already known.

▼ After the Japanese bombing of Pearl Harbor (7 December 1941 the USA launched a crash program to produce enough penicillin to treat all military casualties. One of the main research centers was the Northern Regional Research Laboratory of the US Department of Agriculture. Here the research team is seen in conference in June 1944.

◀▲ Alexander Fleming at St Mary's Hospital, London, where he discovered penicillin in 1928. Inset is the culture of staphylococci which became accidentally infected with a mold *(Penicillium notatum)*: around the mold the bacterial colonies have disappeared.

▶ In the early days of large-scale penicillin manufacture in the USA it was made by a dramatic scaling up of the laboratory method developed by the Oxford workers. Thousands of vessels were inoculated with the mold, fermented for 12 days at 23°C, and penicillin was extracted with solvents from the resulting broth. It was a cumbersome method but at the time the only one known to be reliable. Later, much more efficient deep-culture methods – akin to beer brewing – were developed.

I should like to point out that the possibility that penicillin could have practical use in clinical medicine did not enter our minds when we started our work ...

ERNST CHAIN

No further research on penicillin of any consequence was done until 1939. In that year Howard Florey and Ernst Chain, working at the Sir William Dunn School of Pathology in Oxford, decided to embark on a general study of antibiosis aimed at discovering the scientific factors involved rather than any medical application. By a fortunate chance one of the examples they chose was penicillin, and by May 1940 – when the German armies were sweeping through western Europe – they had established that penicillin was extraordinarily effective in controlling potentially lethal streptococcal infection in mice. But the mold produced only minute quantities of penicillin and to prepare enough of the new drug for even a modest clinical trial seemed almost insuperable, especially in wartime Britain. Florey thus sought, and gained, the support of the pharmaceutical industry in America, where involvement in the war after Pearl Harbor in December 1941 aroused intense interest in penicillin for the treatment of military casualties. In the event, sufficient penicillin became available to treat all military casualties after the D-Day landings in Normandy. By 1950 it was generally available for civilian use, too, and at a very low cost.

Bacteria are not the only microorganisms that can cause disease – in the tropics such deadly and debilitating diseases as sleeping sickness, amoebic dysentry and malaria are caused by trypanosomes, amoebae and plasmodia respectively. Most of them have complex life cycles and successful chemotherapy depends on being able to attack them at a sensitive stage. Malaria, for example, is transmitted by mosquitoes, which can be attacked by insecticides and use of land drainage; the parasite they transmit in biting attacks the blood cells in the body. The traditional remedy was quinine, from the bark of the *Cinchona* tree, introduced into Europe in the 17th century from South America. In 1933 the synthetic product mepacrin (atebrin) was developed

by the Bayer company in Germany and this was followed in 1943 by Paludrine, developed by ICI in Britain in response to the cutting of quinine supplies from Asia by the Japanese, However, malaria can assume many forms and the parasites, like bacteria, can become resistant.

Tissue Culture

In everyday experience tissues survive only as part of a living organism: an amputated limb or a separated shred of skin withers and dies. But in the early years of this century (1905–07) the

World War II and Medicine

World War II provided a major stimulus to clinical research. The need to prevent disease and malnutrition, and to care for the sick and wounded, meant that national coordination of research for strategic ends was necessary. Problems were made more urgent by shortages of drugs. Synthetic substitutes had to be developed as for the use of quinine in malaria treatment. In Britain the Medical Research Council (MRC) took a leading role, and in Germany the research was coordinated by the Deutsche Forschungsgemeinschaft. British biologists shifted their attention to the problem of wound healing. Peter Medawar studied the behavior of skin homografts and with J.Z. Young experimented on the use of a strong solution of fibrinogen in natural plasma as a "glue" for joining the ends of cut nerves. The MRC sponsored research on the pathology and treatment of burns in order to improve healing and treatment by preventing infection. In order

▶ An antimalaria training class. The control of malaria was a major achievement of British military medicine.

▼ World War I had shown that restoration of blood volume was one of the most effective single steps which could be taken to save the life of a wounded person. World War II saw the establishment of a blood transfusion service in Britain with panels of voluntary donors and blood banks. Blood transfusion exemplifies how military priorities had a long-term effect in improving civilian health services. The picture shows blood transfusion for American troops during the invasion of Normandy, 1944.

American embryologist Ross Harrison developed a technique – the so-called hanging drop technique – for growing cells in a nutrient solution outside the animal from which they were derived. The technique has come to be known as tissue culture. His immediate object was to resolve alternative theories about the way in which nerve fibers develop in embryonic tissues. Using nerve tubes from frog embryos he proved that the nerve fibers are outgrowths of nerve cells and not formed independently. Apart from resolving this important question in neurology he showed conclusively that the individual cell is the primary unit in the development of a multicellular organism. This is generally regarded as the greatest simple advance in the study of embryology.

Tissue culture was developed by other research workers such as Alexis Carrel, and was successfully applied to a number of problems in biology and medicine. Meanwhile the Austrian botanist W.F. Haberlandt, also at the beginning of the century, had had some success in growing plant cells in culture but full success in this field was not achieved until 1939 when R.J. Gautheret, in France, succeeded in keeping cultures of carrot tissue alive indefinitely. Later it was discovered that stems from the growing tip of a plant, the meristem, could be used to grow complete plants on a nutrient medium. Nasturtiums were grown in this way in 1946, and after 1955 meristem cultures were increasingly used by horticulturists to propagate a wide range of plants. This technique has two notable advantages. Firstly, the meristem is normally free from infection even if the parent plant carries a virus. Secondly, the speed of propagation is enormously increased. This is not only of direct commercial importance, but enables the breeder to select useful new strains more quickly than by conventional methods.

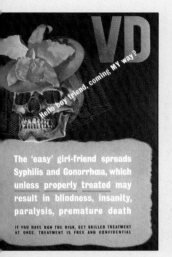

The 'easy' girl-friend spreads Syphilis and Gonorrhœa, which unless properly treated may result in blindness, insanity, paralysis, premature death

IF YOU HAVE RUN THE RISK, GET SKILLED TREATMENT AT ONCE. TREATMENT IS FREE AND CONFIDENTIAL

▲ During World War II the incidence of syphilis and gonorrhoea increased rapidly. From late 1943 penicillin had a rapid impact in controlling the infections. However, immunization against syphilis was not developed.

▼ Hadamar Mental Hospital and Educational Institution was one of the main "euthanasia" centers, where 15,000 psychiatric patients and handicapped children were killed between 1939 and 1945. After the war Heinrich Ruoff, the chief male nurse, was sentenced to death by hanging.

to prevent brain injuries, Hugh Cairns developed a protective helmet, which led to a decrease of injuries among motorcycle despatch riders. Blood transfusion was organized for civilian and military casualties. Research was carried out on the collection, storage and supply of blood, and into blood grouping as a result of the American discovery of the Rh factor. In 1943 it was established that acidification of citrate glucose solution prolonged the storage life of blood. Experiments were carried out on liquid and dried plasma and on sera as substitutes for whole blood.

Although German medical researchers were confronted by similar problems of improving treatment for war casualities, much of the research was affected by Nazi racism. Racial ideology stimulated research on eugenics and human genetics. For population groups deemed to be of high racial value, doctors provided the best possible treatment. But medicine could be lethal for anyone judged to be incurable or racially inferior. Compulsory sterilization was instituted in 1933 for the "mental defectives", schizophrenics and for sufferers from Huntington's chorea. Hitler's declaration of war in 1939 was accompanied by an order that mentally ill patients should be killed. Measures were extended to crippled babies and children, and to the elderly and to the so-called "anti-social". The "euthanasia" of those whose lives the Nazis deemed to be incurable and worthless represented a medical holocaust which preceded and complemented the genocidal killing of millions of Jews, gypsies and other population groups deemed to be racially inferior. It is estimated that 100,000 were killed as part of the "euthanasia" program.

The medical holocaust provided a stimulus to the development of human genetics, biochemistry and ethology, although the extent to which research was explicitly linked to Nazism varied. The Nazis treated those persecuted in concentration camps as racially inferior, as if they were laboratory animals. Experiments on sterilization by exposure to X-rays were linked to plans for eliminating unwanted populations. New vaccines and drugs were tested, as for typhoid at Buchenwald. Low-pressure and low-temperature experiments were conducted on prisoners at Dachau to discover how long pilots could survive in conditions of extreme cold or low pressure. (Doctors who participated in such work later assisted United States air force and space programs.)

In general the Nazi medical holocaust represents a cautionary case of the abuse of social and ethical responsibilities of doctors and researchers.

Artificial organs

The human body can survive a remarkable degree of mutilation – the loss of limbs, eyes, stomach, for example – but certain organs are essential. These include the lungs, kidneys, heart and brain. If the function of these is severely impaired, by either disease or accident, death is inevitable. For most of human history this outcome had to be accepted with resignation but during the 20th century a good deal of progress has been made in developing devices which reproduce the function of lost organs. Among the first of these was the tank respirator, the so-called iron lung, which was invented in 1929 by an American engineer, Philip Drinker. It was designed primarily for the use of polio victims whose lungs were intact but did not function because of failure of the nerves supplying them. It consisted of a sealed chamber in which the patient was enclosed from the neck down: pressure changes with the same rhythm as normal breathing forced air in and out of the lungs, allowing indefinite – but sadly restricted – survival.

The principal role of the kidneys is to excrete toxic nitrogenous waste materials from the blood; malfunction causes ill health and, in severe cases, death. In 1943 a Dutch physician, W.J. Kolff, devised the first machine which could take over the function of the kidneys. In this, the patient's blood was circulated through a cellophane filter immersed in water: the impurities diffused through the film, leaving the blood corpuscles and proteinaceous matter to go back into the bloodstream. However, Kolff's machine was of limited application: it could keep patients alive for a short time while damaged kidneys recovered but could not be used indefinitely. Not until the 1960s, with the development of the dialysis machine by B.H. Scribner, was this problem solved.

Datafile

This period saw great progress in the new science of experimental nuclear physics. Investigations of cosmic rays revealed the existence of the positively charged electron which had been predicted theoretically in 1928. In 1932 fast-moving protons were used to break up an atomic nucleus for the first time. The splitting of uranium atoms enabled the creation of the first nuclear weapons, which was achieved by the Manhattan Project.

▼ The 1920s and early 1930s saw both an expansion in the provision of university education in Italy and notable progress in many areas of science. Unfortunately, while student numbers continued to rise, many of the ablest academic scientists left the country to escape fascism.

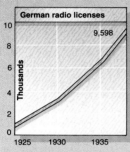

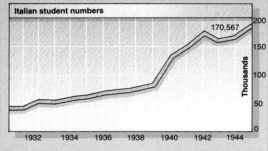

▲ Radios were enormously popular in Germany. The number of licenses issued grew from one million in 1925, the first full year for which a license was required, to nearly 10 million in 1938. This immediate access to people's homes was a valuable weapon in the propaganda campaign of the Nazis.

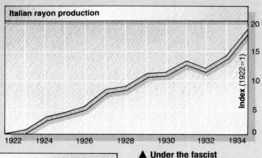

▲ Under the fascist government which took office in 1922 all indexes of industrial production in Italy increased significantly in the following years, particularly those of building, gas and electricity and the chemical industry. Rayon production, shown here, increased by a further 250 percent between 1934 and 1937.

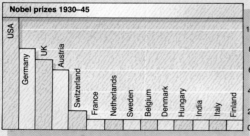

▲ In 1935 Frédéric and Irène Joliot-Curie were awarded the Nobel Prize for Chemistry for their synthesis of new radioactive elements. In physics, prizes honored the founders of quantum mechanics, Werner Heisenberg (1932), Paul Dirac and Erwin Schrödinger (both 1933). Nobel prizes were not awarded in 1940–42).

▶ The foundations of modern physics and chemistry were laid in the first 30 years of the century and were published almost entirely in German. But by the outbreak of war in 1939 an entire generation of German speaking scientists had fled the Nazi oppression of "Jewish" science; enriching science elsewhere.

Nobel prizewinners forced to leave positions in Nazi Germany

Name	Country of birth	Left	Settled in	Nobel award
Physics				
Hans Bethe	Germany	1933	USA	1967
Felix Bloch	Switzerland	1933	USA	1952
Max Born	Germany	1933	UK	1954
Albert Einstein	Germany	1933	USA	1921
James Franck	Germany	1933	USA	1925
Dennis Gabor	Hungary	1933	UK	1971
Otto Stern	Germany	1933	USA	1943
Eugene Wigner	Hungary	1933	USA	1963
Gustav Hertz	Germany	1935	—	1925
Viktor Hess	Austria	1938	USA	1936
Erwin Schrödinger	Austria	1938	Ireland	1933
Chemistry				
Fritz Haber	Germany	1933	(d.1934)	1918
George de Hevesy	Hungary	1934	Sweden	1943
Gerhard Hertzberg	Germany	1935	Canada	1971
Peter Debye	Netherlands	1940	USA	1936
Medicine				
Boris Chain	Germany	1933	UK	1945
Hans A. Krebs	Germany	1933	UK	1953
Max Delbrück	Germany	1937	USA	1969
Otto Loewi	Germany	1938	USA	1936
Otto Meyerhof	Germany	1938	USA	1922

In the popular estimation the outstanding achievements in atomic physics in the inter-war years – leading on to the atom bomb and the development of nuclear power – have greatly overshadowed other developments in the physical sciences. However, this was not the way in which contemporary observers viewed the scene. While new discoveries about the atom were interesting and had considerable appeal they were not generally seen as having any great relevance to everyday life. The reality was that progress was being made over a broad front – including such fields as metallurgy, geophysics, meteorology and oceanography – on both theoretical and practical levels.

Advances in the 1930s

Classical studies in electricity were concerned with the flow of electricity in conductors, but from the turn of the century devices depending on the flow of electrons – exemplified by J.A. Fleming's thermionic tube (valve) of 1900 – became increasingly important. In this field a major development was the electron microscope, in which an electron beam, focused electromagnetically, takes the place of the light beam and glass lens of the conventional microscope. Although it has certain inherent disadvantages these are far outweighed – in certain contexts – by the far higher magnification obtainable. The first was built by M. Knoll and E. Ruska in Germany in 1932. In this, the electron beam was transmitted through an extremely thin section of the specimen to be observed. In 1938 a scanning electron microscope was introduced which made it possible to examine the surface configuration of thick specimens.

As electronic devices grew in complexity, so too did the associated circuitry, which presented particular difficulties in the face of a trend towards miniaturization. An important development here was the invention of the printed circuit by P. Eisler in 1943: in this, as its name implies, the wiring connecting major components was prepared by using conventional printing techniques to delineate them in what was essentially a conducting ink.

It is a commonplace in science and technology that principles are often discovered long before they are put to practical use – radar is a case in point. Another is the liquid crystal display, which eventually became familiar in many forms, such as digital watches and miniature television sets. The basic principle of this important electro-optical device was enunciated in 1934 by J. Dreyer in Britain, though it was not until after the war that it became widely used.

In general accounts of the history of science and technology the role of chemistry tends to be

SPLITTING THE ATOM

underestimated and for this there seem to be two main reasons. The first is that the number of specific chemical compounds is numbered literally in millions and few have trivial, common-place names, so that the number familiar to the ordinary citizen is quite small. Examples that spring to mind are ozone, methane, chlorophyll, alum, DNA and chloroform. But chemists themselves must necessarily take refuge in an esoteric jargon which defines precisely how the various atoms are arranged in the molecule of a particular substance. This creates great difficulties in popular exposition. The second reason is that chemistry is very much an all-pervasive science: very few advances have no chemical content. The chemistry therefore tends to be taken as read and attention is focused on the ultimate scientific or technological novelty.

Most new inventions tend to be identified solely with the fields to which they contribute, ignoring the means by which they are reached. This is true, for example, of many of the medical advances of the 1930s and 1940s. The sulfonamides were possible only because the chemists of I.G. Farben had the professional expertise to synthesize a wide range of substances which do not exist in nature. Even penicillin – a discovery rather than an invention, because it is a natural product – could never have been extracted and manufactured without the assistance of skilled chemists, both pure and applied. The same is true of all drugs, from aspirin to cortisone, from paracetamol to vaccines.

Equally, it must be remembered that all plastics are chemical products. These include such man-made fibers as polyamides, polyesters, and acrilan, and moldable plastics such as polythene, PVC and perspex (lucite). Even rubber, a natural product, has to be chemically processed – as in vulcanizing – before it can be put to effective use. Other products of the chemical industry which are now indispensable are a wide range of dyes, paints, agrochemicals and explosives for both civil and military use.

Proverbially nothing comes from nothing and the chemical industry had to recognize this in respect of its raw materials. The heavy chemical industry – making traditional products such as sulfuric acid, soda and bleach – continued to expand using as raw materials minerals such as salt, limestone, coal and pyrites. But the most interesting new products were in the field of organic chemistry, representing a vast and disparate range of compounds, but united in all containing a substantial proportion of the element carbon. Inevitably every organic product had to derive from an equivalent quantity of carbonaceous raw material. Initially the basic raw material in Europe was coal tar, abundantly available as a by-product

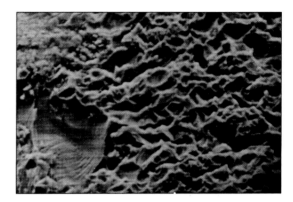

► Etched surface of aluminum, as revealed by a scanning electron microscope built in 1952.

▼ Electrons can behave not only as particles but as waves and these can be focused by magnetic coils. This principle is applied in the electron microscope: here E.A.F. Ruska, pioneer of electron microscopy, is seen (left) with an early instrument.

of manufacturing coal gas. The tar was distilled to yield a variety of chemical intermediates – benzene, tolvene, xylene, and naphthalene to name only a few. This dependence remained unchanged until the 1950s. But in the USA the situation was very different. There, natural gas had been used in substantial quantities since the 1880s and there was no comparable coal-gas industry. From the 1920s onward petroleum began to be a major raw material, and was established as such by the time of World War II.

Polar Research and the Second International Polar Year

The Earth's polar regions have been objects of intense interest to science for over 200 years and from the early 19th century onward there was a steady flow of scientific voyages to both the Arctic and the Antarctic. In the 19th and early 20th centuries it was possible to attract considerable interest to this area of scientific research because of its apparently heroic dimension and the expansionary ambitions of major powers. The research itself intensified with technological developments. Much was learnt from the indigenous people, the Inuit: using their kayak boats and dog teams it became possible to reach areas previously inaccessible. When motorized land vehicles and aircraft became available in the 1920s it was possible to maintain permanent scientific stations and observatories.

Polar research provided a strong impetus for the growth of existing environmental disciplines, such as oceanography, meteorology, terrestrial magnetism and geophysics, but scientists also became aware that the polar regions were themselves of importance in the world's oceanic and climatic systems and had to be understood in their entirety: this provided a powerful incentive to international cooperation in research.

The size of the polar regions had long encouraged some international cooperation: in 1882 11 nations had cooperated in the first International Polar Year to build circumpolar observatories. In 1932, 26 nations cooperated in the second International Polar Year. The thrust of their research was to construct a more detailed picture of the Arctic environment over a period of 12 months in all northern countries: Canada, the United States, Norway, Sweden, Finland, Iceland and the Soviet Union. Two stations were also established by Norway in Antarctica. Plans were made to coordinate measurements in four environmental disciplines: terrestrial magnetism, meteorology, atmospheric electricity and the aurora phenomena or "northern lights". To understand these continually changing phenomena, researchers stressed the need for making simultaneous measurements at all observing sites. A network of stations was constructed around the Arctic and the development of telephones, radio aparatus, photographic equipment and self-registering instruments contributed to the precise coordination of the observatories. The colonization of the Arctic Regions by the scientific community was also a financial battle, taking place at a time of great depression in the global economy. The International Polar Commission played a key role in raising the prerequisite funds through grants from charitable organizations.

This network of Polar observatories meant that more Arctic phenomena could be comprehensively scrutinized for the first time. More generally, measurements of climate could be used to extend the global models of climate developed in the more temperate regions of the earth. Other polar projects outside IPY also contributed to climatic research. Workers in the Antarctic seas, for example, were also suggesting intricate relationships between sea life and the circulation of ocean currents.

▼ **Photography played an important role in all polar science by portraying expeditions as heroic voyages and providing picturesque visual imagery of seascapes. Silhouetted figures of the 1938 Soviet Papanin expedition are shown here on their return journey aboard the icebreaker *Taimyr*. The expedition's principal research had been to carry out meteorological research while drifting on a Polar ice floe in the East Greenland Current. Yakov Khalip (1908–80) had been appointed photographer to the expedition and for his crucial contribution was awarded a medal by the Soviet press.**

Geophysics, cosmic rays and heavy elements
Like religious and political parties, science steadily generates splinter groups. In physics, one of the most viable of these has been geophysics, dealing with phenomena in the Earth's crust, such as continental drift. In 1929 the Japanese geologist Motonori Matuyama discovered, from the remanent magnetism in certain rocks, that in the course of geological time the Earth's magnetic field has reversed. It is now known that this reversal has taken place many times over the last five million years or so. This is a total north–south reversal, quite different from the local wandering of the north magnetic pole which has been known since the 16th century. The cause of these reversals is still unclear but may be due to fluctuations in convection currents in the liquid core of the Earth, about which little is known. Advances in seismology gave new information about the more easily accessible crust of the Earth. Until 1935 the violence of earthquakes had been measured in qualitative terms, but in that year C.F. Richter devised the absolute scale named after him. It is based on the logarithm of the observed amplitude of associated shock waves.

In 1910 T. Wulf discovered that radiation in the atmosphere was higher than at ground level, suggesting that it must be of extra-terrestrial origin. In 1911–12 V.F. Hess, in Austria, made a number of balloon ascents to measure the effect at heights greater than 5,000m (16,400ft). At first it was suspected that the origin was the Sun, but measurements during an eclipse in 1912 disproved this and in 1925 the radiation was named cosmic rays by the American physicist R.A. Millikan. In 1932 the French physicist A. Piccard ascended 16,000 meters (52,500ft) into the stratosphere to observe the rays.

In 1933 investigations of radiation from outer space took a new turn when Karl Jansky in the USA detected a source of radio frequency radiation in the direction of the Milky Way. Surprisingly, this excited little interest at the time but after the war it gave rise to the major new field of radio astronomy, exemplified by the huge steerable dish built in England at Jodrell Bank. Though designed to further the new science, one of its first tasks was to track the early artificial satellites.

The four elements of the ancient world were fire, earth, air and water, but in 1931 the latter was found still to have a secret to reveal. The American chemist Harold Urey, at Columbia University, had predicted, on the basis of certain regularities in the structure of atomic nuclei, that hydrogen – the lightest element with a mass of one – should exist also in the form of isotopes with masses two and three respectively. Using a highly sensitive spectrograph he identified in the spectrum of ordinary hydrogen a very faint line corresponding to an admixture of one part in 4,500 (later amended to 6,500, near the modern figure) of hydrogen-2. This was later confirmed using hydrogen enriched by fractionally distilling it in liquid form. Corresponding to this heavy hydrogen (deuterium) there must be a heavy water. This was subsequently prepared by the electrolysis of ordinary water, when the normal water is decomposed faster than heavy water.

▲ In 1925 radioactivity in the upper atmosphere was attributed to cosmic rays falling on the Earth from outer space. To investigate them in the 1930s scientists took themselves and their instruments aloft with the aid of huge balloons, such as this Soviet balloon which is about to rise into the stratosphere.

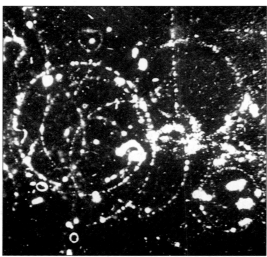

▶ Atomic particles are far too small to be visible and their presence must be inferred indirectly. In 1911 C.T.R. Wilson invented the Cloud Chamber in which the path of a single charged particle is revealed by a line of tiny water droplets like the vapor trail of an aeroplane. This Russian photograph (1927) shows a complex pattern of tracks.

The Struggle of Science in Italy

Between the 16th and the early 19th centuries Italians made many notable contributions to science, but by the late 19th century Italian science had become stagnant. The political unification of Italy in 1870 had failed to restore a position of prominence and, if anything, the gap widened in the 1890s, when the pace of development in pure and applied research set by other European countries – especially Germany – quickened. It was only on the eve of World War I, in the context of mounting international tension, that the goverment began to heed the calls of Italian scientists for the reform of the country's scientific institutions. However, scarcely any action was taken before the 1920s, and even the creation of the Consiglio Nazionale delle Ricerche (National Council of Research) in 1927, under the chairmanship of Guglielmo Marconi, yielded a disappointing harvest. The resources were too limited and too thinly spread to generate substantial research programs.

In the early 20th century, mathematics was (as in the 19th century) the only sector in which Italy retained a truly international reputation. The situation was less favorable in other sectors. Inadequate facilities, and a tendency to focus on rather narrow experimental problems, detached Italian physics from the profound theoretical developments changes that led to the emergence of relativity, quantum mechanics and nuclear physics. Galileo Ferraris created an important school of electrical engineering in Turin (the forerunner of the "Galileo Ferraris" National Institute of Electrical Engineering opened in 1935), and Augusto Righi conducted brilliant experiments on electricity in Bologna, but neither of them succeeded in building a research school. The decline of this individualistic tradition was eclipsed by the emergence of a new breed of capable "academic" managers in the 1920s. The most remarkable product of this change was the creation of a research center attached to the University of Rome, where Orso M. Corbino brought together an outstanding team, including Enrico Fermi, Emilio Segrè, Franco Rasetti and Edoardo Amaldi. Their work on atomic physics flourished until 1935, but by then not even their reputation could secure the funds that were necessary to continue their research. This, with the growing influence of fascism, triggered a diaspora from which Italian physics did not recover until the 1950s.

The reorganization of science after World War II was slow and fraught with financial and organizational difficulties. While the Consiglio Nazionale delle Ricerche retained its central role in sponsoring research, other national and private agencies were created, with responsibilities that often overlapped. Applied research achieved the most visible results. Giulio Natta, for example, with the support of industrial as well as governmental funds, developed the study of polymers for which he was awarded a Nobel prize in 1963. Physics began to gather new momentum in the late 1950s, chiefly through the launching of European cooperative projects in the nuclear field, such as CERN, which was strongly supported by Italy.

Splitting the atom

In retrospect, it is a little surprising that the atom-splitting experiment of J.D. Cockcroft and E.T.S. Walton at Cambridge in 1932 attracted the attention it did. So far as the general public was concerned the slow unraveling of the structure of atoms was still essentially an erudite but harmless pursuit for academic scientists. There was no general notion of the potential possibilities of atomic power, even though the decision to attempt the construction of an atomic bomb lay less than ten years ahead.

Such then was the climate of opinion in which Cockcroft and Walton's classic experiment was performed. For an explanation of why it captured the public imagination we must look to traditional views about the nature of atoms. Despite the accumulated evidence that atoms had structure, the old belief that they were indivisible died hard. News that two Cambridge scientists had in fact deliberately split an atom thus had the effect of undermining dogma – always newsworthy in whatever context. Was the alchemist's dream of converting lead to gold feasible after all?

Put in its simplest terms, the experiment was a microversion of the familiar coconut shy at the fair, when a powerfully hurled wooden ball breaks open a larger nut. In the Cockcroft/Walton experiment the ball was a positively charged hydrogen atom (proton) – electrically accelerated to a high speed – and the coconut was an atom of lithium. On traditional theory the mass of the projectile and the target should have exactly equaled that of the two new particles: in fact,

▼ The discovery of artificial radioactivity by the French physicists J.F. Joliot and his wife opened up new avenues in atomic physics. Joliot (right) is seen here in his laboratory in the early decades of atomic physics, the dangers of exposure to radioactivity were not entirely appreciated. Marie Curie died of leukemia in 1934, presumably caused by an excessive intake of radioaction; so too did her daughter Irène (who married J.F. Joliot in 1926). Even after World War II servicemen were exposed to radiation during open-air testing of atomic bombs.

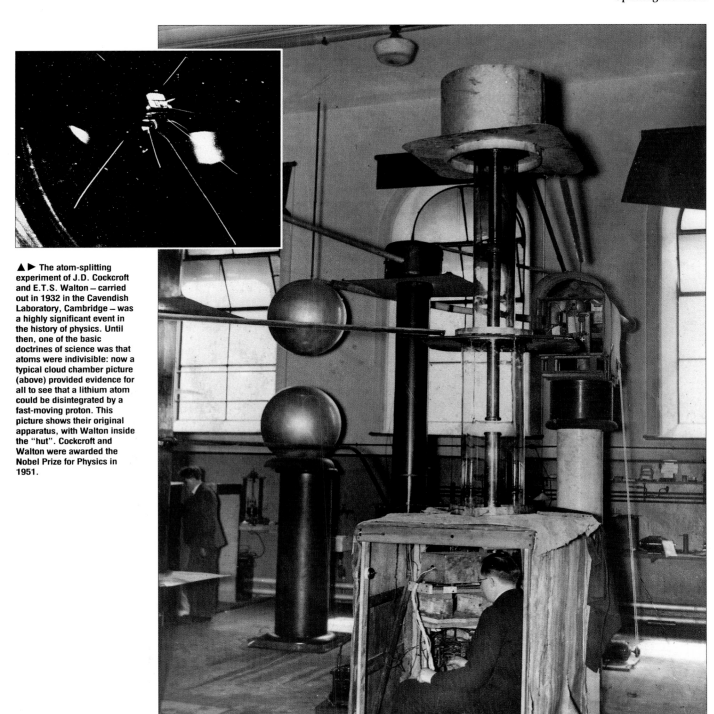

▲ ▶ The atom-splitting experiment of J.D. Cockcroft and E.T.S. Walton – carried out in 1932 in the Cavendish Laboratory, Cambridge – was a highly significant event in the history of physics. Until then, one of the basic doctrines of science was that atoms were indivisible: now a typical cloud chamber picture (above) provided evidence for all to see that a lithium atom could be disintegrated by a fast-moving proton. This picture shows their original apparatus, with Walton inside the "hut". Cockcroft and Walton were awarded the Nobel Prize for Physics in 1951.

In my own experimental work, we used to hold up a fluorescent screen in our lab, and we judged the intensity of the radiation to which we were exposed by the intensity of the glow. If it was too bright we added another millimeter of lead shielding. It is remarkable that we are alive and well today!

J.D. COCKCROFT

there was a slight discrepancy exactly accounted for, in terms of Einstein's theory, by the amount of energy released. Thus although there was a long way to go, and the road ahead was not clear, the possibility – and no more – of deliberately converting mass to energy had been established. For the moment, however, this was ignored: the pioneers were intent not on providing a new source of energy but with satisfying their curiosity about the ultimate structure of matter.

The essence of the new method for investigating atomic structure was to provide increasingly faster and more energetic particles – on the principle that a rifle bullet will penetrate further into a target than a pea from a peashooter. The velocity a particle acquires from an electric field is determined both by the potential difference and the length of the path it travels. To increase the latter, E.O. Lawrence and D.H. Sloan in the USA conceived the ingenious idea of using a combined electric and magnetic field to make the particle travel in a spiral, shaped like the hairspring of a watch. In this way, a long path can be contained within a relatively small space. Such a device came to be known as a cyclotron and by 1939 Lawrence had built one at Berkeley, California, with a power more than 50 times greater than that of Cockcroft and Walton. With a 200-tonne magnet, this was already an expensive machine but Lawrence even then had a far more ambitious

▼ The first controlled self-sustaining nuclear reaction – capable of being stopped and started at will – began on 2 December 1942 in a device operated in Chicago by the Italian physicist Enrico Fermi. Called simply a "pile" for security reasons, it was an historic piece of equipment, the ancestor of all atomic weapons and power plants. A.H. Compton, who designed the pile with Fermi, sent a dramatic message to the steering committee of the Manhattan Project at Harvard: "The Italian navigator has just landed in the New World."

machine on the drawing board. This was designed for a 5m (16ft) magnet weighing nearly 4,000 tonnes and containing 300 tonnes of copper wiring: its power was no less than 200 MeV. In the event this was not at first used for research, as intended, but was diverted to the Manhattan Project to separate uranium isotopes for the first atomic bombs.

The Manhattan Project

The advent of atomic power was undoutedly one of the most – perhaps the most – significant events in the history of civilization. It not only stopped a world war in its tracks, but powerfully influenced the subsequent course of history. The bridge between the old world and the new was the Manhattan Project, set up initially to manufacture atomic bombs but which incidentally established the basis of generating nuclear power for peaceful purposes. It was an enormous and highly complex enterprise, both technically and managerially, but four different phases can be discerned: the scientific, the political, the technological and the military.

Scientifically the story begins before the beginning of the century, with the discovery of the X-rays by W.K. Röntgen and radioactivity by A.H. Becquerel, but all that need concern us here is the state of the art in the crucial months before the outbreak of World War II. By then a handful of European scientists – notably Fermi, Hahn, Strassmann, Meitner, Frisch, the Joliots, Bohr and Wheeler – had established that if uranium nuclei are split by bombardment with neutrons, not only is energy released but also secondary neutrons more numerous than those used for the original bombardment. Thus under appropriate conditions a chain reaction is possible, fission spreading through a mass of uranium with enormous release of energy. It was further reported – just two days before the outbreak of war – that fission was more readily achieved with the rare

uranium-235 isotope than with the uranium-238, which comprises 99.3 percent of natural uranium. Early in 1940 two refugee scientists in Britain, Otto Frisch and Rudolf Peierls, presented a memorandum to the British government persuasively arguing that a critical mass of uranium-235 – perhaps no more than 10kg (22lb) – could be the basis of a bomb equivalent to several thousand tonnes of TNT.

The matter then entered the political arena. On the recommendation of the MAUD Committee the British government set up a bomb project under the code name Tube Alloys. At the same time the MAUD Report was communicated to the US government – then interested in atomic energy primarily as a source of power for submarines – and thereafter the center of gravity shifted to the other side of the Atlantic. On the eve of the Japanese bombing of the US Pacific Fleet at Pearl Harbor (7 December 1941) the US government approved a bomb project, code-named the Manhattan Project. For a year it was an British-American venture then, ironically, Britain was excluded, rejoining as a junior partner late in 1943. By 1945 the Manhattan Project was demanding as much funding as the entire American motor industry.

Technologically, the project bristled with difficulties. Chemically U-238 and U-235 are identical so that normal metal-separating processes were inapplicable. Physical methods were therefore necessary but the difference in weight between the two isotopes (1.25 percent) meant that hundreds of fractionations were necessary. Three main techniques were all pressed into service: gaseous diffusion, thermal separation and electromagnetic separation. A secondary project was to manufacture plutonium, an artificial element not known in nature. It is a product of bombarding uranium with neutrons and it too can be the starting point for an explosive chain reaction. It can be created only slowly, but being a specific element it can be separated by chemical methods: the difficulty was that being unknown, its chemical properties had to be determined from the start. In the event the chemistry was determined by the use of plutonium no bigger than a pin's head: from this a full-scale extraction plant had to be designed and built. The whole operation was complicated by the fact that many highly radioactive substances had to be safely handled. Lastly, there was the problem of detonation. To achieve criticality effectively the two subcritical masses comprising the bomb had to be united in less than one millionth of a second. This was eventually achieved by surrounding a marginally subcritical mass with a conventional explosion: on detonation the core was compressed to critical volume.

Finally there was the military input. This was led by General Leslie Groves, while Robert Oppenheimer was in charge of the scientific side. Here the constraint was not merely to produce a bomb that would detonate but one that could be delivered by a B-29 bomber. In the event, a successful trial was conducted in New Mexico on 16 July 1945 and bombs were dropped in Hiroshima (uranium) on 6 August and Nagasaki (plutonium) on 9 August. On 15 August Japan's surrender was announced and World War II was ended.

◄ When the Manhattan Project was set up in 1942 to develop an atomic bomb, the American physicist Robert Oppenheimer was called on to play a key role, as director of the Los Alamos Laboratory where much of the basic research was done. The speedy and successful outcome of the venture was a tribute to both his deep scientific insight and his powers of administration.

▲ For the first test, the atomic bomb was located on top of a steel tower, here seen being inspected by Robert Oppenheimer (left) and General L. Groves, respectively in charge of the scientific and military aspects of the project. After the explosion the tower and the ground below it were fused.

◄ The moment of truth: the first atomic bomb exploded in New Mexico on 16 July 1945. Until that moment there was no guarantee of success, for the nature of the project was such that intermediate tests were impossible. Many of the critical calculations were based on laboratory experiments using minute quantities of chemicals whose properties had previously been little, if at all, studied.

119

THE STRUCTURE OF THE NUCLEUS

The 1930s saw the birth of nuclear physics – a new science that was to have profound consequences. The structure of the atom, with electrons surrounding a tiny, central nucleus, became clear around 1911–13, but it was only with an understanding of the structure of the nucleus itself that scientists could harness the great energies locked at the heart of the atom.

The discovery of the nucleus showed that most of an atom's mass and all its positive electric charge are concentrated in a small central region. And by 1919 Ernest Rutherford had found that the nuclei of several elements contain positively-charged particles of matter identical to the nucleus of hydrogen, the lightest atom. He argued that these particles are constituents of all nuclei, and he named them "protons", for the first nuclear particles. Thirteen years later the picture was completed with the British physicist James Chadwick's discovery of the neutron – an electrically neutral particle, only slightly heavier than the proton.

Together protons and neutrons form the atomic nuclei of all elements (except hydrogen, where the nuclei are single protons). Scientists now know that they are bound together by the so-called strong nuclear force. In any atom the number of electrons – which determines the chemical properties of the element – exactly balances the number of protons in the nucleus. The role of the neutrons is to dilute the repulsive electric force between the protons, and so prevent the nucleus from flying apart. In larger nuclei, an increasing number of neutrons is necessary to counteract the electric force, so that in the heaviest nuclei the number of neutrons greatly exceeds the number of protons.

Only certain configurations of protons and neutrons prove to be completely stable, however. Others give rise to nuclei that are unstable, or, in other words, radioactive. They change to more stable structures through the spontaneous emission of radiations – alpha particles (helium nuclei), beta particles (electrons) and gamma rays (high-energy photons). And in some instances a large nucleus, such as uranium, can break into two more or less equal fragments and a few neutrons in a process known as fission.

The energies released in these transmutations are millions of times greater than those involved in chemical reactions. This is due to the strength of the nuclear binding force. In the late 1930s, after the discovery of fission in uranium bombarded by neutrons, physicists realized that they could set up a chain reaction, with the neutrons released in one fission producing another and so on. When controlled, this chain reaction leads to a useful source of power; when uncontrolled it leads to a devastating explosion.

Physicists soon saw the potential uses and misuses of this source of nuclear energy. Their greatest dreams have come to life with the development of reactors to generate nuclear power; their worst nightmares are realized with the proliferation of nuclear weapons.

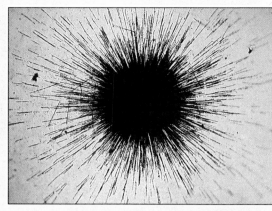

► Alpha particles stream out from a speck of radium salt, about a tenth of a millimeter across. The particles – stable clusters of two protons and two neutrons (helium nuclei) – leave trails in a special photographic emulsion, which appear here as dark lines. Alpha particles were used in much of the early research on the atomic nucleus, as they are energetic charged projectiles which can closely approach and penetrate other nuclei.

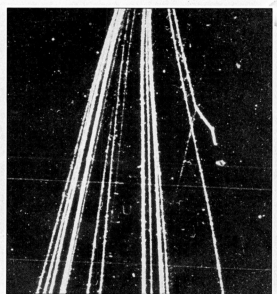

► Using alpha particles as "projectiles", physicists discovered that nuclei contain protons. Here alphas leave tracks in a detector called a cloud chamber, photographed by Patrick Blackett at Cambridge in 1925. The alpha particle (2 protons, 2 neutrons) at the far right collides with a nucleus of nitrogen (7 protons, 7 neutrons) in the air inside the detector, and is captured. The nucleus absorbs two of the neutrons and one of the protons, thereby becoming a nucleus of oxygen-17 (8 protons, 9 neutrons), but it spits out the remaining proton. The proton continues, to make the fainter left-hand branch of the forked track, while the oxygen nucleus goes only a little way, leaving the short thick, kinked track.

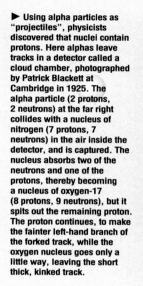
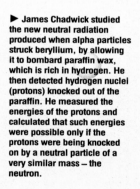

► James Chadwick studied the new neutral radiation produced when alpha particles struck beryllium, by allowing it to bombard paraffin wax, which is rich in hydrogen. He then detected hydrogen nuclei (protons) knocked out of the paraffin. He measured the energies of the protons and calculated that such energies were possible only if the protons were being knocked on by a neutral particle of a very similar mass – the neutron.

Hydrogen nucleus

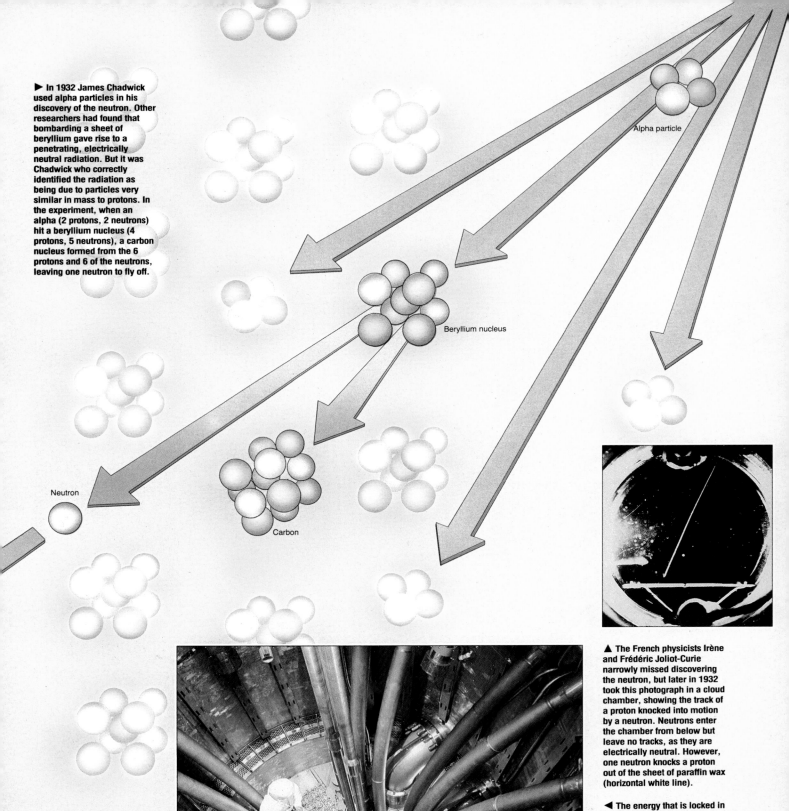

► In 1932 James Chadwick used alpha particles in his discovery of the neutron. Other researchers had found that bombarding a sheet of beryllium gave rise to a penetrating, electrically neutral radiation. But it was Chadwick who correctly identified the radiation as being due to particles very similar in mass to protons. In the experiment, when an alpha (2 protons, 2 neutrons) hit a beryllium nucleus (4 protons, 5 neutrons), a carbon nucleus formed from the 6 protons and 6 of the neutrons, leaving one neutron to fly off.

Alpha particle

Beryllium nucleus

Neutron

Carbon

▲ The French physicists Irène and Frédéric Joliot-Curie narrowly missed discovering the neutron, but later in 1932 took this photograph in a cloud chamber, showing the track of a proton knocked into motion by a neutron. Neutrons enter the chamber from below but leave no tracks, as they are electrically neutral. However, one neutron knocks a proton out of the sheet of paraffin wax (horizontal white line).

◄ The energy that is locked in binding protons and neutrons together in the nucleus is released in a controlled fashion within the core of a nuclear reactor. Here the large nuclei of uranium-235 (92 protons and 143 neutrons) absorb neutrons and break into two fragments in a process called fission, which releases energy. The fission also releases a few neutrons, which can in turn be absorbed by other uranium nuclei in a chain reaction.

THE UNIVERSAL PANACEA

Time Chart

	1946	1947	1948	1949	1950	1951	1952	1953
Nobel Prizes	• *Chem*: J.B.Sumner, J.H. Northrop, W.M. Stanley (USA) • *Phys*: P.W. Bridgman (USA) • *Med*: H.J. Muller (USA)	• *Chem*: R. Robinson (UK) • *Phys*: E.V. Appleton (UK) • *Med*: C.F. Cori, G.T. Cori (USA), B.A. Houssay (Arg)	• *Chem*: A. Tiselius (Swe) • *Phys*: P.M.S. Blackett (UK) • *Med*: P.H. Müller (Swi)	• *Chem*: W.F. Giauque (USA) • *Phys*: H. Yukawa (Jap) • *Med*: W.R. Hess (Swi), E. Moniz (Port)	• *Chem*: O. Diels, K. Alder (FRG) • *Phys*: C.F. Powell (UK) • *Med*: P.S. Hench, E.C. Kendall (USA), T. Reichstein (Swi)	• *Chem*: E.M. McMillan, G.T. Seaborg (USA) • *Phys*: J.D. Cockcroft (UK), E.T.S. Walton (Ire) • *Med*: M. Theiler (SA)	• *Chem*: A.J.P. Martin, R.L.M. Synge (UK) • *Phys*: F. Bloch, E.M. Purcell (USA) • *Med*: S.A. Waksman (USA)	• *Chem*: H. Staudinger (FRG) • *Phys*: F. Zernike (Neth) • *Med*: F.A. Lipmann (USA), H.A. Krebs (UK)
Technology	• First vinylite record disk released (USA) • Xerography invented by C. Carlson (USA) • ENIAC, the first automatic computer, is developed (USA) • First Soviet nuclear reactor begins to work • Invention of ceramic magnets, later used in microwave and radar technology (Neth)	• Development of the reflecting microscope • US airplane first flies at supersonic speeds • D. Gabor develops the basic principle of holography (UK) • Introduction of the printed circuit board by J.A. Sargrove (UK)	• Invention of the transistor by scientists at the Bell Telephone Company (USA) • D. Gabor develops holography (UK) • Cesium atomic clock developed, accurate to 1 second in 1000 years • E. Land develops camera with instantly processed film (USA)	• First atomic bomb test in the USSR • Aug: BINAC, the first electronic stored-program computer in the USA, goes into operation • Cape Canaveral rocket test-site opened (USA) • First multi-stage rocket tested (USA)	• Commercial color television begins in the USA • Development of the Vidicon, the first TV camera tube to use the principle of photoconductivity (USA)	• J.W. Mauchly and J.P. Eckert build UNIVAC I, the first computer to be commercially available and the first to store information on magnetic tape (USA) • First 3-D movies shown (USA) • Dec: Electric power is produced from atomic energy at Arcon, Idaho (USA)	• 3 Oct: first British atomic bomb tests at Monte Bello Islands, Western Australia • 6 Nov: USA explodes the first hydrogen bomb, at Eniwetok Atoll in the Pacific Ocean • First pocket-sized transistor radio marketed by Sony (Jap)	• 29 Aug: USSR explodes its first hydrogen bomb • Michelin (Fr) and Pirelli (It) introduce radial-ply tyres • MASER (Microwaves Amplified by the Stimulated Emission of Radiation) developed by C.H. Townes (USA)
Medicine	• Publication of B. Spock's *Baby and Child Care* (USA) • Benadryl becomes commercially available as a relief from the symptoms of hayfever	• J.C. dos Santos perform the first endarterectomy (Port) • "The Doctors' Trial": trial by an Allied court of doctors responsible for medical atrocities in Nazi concentration camps during WWII • Publication of F. Dunbar's *Mind and Body: Psychosomatic Medicine*	• Publication of A.C. Kinsey's *Sexual Behavior in the Human Male* (USA) • World Health Organization (WHO) set up • Preparation of the antibiotics aureomycin and chloromycetin	• P. Hench discovers cortisone (compound E) as a treatment for rheumatism • S.A. Waksman isolates the antibiotic neomycin • The first synchotron is demonstrated, to be used to treat cancer (UK)	• Miltown, a meprobamate, comes into wide use in the USA as a tranquillizer • Antihistamines become popular remedy for colds and allergies • Invention by J. Gibbon of the first heart-lung machine (USA)	• Introduction of methotrexate, a drug for treating uterine cancer (Hong Kong) • J. Andre-Thomas devises a heart-lung machine for heart operations	• First successful sex-change operation performed by K. Hamburger (Den) • D. Bevis develops amniocentesis – drawing off amniotic fluid from the womb to test for abnormalities in the fetus (UK) • Discovery of a new antitubercular drug, isoniazid (USA)	• First successful open-heart operation performed by J. Gibbon (USA) • Publication of A.C. Kinsey's *Sexual Behavior in the Human Female* (USA)
Biology	• J. Lederberg and E.L. Tatum discover sexuality in bacteria (USA) • L. Pauling suggests how enzymes act as catalysts for biochemical reactions • C. Leblond uses radioautography to study protein synthesis (Can)	• K. von Frisch discovers that bees use the polarization of light for orientation	• A. Tiselius develops electrophonesis, a technique for studying blood proteins (Swe) • Research into mainstream genetics virtually outlawed in the USSR under the influence of biologist T.D. Lysenko • A.E. Mirsky discovers RNA in chromosomes (USA)	• L. Pauling discovers that sickle-cell anemia has an abnormal hemoglobin (hemoglobin S) (USA) • M. Barr identifies the tiny black dots (Barr bodies) found in the cell nuclei of females (Can)	• First embryo transplants for cattle are performed • Deep-sea expedition in the *Galathea* to investigate fauna (Den)	• R.B. Woodward synthesizes the steroids cortisone and cholesterol (USA) • F.M. Burnet proposes clonal selection theory in immunology (Aus)	• A.D. Hershey and M. Chase prove DNA is the genetic information carrier (USA) • Modern theory of transmission in nerve cells proposed by A.L. Hodgkin and A.F. Huxley (UK) • J.E. Salk first tests his polio vaccine (USA)	• H.B.D. Kettlewell shows industrial melanism in moths (UK) • F. Crick (UK) and J. Watson (USA) present their double-helix model for the structure of the DNA molecule
Physics	• F. Bloch (Swi) and E.W. Purcell (USA) independently develop nuclear magnetic resonance • E.V. Appleton and D. Hey discover that the sun emits radio waves	• Publication of P.M.S. Blackett's theory that all massive rotating bodies are magnetic • Pi-meson discovered by C.F. Powell	• Shell theory of atomic nucleus proposed by M. Gopper-Meyer (USA) • Publication of L. Jánossy's *Cosmic Rays and Nuclear Physics* (Hung)	• L.J. Rainwater begins to work on the notion that the atomic nucleus may not be spherical (USA)	• Confirmation of the existence of "V" particles (USA/Fr) • New calculations of the speed of light obtained through radio waves (USA/UK)	• Field-ion microscope developed by E.W. Mueller (USA) • Second British plutonium pile, at Windscale, Cumberland, in operation	• D.A. Glaser devises the bubble-chamber (USA) • Rapid extension of the use of radio-isotopes in scientific research, medicine and industry	• M. Gell-Mann introduces the concept of "strangeness" to quantum mechanics • Breeding of atomic fuel achieved at the nuclear reactor at Arcon, Idaho (USA)
Chemistry	• U. von Euler isolates noradrenalin (Swe)		• E.L. Smith (UK) and E.L. Rickes (USA) independently isolate vitamin B12 • R.A. Alpher, M.A. Bethe and G. Gamov propose a scheme to explain the evolution of chemical elements in the early universe	• D.C. Hodgkin uses a computer to work out the structure of penicillin	• US Atomic Energy Commission separates plutonium from pitchblende concentrates • G.T. Seaborg discovers californium (USA)	• L. Pauling proposes that some protein molecules are helical (USA) • Krilium is developed from acrylonitrile for use in fertilization	• Discovery of Acrilan, an acrylic fiber (USA) • G.T. Seaborg discovers einsteinium (USA) • A contraceptive tablet of phosphorated hesperidin is made	• K. Ziegler (Ger) develops the first catalyst combining monomers into a polymer in a regular fashion, thus making an improved polyethylene. G. Natta (It) uses this idea to develop the first isotactic polymers
Other	• Radiocarbon dating introduced by F.W. Libby (USA)	• H. Hartley adds titanium to an iron-graphite mixture to produce an extra-strong type of cast iron (UK)	• Publication of H. Bondi and T. Gold's *The Steady State Theory of the Expanding Universe* (UK)	• Discovery by USSR of its first offshore oil field, in the Caspian Sea	• Apr: J. von Neumann and colleagues, using the ENIAC computer, make the first computerized 24-hour weather predictions (USA)	• E.M. Purcell detects radiation from interstellar hydrogen (USA)	• W. Baade discovers that galaxies are about twice as distant as had been previously believed, due to an error in the Cepheid luminosity scale	• A cosmic ray observatory is established on Mount Wrangell, Alaska (USA)

1954	1955	1956	1957	1958	1959	1960
● *Chem*: L.C. Pauling (USA) ● *Phys*: M. Born (UK), W. Bothe (FRG) ● *Med*: J.F. Enders, F.C. Robbins, T.H. Weller (USA)	● *Chem*: V. du Vigneaud (USA) ● *Phys*: W.E. Lamb (Jr), P. Kusch (USA) ● *Med*: A.H.T. Theorell (Swe)	● *Chem*: C.N. Hinshelwood (UK), N. Semenov (USSR) ● *Phys*: W.B. Shockley, W.H. Brattain, J. Bardeen (USA) ● *Med*: D.W. Richards (Jr), A.F. Cournand (USA), W. Forssmann (FRG)	● *Chem*: A.T. Todd (UK) ● *Phys*: Tsung-Dao Lee, Chen Ning Yang (China) ● *Med*: D. Bovet (It)	● *Chem*: F. Sanger (UK) ● *Phys*: P.A. Cherenkov, I.Y. Tamm, I.M. Frank (USSR) ● *Med*: J. Lederberg, G.W. Beadle, E.L. Tatum (USA)	● *Chem*: J. Heyrovsky (Czech) ● *Phys*: E. Segrè, O. Chamberlain (USA) ● *Med*: S. Ochoa, A. Kornberg (USA)	● *Chem*: W.F. Libby (USA) ● *Phys*: D.A. Glaser (USA) ● *Med*: F.M. Burnet (Aus), P.B. Medawar (UK)
● Bell Telephone Company develops photovoltaic cell capable of converting the Sun's radiation into electricity (USA) ● First use of silicon instead of germanium in the transistor by Texas Instruments (USA)	● 18 Jul: First use of atomically-generated power in the USA ● First optical fibers produced by Narinder Kapary (UK) ● First artificial diamonds for industrial use (USA) ● Ultra high frequency (UHF) waves are produced (USA) ● First test run of C. Cockerell's hovercraft (UK)	● 17 Oct: Opening of Calder Hall, the world's largest nuclear power station (UK) ● Bell Telephone Company develops a videophone (USA) ● FORTRAN, the first computer programming language, developed (USA) ● A. Poniatoff demonstrates the first video tape-recorder (USA)	● 4 Oct: USSR launches *Sputnik I*, and on 3 November launches *Sputnik II*, carrying a dog, for studying living conditions in space ● Philips introduce the Plumbicon TV camera tube, as an improvement on the Vidicon (UK)	● USA launches satellites *Explorer I* (31 January), to study cosmic rays, *Vanguard I* (17 March) rocket, to test solar cells, and *Atlas* (18 December), to investigate radio relay ● 15 May: USSR launches *Sputnik III* ● Launch of nuclear-powered icebreaker *Lenin* (USSR)	● 2 Jan: Launch of *Lunik I. Lunik II* reaches the Moon on 12 September and *Lunik III* launched on 4 October to photograph the Moon (USSR) ● First commercial Xerox copier is introduced ● G.M. Hopper invents computer language COBOL (USA) ● 25 Jul: Launch of first nuclear passenger-cargo ship *Savannah* (USA)	● First laser constructed by T.H. Maiman (USA) ● US nuclear submarine *Triton* completes first circumnavigation of the globe under water ● 1 Apr: Launch of *Tiros I*, the world's first weather satellite (USA) ● 12 Aug: Launch of the first communications satellite, *Echo*, by J.R. Pierce (USA)
● First suggestion of a connection between smoking and lung cancer ● J.E. Salk begins to inoculate schoolchildren with his antipolio serum (USA) ● G.G. Pincus introduces the oral contraceptive pill after field trials in Haiti and Puerto Rico (USA) ● First successful kidney transplant (USA)	● Invention of the Holtzer Shunt to drain off excess fluid from the brains of hydrocephalic infants	● M. Shiner perfects a biopsy capsule to collect tissue from the intestine as an aid to diagnosis of intestinal diseases (UK) ● US physicians pioneer hemodialysis – cleansing the blood on an artificial kidney machine	● After workers manufacturing 2,4,5-T develop skin disease, it is recognized that dioxin frequently contaminates such herbicides (FRG) ● High-speed dental drill developed (USA) ● A. Isaacs and J. Lindenmann discover interferon ● A.B. Sabin produces an oral polio vaccine (USA)	● I. Donald pioneers the use of high-frequency sound waves to examine unborn fetuses (UK) ● J. Enders prepares an effective vaccine against measles (USA) ● Von Ardenne invents a tiny radio transmitter for gastro-intestinal investigations (GDR)	● Inauguration of the world's first bone-marrow bank by the Vienna Cancer Institute (Aut) ● A new antibiotic is isolated, Cephalosporin C, which is effective against penicillin-resistant bacteria (UK)	● First heart pacemaker developed (UK) ● G.N. Robinson discovers the antibiotic methicillin
● J. Hin Tjio and A. Levan demonstrate that humans have 46 chromosomes rather than 48 as was previously believed ● Hybrid wheats are produced by E.R. Sears that are more resistant to disease and drought (USA)	● C.R. De Duve identifies lysosomes (Bel) ● S. Ochoa isolates an enzyme, from the bacterium *Aztobacter vinelandii*, capable of catalyzing the formation of RNA from nucleotides (Sp/USA)	● P. Berg discovers transfer RNA (USA) ● G.E. Palade discovers ribosomes (USA) ● A. Kornberg isolates an enzyme, from the bacterium *Escherichia coli*, capable of catalyzing the formation of DNA from nucleotides	● Giberellin, a growth-producing hormone, is isolated ● M.S. Meselson and Stahl verify Crick and Watson's ideas on replication (USA)	● I. Darevsky discovers in Armenia an all-female lizard species that reproduces parthenogenetically (USSR) ● J. Dausset discovers the human histocompatibility system (Fr)	● G.E. Hutchinson points out that two different species cannot occupy the same ecological niche ● C. Ford devises a method of making human chromosomes visible and sorting them into pairs (UK)	● S. Moore and W.H. Stein determine the sequence of all 124 amino acids in ribonuclease ● J.C. Kendrew and M.F. Perutz elucidate the three-dimensional structure of the protein myoglobin
● Chen Ning Yang and R. Mills establish the basis for modern quantum field theory by developing the Yang-Mills gauge-invariant fields (USA) ● 29 Sep: *Centre Européen de Recherche Nucléaire* (CERN) founded	● A. Einstein pleas on radio for a halt to the arms races (USA) ● E.G. Segrè and O. Chamberlain discover the anti-proton	● Detection of the neutrino, a particle of no electric charge (USA) ● Discovery of the anti-neutron by Cork, Piccioni, Lambertson and Wenzel (USA)		● Discovery by L. Esaki of the tunnel diode (Jap)	● L.W. Alvarez and colleagues complete a bubble-chamber for detecting particles produced by the Cosmotron accelerator (USA) ● L.W. Alvarez discovers the neutral *xi*-particle (USA)	● Standard meter redefined in terms of the wavelength of the krypton spectrum ● R.L. Mossbauer develops gamma rays of accurately defined wavelength: used to detect gravitational red shift (FRG)
● Polypropylene invented by G. Natta (It) ● Paraquat developed by ICI (UK)	● F. Sanger establishes the amino acid sequence of insulin (UK)	● Choh Hao Li and colleagues isolate human growth hormone ● D. Hodgkin discovers the composition of Vitamin B12 using a computer (USA)	● Nobellium is discovered (Swe)			● Chlorophyll synthesized independently by M. Strell (FRG) and R.B. Woodward (USA) ● K.H. Hofmann synthesizes pituitary hormone
● Composite photograph of the night sky completed by Lisk Observatory, California (USA)	● A. Dolfus ascends 7.25 kilometers above the Earth to make photoelectric observations of Mars	● H. Friedman announces that solar flares are a source of X-rays (USA) ● B.C. Heezen and M. Ewing discover the Mid-Oceanic Ridge (USA)	● 1 Jul: International Geophysical Year begins (until 31 December 1958), concentrating on Antarctic exploration, oceanographic and meteorological research	● E.M. Parker shows that a solar wind of particles thrown out by the Sun makes comets' tails point away from the Sun	● First International Congress of Oceanography in New York	● H.H. Hess develops the theory of sea-floor spreading (USA) ● T.A. Matthews and A. R. Sandage deduce the existence of quasars

Datafile

Postwar technology advanced rapidly in two areas which have had lasting significance. The development of the atomic bomb within the Manhattan Project led to the tension of the Cold War and the nuclear arms race, and to the construction of nuclear power stations. The invention of the semiconductor transistor and the consequent miniaturization of electronics resulted in powerful computers, satellites and color televisions.

▼ The technical advances in aircraft design and navigation by radar made during World War II coupled with the huge production capacity set up to service the military requirements resulted in a rapid postwar increase in the field of civil aviation. Jet aircraft were the most successful.

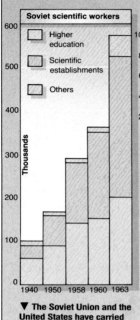

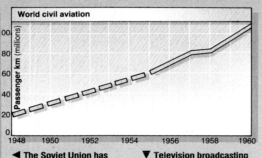

◄ The Soviet Union has consistently invested in science as a means of creating a socialist society. After World War II the pool of "scientific workers" grew enormously and notable advances were achieved (eg in space technology). Ideology, however, hampered creativity. The Soviet tally of Nobel prizes is only seven.

▼ Television broadcasting began before World War II but only in the affluent 1950s did sales of receivers became substantial and the medium spread round the world. The success of black-and-white TV demonstrated the potential for color. Effective color broadcasting began in the USA, in 1954; it spread to Japan and Europe in the 1960s.

▼ The Soviet Union and the United States have carried out over 90 percent of all nuclear tests. The USA is responsible for nearly two-thirds of the world total. The initial acceleration in the frequency of testing was followed by two years of relatively amicable relations during which time Harold Macmillan visited Moscow, and Khrushchev visited the USA, the first such visit by a Soviet leader. But soon after, in 1962, the stress of rapidly increasing testing, the space race, and the Cuban crisis almost led to war.

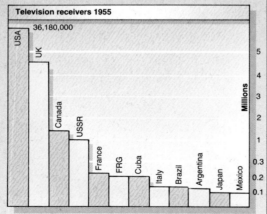

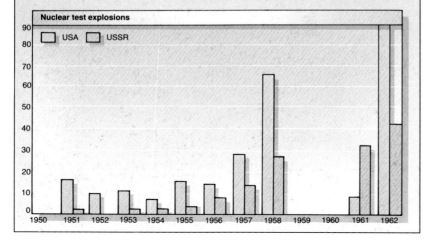

During the postwar years technology increasingly changed the pattern of daily life, but unquestionably one issue transcended all others: the bitter and expensive military rivalry between the United States and the Soviet Union into which the democracies of Western Europe and the Soviet-dominated countries of Eastern Europe inevitably became drawn. It was a rivalry largely fought out on a technological battlefield. On the one hand there was the proliferation of nuclear weapons of ever increasing power, on the other the development of means of delivering them from more and more distant bases. Save in the unquantifiable terms of national security and industrial spin-off, the enormous expenditure was non-productive and sufficient, through economic repercussions, to affect the social pattern.

Postwar technological developments

Inevitably, the early postwar years were largely ones of reconstruction rather than innovation. Many prewar activities had been virtually stopped in their tracks and simply picked up where they had left off. In the United States, for example, there had been a wartime ban on the construction of television transmitters and receivers, which was not lifted until 1946. Regular transmissions in color began in 1950.

The immediate postwar automobiles were mostly similar to their prewar counterparts. Chrysler broke new ground by introducing disk-brakes in 1949, though these had been envisaged by Lanchester at the beginning of the century. Radial-ply tires, so constructed that they provide a better grip on the road, were introduced in 1953. In automobiles themselves there was, however, a significant trend toward smaller models, in keeping with reduced public spending power. The German Volkswagen, the people's car, was made in very small numbers before the war but subsequently reappeared as the immensely popular "Beetle" and sold in millions over a period of 40 years. In 1949 the French automobile company Citroën launched their equally successful 2CV model: five million were sold over the next 30 years and it was still popular when discontinued in 1987. Mechanization in agriculture, forestry, and other land uses is reflected in the British Land Rover vehicle, introduced in 1948, employing a four-wheel drive adopted from the American military Jeep.

Motorcycling, too, entered a new phase with the appearance of a variety of light low-power models. In Italy, the Vespa scooter appeared in 1946 and a million had been sold ten years later. In Japan Soichiro Honda, in 1947, laid the foundations of an immense international motorcycle industry by fitting very small engines into the frames of ordinary pushbikes.

THE ADVENT OF ELECTRONICS

As was to be expected, some of the greatest changes occurred where developments pursued for purely military purposes suddenly became available for civilian use. The impact was rapid for two reasons: first, because the research and development was already done; and second, because manufacturers, lacking large government contracts, urgently needed civilian outlets in order to stay in business.

The aircraft industry was case in point: there was a great manufacturing capacity but a dearth of contracts. This facilitated an immense and rapid expansion of civil aviation, assisted also by the availability of advanced radar for navigation and air traffic control. This heralded a revolution in the pattern of travel, exemplified by that across the Atlantic.

There the great liners had vied with each other in prewar days in providing high standards of comfort and speed. In 1952 the existing fleet was joined by the USS *United States*, built at the then enormous cost of 75 million US dollars, and innovative to the extent of making extensive use of light aluminum alloys in her superstructure. But she was obsolescent when launched, for already civil aviation could offer a crossing in one-tenth of the time. By 1957 more passengers crossed the Atlantic by air than did so by sea and by the late 1960s more than 97 percent of trans-Atlantic passengers went by plane. The same sort of change occurred worldwide and the factor of speed opened up an entirely new market.

In prewar years the chemical industry had invented many new products in the field of polymers but again military needs had diverted these from the public domain. Afterward, however, they flooded the market. As fibers, nylon and Dacron (Terylene) were powerful recruits to the textile industry. Polythene, originally regarded as a plastic for limited specialist use in the electrical industry, proved to be a useful material for a great variety of purposes, and production came to be measured in hundreds of thousands of tons. It became even more widely used after K. Ziegler, in 1953, devised a low-pressure process to replace the original high-pressure one. In Italy, Giulio Natta applied Ziegler's process to the polymerization of propylene, thus opening up a big market for polypropylene.

Development of the transistor

If radios are to operate with alternating current, the normal mains supply, this must be rectified; that is, made unidirectional. In the earliest days "cat's whisker" devices were used for this purpose, utilizing the fact that certain crystals (such as galena, lead sulphide) allow current to pass in only one direction. For virtually the whole of the first half of the 20th century, however, these were

entirely replaced by thermionic tubes (valves) which can both rectify and amplify a current. They had the disadvantage, however, that they were large, used large amounts of electricity, and took time to warm up before becoming effective.

During the early 1930s, at the Bell Telephone Laboratories in the United States, W.H. Brattain had begun detailed studies of the properties of semiconductors; that is materials that have an electrical resistance between that of conductors (low resistance) and insulators (high resistance). His work had shown that surface effects in a semiconductor could lead to current rectification. Such rectifiers had certain advantages over thermionic tubes and indeed silicon rectifiers were used in radar systems during World War II.

After the war, Brattain resumed the research together with J.Bardeen. They discovered that with two contacts on a piece of germanium, like

▼ The Space race epitomized the new prestige of science after World War II, and excited a whole generation of schoolchildren.

▼ The transistor's ability to control a large current with a small one has made it a vital component in many types of circuitry, both as an amplifier and as a switch. Equally important has been the fact that the transistor's basic structure lends itself to miniaturization. The first transistor, shown here, was built from the semiconducting material germanium. Now the devices are most commonly built with silicons: many thousands of transistors can be constructed on the same small "chip".

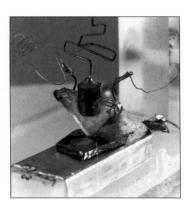

▲ The introduction of the long-playing record with a "microgroove", in 1948, allowed up to 30 minutes playing time per side. Ten years later, stereo recordings were introduced, with opposite sides of the groove (seen here) containing separate patterns for sound coming from left and right.

► The first electronic digital computers, built in the mid-1940s, were huge machines based on thermionic tubes (valves), often designed for specific applications. The American Whirlwind, shown here, was built to solve problems associated with the performance of an aircraft in flight. In the late 1950s transistors took over from valves, leading to a second generation of smaller and faster computers.

two cat's whiskers, they could control the current flowing through the semiconductor and on 23 December 1947 they demonstrated that their device could behave as an amplifier. Its behavior depended on regions devoid of electrons forming in the surface layer of the semiconductor, beneath the contacts. Because the device worked by transferring current across a resistor they called it a transistor. The original version – a point-contact transistor – had limitations: it was electrically "noisy" and could control only low power inputs. The improved junction transistor was developed soon afterward.

The versatility and miniaturization made possible by the transistor were phenomenal and gave rise to a new multibillion dollar industry for the manufacture of silicon chips. It ranks as one of the most important inventions of all time, yet the research program which gave birth to it demanded only quite simple apparatus: the most expensive is said to have been an oscilloscope.

The first electronic computers

A fair way of assessing the importance of a technological development is to consider the consequences of being deprived of it. A general and prolonged breakdown in basic utility services such as water, electricity, telecommunications, public transport, or sewerage quickly brings chaos and the ordinary pattern of life cannot be sustained. In the late 20th century another technological innovation came to be regarded among those essential to sustaining the normal pattern of life, namely the computer. It was originally – as its name implies – just a calculating device but it developed into a highly sophisticated means of storing and retrieving information. It is now so much a part of everyday life that it is hard to realize that in its modern electronic form it had no prewar counterpart and that the cheap "desktop" computers of the late 20th century have a far greater capacity than the huge prototypes of the 1940s. The world market for computers of all kinds was estimated to have reached £100 billion annually by the end of the 1980s – a remarkable achievement for such a young industry.

There is a case to be made for regarding the ASCC (Automatic Sequence Controlled Calculator) of 1944, built in the USA by IBM Inc., as the original prototype. Although basically an electronic apparatus, it contained many mechanical devices akin to those found in a long line of earlier machines going back to the French mathematician Blaise Pascal's adding machine of 1642 and the German Gottfried Leibniz's much more sophisticated device built 30 years later. ASCC was a veritable dinosaur – it weighed five tonnes, was 16m (52 ft) long, and contained 800km (500 mi) of electric wiring. By present-day standards its ability to multiply two 11-digit numbers in three seconds is remarkably unimpressive: computer operation times are now measured in millionths of seconds. The successor to ASCC was ENIAC (Electronic Numerical Integrator and Calculator) which, apart from some switches used for circuit control, was fully electronic. It was built at the University of Pennsylvania by J.P. Eckert and J.W. Mauchly and was

originally intended for wartime use to make computations for ballistic tables. In the event, however, it was not finished until 1946. It, too, was a formidable machine, twice the size of ASCC. It contained no less than 18,000 thermionic valves and at maximum output consumed 100 kilowatts of electricity – to disperse the resultant heat was a problem in itself. To program ENIAC necessitated setting switches and plugging in connections by hand, while ASCC was programmed with punched tapes. Each method had its advantages, but both were basically slow and tedious.

For the next advance, inventors again turned to the 19th century, and in particular to the analytical engine built by the English inventor Charles Babbage in the 1830s. Rather surprisingly, an important publicist for this was Lady Lovelace, a daughter of Lord Byron, who stressed that many computations involved repeated repetition of a given sequence of operations: time and trouble could therefore be saved if the machine were primed to do this automatically. In effect, the computer should have a built-in memory.

Even earlier, in the later 17th century, Leibniz had suggested that for mechanical calculation it would be simpler to use a binary notation rather than the traditional decimal one. This means in effect that all numbers are expressed in terms of only two digits, 0 and 1, instead of ten, 0–9. For an electronically operated computer this is particularly attractive because it corresponds to the two electrical modes, off and on.

These two concepts were incorporated by John von Neumann, also of the University of Pennsylvania, in a much more advanced machine, EDVAC (Electronic Discrete Variable Automatic Computer). In this the "memory" consisted of sonic pulses maintained in a column of mercury. A more sophisticated memory device followed in the early 1950s: this depended on the fact that the direction of magnetization in materials known as ferrites could be almost instantaneously reversed according to the direction of flow of electricity in a linked circuit.

By this time the computer was, as it were, getting ahead of itself: it could do the sums faster than punched tape could provide instructions. The next advance was to record the program of operations on a magnetic tape, similar to the tapes used in sound recording. This was incorporated in the Remington-Rand UNIVAC in 1956. This is noteworthy as having been the first electronic computer to be available commercially: the relatively few earlier ones had been custom-built at great cost for special purposes. For this reason, the general public was almost completely unaware of developments in the computer world. This situation changed very considerably in 1952, when the original UNIVAC, designed to process US census information, successfully predicted the election of Dwight Eisenhower as US president.

One further invention was crucial to the development of the computer for general use: the transistor. As in the electronic engineering field generally, this allowed both a great degree of miniaturization and a great saving in electricity.

Magnetic recording

The principle of magnetic recording is very simple and was in fact demonstrated as long ago as 1898 by a Danish engineer Valdemar Poulsen; his Telegraphone won the Grand Prix at the Paris Exposition of 1900. Current from a microphone modulated the field of an electromagnet which in turn imposed a variable degree of magnetization on a moving steel wire. The magnetically stored message on the wire was reproduced by reversing the process: the concept is thus very similar to the modulation of an electric current by sound signals in a telephone system. The system had two features to commend it. First, as there was no mechanical contact with the wire the recording was not impaired by repeated playing. Second, an existing message could be erased in a magnetic field and the wire used to record something quite different.

Despite these advantages, magnetic recording never became widely used until after World War II. Then the original wire was replaced by plastic tape coated with magnetized ferromagnetic powder. At first, magnetic recording was used largely for its original purposes: to record and reproduce sound. In this there were two main lines of development. The first was use by individuals, for purposes as diverse as dictating letters in the office or recording radio programs for future replay. The second was the sale of prerecorded material, mainly music. Hand in hand with this went the marketing of a wide range of audio-cassette players, ranging from a few dollars upward.

▶ The broadcasts of color television after World War II were based on an electromechanical system developed during the 1930s. One of its big disadvantages, however, was that it produced images in the three primary colors sequentially and was totally incompatible with black-and-white receivers. In the early 1950s in the USA, RCA developed an all-electronic system, seen here under test, which overcame the problem of incompatibility. The technique, still in use in the USA today, was to transmit two images simultaneously — one containing detailed information about brightness, the other containing coarser information about color. Color receivers can respond to both parts of the signal, while black-and-white receivers respond only to the part containing the detailed brightness information.

Color has brought a new dimension to television, turned the gray-scale of early black and white video into the rainbow of the 1960s

A.C. NIELSEN, JR

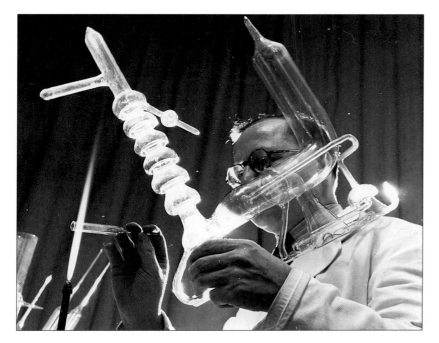

▲ The development of television relied on more than advances in electronics. Here a technician perfects an intricate system of glass tubing, the basis of a pump needed to produce the high vacuum inside the tubes of TV cameras. A camera tube contains a beam of electrons that scans a specially coated surface and produces a signal at places where light has fallen on the surface. The vacuum is necessary to keep the electrons in a tight beam.

▶ As with computers, so with television. By the 1960s the replacement of electronic valves by transistors led to smaller products, and in the case of televisions, to portable sets. In 1959 the Japanese company Sony was the first to develop a set based fully on the solid-state electronics centered on transistors. The set, shown here, had a 20cm (8in) screen and weighed only 6kg (13lb).

Color television

The principles of modern all-electronic television were well established by the mid-1930s and regular public broadcasts were made on both sides of the northern Atlantic, though to relatively small audiences. Definition was not good, the picture was liable to flicker, and camera techniques were primitive, but nevertheless an acceptable service was offered. This was brought to an abrupt halt on the outbreak of war in Europe and afterward was resumed roughly where it had stopped in 1939. Some considerable improvement in quality was effected by applying wartime advances in electronics but one basic feature remained unchanged: the picture was in black and white. This was not due to any particular technical difficulties, for the problems of color television are essentially the same as those of color photography, which had long been mastered. In essence, the picture transmitted must be broken down into red, green, and blue images which must then be recreated and superimposed on the screen of the receiver.

Rather surprisingly, in view of the decisiveness with which photomechanical TV systems were abandoned in the 1930s, the first method to go into service (devised by Peter Goldmark in the USA in 1951) involved a whirling disk fitted with color filters placed in front of the camera lens. By 1953, however, RCA had perfected an all-electronic system. In this, the beam of light transmitted through the lens of the camera is split up by color-selective mirrors into red, green and blue components. The three images are then turned into a picture signal, for transmission in two modes. One mode, luminance, relates to the brightness of the image. The other, chrominance, relates to its color. At the receiving end the chrominance signals are picked up by three electron guns, whose beams scan the screen, activating very large number of dot-size phosphors which glow with red, green and blue light respectively. Because the luminance and the chrominance systems are

separated, color transmissions can also be picked up by black-and-white receivers.

By 1960 television was in the Western world no longer a novelty but a familiar piece of domestic equipment. The total number of sets was then around 100 million, some 85 percent of them in the USA. By 1970 this had more than doubled, to about 230 million. In the early 1980s color television had far outstripped black-and-white. By then virtually every home in the USA (98 percent) owned at least one television set, of which four-fifths were color sets. These figures relate, of course, to domestic receivers and take no account of the large number of television systems – many on closed circuit – which were used for special purposes. These included antitheft sets in shops, demonstrations of surgical operations to students, and a variety of earth scans from satellites.

Jet Flight

The basic principle of aircraft flight is very simple: a forward thrust creates an airflow over an aerodynamically shaped wing such that the pressure above it is less than that below, thereby creating lift. The thrust moves the aircraft forward and the lift stops it falling to the ground.

For the first three decades of aviation thrust was provided by a propeller or – in the case of the largest pre-World War II aircraft – a bank of propellers. For smaller aeroplanes the propeller has remained dominant but by the late 1920s it was increasingly being recognized as far from ideal. In particular, its engine requires a multiplicity of moving parts – crankshaft, reciprocating pistons, camshafts, connecting rods, and so on – and the engineering problems increase rapidly as higher speeds are sought. In the simple jet engine, air is drawn in at the front and compressed – which in itself heats it considerably – and then further heated by injecting fuel. The hot, high-pressure gas streams out through a nozzle at the rear,

causing a forward thrust, as in a rocket. The latter's fuel contains its own oxidant, so that it can operate in airless space, while the jet engine requires air to burn its fuel.

In the realm of fighter aircraft, superior speed is of paramount importance and even in the 1920s the Royal Aircraft Establishment in Britain was experimenting with the possibility of using a gas turbine instead of a piston engine to turn the propeller. But jet propulsion – pioneered by Frank Whittle in the UK – eliminates the propeller as well. He completed his first successful prototype engine in 1937 and his first jet-propelled fighter aircraft flew in May 1941. Meanwhile the Germans had been experimenting on similar lines, under the guidance of H. von Ohain and they were in fact first in the air, in August 1939, with the Heinkel He 178. However, there remained many technical problems to solve and in the event jet aircraft did not become operational until the last year of the war, with the appearance of the Gloster Meteor and the Messerschmidt Me 262. The USA developed the Bell P-59 but it never saw active survice.

After the war the development of jet fighters continued. In 1947 the USA introduced the F-86 Sabre, which had wings swept back, to avoid the shock wave from the nose of the aircraft. Some 10,000 were built. The Soviet reply, which also appeared in 1947, was the MiG-15. The first really successful French jet fighter plane, the Dassault Mystère, was introduced in 1955.

With the return of peace in 1945, the great future of jet propulsion necessarily lay in civil

▼ The 1950s saw a revolution in commercial aviation, with jet propulsion, developed during World War II for combat aircraft, taking over from engines driven by propellers. The first commercial jet aircraft was the British De Havilland Comet, shown here during construction. It appeared in 1952, but had a chequered career, including two serious accidents soon after its entry into commercial service.

aviation. The world's first jet air transport was the British de Havilland Comet. Introduced in 1952, it had to be withdrawn in 1954 when two aircraft disintegrated in mid-air. The fault was found to lie in metal fatigue but immediately after it was reintroduced in 1958, with redesigned fuselage, it was rendered obsolescent by the American Boeing 707 and Douglas DC8, which were roomier, faster and had a longer range. In the late 1950s Sud-Ouest in France introduced the first of the short/medium range Caravelles, which were very successful. Meanwhile the Soviet Union had been pursuing a largely independent line of development with the Tupolev 114. First flown in 1957, it was then the world's largest long-range air transport aeroplane, but serious technical problems arose, particularly with the Kuznetsov turboprop engines.

A major problem in the design of such aircraft is aerodynamic heating. This led to the development of sandwich-type skins, in which aluminum was to some extent replaced by the more resistant steel or titanium.

Concurrently, there had also been advances in the shape of propeller-driven aircraft powered by gas turbines. Among the most successful of the so-called turboprop aircraft was the Vickers Viscount, the prototype of which flew in 1950, and went into scheduled service in 1953. Another variation on the turboprop design is the turbofan unit, in which part of the sucked-in air bypasses the compression chamber, resulting in cooler and quicker running. Such engines have proved very economical and quiet for the new generation of wide-fuselage "jumbo" jets that emerged in the 1970s. The supreme technological achievement in civil aviation – though unfortunately not the most successful economically – was Concorde, the first supersonic airliner. Developed jointly by France and Britain, it made its maiden flight in 1969 and went into regular service in 1976. Fears about damage from its sonic booms, and deliberate obstruction by certain countries backward in this field of technology, limited the routes on which it could operate and militated against its commercial success.

Rocket development, satellites and space flight

In 1942, when Germany successfully launched the first V2 long-distance rocket from its rocket research establishment, at Peenemünde on the Baltic coast, Walter Durnberger – the station commander – remarked that this had proved that rocket propulsion was suitable for space travel. In the immediate postwar years, however, this possibility was not directly pursued. The advent of the atomic bomb now completely altered military strategy in respect of missile delivery. The manned bomber had become too vulnerable for attack, particularly in view of the development of the proximity fuse, and to rely on interceptor aircraft for defense called for larger numbers to be kept flying at all times to avoid the risk of large-scale destruction on the ground.

The answer was the long-range rocket, which was launched from a strongly fortified base and traveled to its destination in a high trajectory. Both the United States and the Soviet Union pur-

FEBRUARY 1951   35 CENTS

POPULAR MECHANICS MAGAZINE

WRITTEN SO YOU CAN UNDERSTAND IT

See page 118

▲ Aircraft with vertical or short take-off and landing (V/STOL) are vital in places where there is limited space on the ground, particularly during combat. The British engineering company Rolls Royce tested the first pure jet aircraft of this type, the "Flying Bedstead", in 1954. This was the forerunner of the first V/STOL aircraft to go into service, the highly successful Hawker Siddley Harrier, or "Jump Jet", introduced in the 1970s. Commercial STOL aircraft operate in places such as the Docklands Airport in London.

◄ As commercial air travel expanded during the 1950s there seemed to be no limit to future possibilities. Air travel did indeed increase dramatically, with the transport of people to holiday destinations by air becoming a business in its own right. This optimistic view of 1951, of the private helicopter replacing the family car, never did come to pass, however.

sued this new policy energetically, the former with the help of German rocket experts led by Wernher von Braun, and the latter with the aid of captured rockets and designs.

This led to the development of rockets of increasing power and range, culminating in the Intercontinental Ballistic Missiles (ICBMs) of the 1960s. Progress in the field was dramatically highlighted on 4 October 1957, when the Soviets launched Sputnik I, the first artificial satellite. This was a tiny spacecraft, no more than 60cm (23 in) in diameter and weighing 84kg (185lb). It carried two radio transmitters and emitted a steady bleep as it circled the Earth every 90 minutes.

The Impact of Ideology: Science in the Soviet Union

Before the Bolshevik Revolution in 1917 Russian science had largely been developed in a few elite establishments which were generously supported by the state. They formed a hierarchical pyramid, with the Imperial Russian Academy of Sciences (founded in 1724) at the apex. State support for science was part of the general drive to westernize the Russian Empire which had been initiated by Czar Peter the Great in the 18th century. At the end of the 19th century, when industrial development accelerated, applied science and technological research emerged.

The creation of comparatively advanced research facilities in some institutes of the Imperial Academy and at the universities of St Petersburg and Moscow gave Russian science a leading position in several important fields. Dmitri Mendeleyev, for example, proved that the chemical properties of elements are periodic functions of their atomic weights (periodic law). On the eve of World War I Russian science was well on the way to being integrated with European science.

Many prominent scientists emigrated during the revolution and the subsequent Civil War (1918–21). Needing urgently to restore and develop Russian industry, the Soviet government under Lenin considered the development of education and science essential conditions for fulfilling the task of building socialism. Facilities for research and education expanded in the 1920s, several new academies were established on geographic and specialist lines and numerous new research and educational institutes were created. Soviet scientists attempted to restore international links and cooperation with foreign institutions.

This trend ended unexpectedly in 1929 when Stalin inaugurated a program of rapid industrialization and the collectivization of agriculture. Suspicious of the foreign links of Soviet scientists and technologists, he organized several show trials of prominent engineers and scientists in 1929–31 which resulted in the imprisonment, exile and execution of dozens of innocent scientists. Travel abroad was forbidden and "socialist" science was declared principally different from "bourgeois" science.

Some Soviet scientists tried in those years to combine Marxist ideology with scientific theory. In 1935–37 many pseudoscientific ideas received government support. The most notorious product was Trofim Lysenko's pseudobiology which promised the transformation of plants, animals and even humans by neo-Lamarckist methods. (ie, genetic properties could be changed by environmental conditioning). The repression of science became most intense in 1937–40 when many of the best Soviet scientists were shot or perished in Arctic labor camps.

Stalin's terror affected Soviet military research and contributed to the defeats suffered by the Red Army in 1941 and 1942. However, the total mobilization of all technical and scientific potential for the war effort improved the situation in 1943 and 1944 and Soviet military technology gained new strength as prominent engineers like Anatoli Tupolev and Sergei Korolev were released from jail. Scientists like Pyotr Kapitza, Nikolai Semyonov and Lev Landau (all future Nobel laureates) and others devoted themselves to the war effort. In 1941 Kapitza suggested the idea of an atomic bomb but a research project did not begin until in 1943.

After World War II Stalin realized that the USSR lagged behind the USA and the UK in some key technologies relevant to atomic weapons – missiles, jet aircraft, radar, etc. Generous support was given to research and scientists again became respected members of the intellectual elite. Igor Kurchatov and Andrei Sakharov contributed to the creation of a Soviet atomic bomb in 1949 and the hydrogen bomb in 1953. The Academy of Sciences again received strong support from the government. It was able to increase its membership substantially and expand its network of research institutes.

From 1954 to 1961 the USSR led the field in the practical use of atomic energy, building the first atomic power station and the first atomic-powered icebreaker. It also excelled in space. It launched the first satellite in 1957, the first manned space flight in 1961. But the old pseudoscientific trends also enjoyed government support and in organic chemistry, biology and computer science the development of normal research remained impossible until the late 1950s.

A rapidly growing research infrastructure, the creation of isolated "scientific towns" like Akademgorodok in Siberia and the comparatively high salaries paid to research workers attracted

▲▼ Soviet science and technology, epitomized by the Sputnik satellites (above) and machines such as this particle accelerator (below), flourished in the 1950s.

nearly 400,000 young graduates to a scientific career between 1950 and 1960. Imported Western equipment improved the technical base of research. The quality of Soviet research in modern fields like molecular biology, superconductivity, the use of powerful accelerators, fusion, laser and computer technology approached international levels.

Although the politically motivated isolation of the scientific community had been breached, political interference into scientific life continued to blight Soviet science. The Communist Party controlled all important appointments in the research institutes and universities. Repressive measures against political dissidents after the Soviet invasion of Czechoslovakia in 1968 had a negative impact on science.

In the late 1970s there was a period of stagnation in scientific and technological research. As a result the USSR missed the electronic and biotechnological revolutions which took place in Western science. In the late 1980s President Mikhail Gorbachev's policy of *perestroika* ("restructuring") had a limited impact on Soviet science because of the shortage of funds and foreign currency to subsidize new trends in research and development.

Its immediate significance was not that it was the first satellite but that it clearly signaled to the world that the Soviet Union had developed a rocket powerful enough to rain H-bombs on the United States. The Americans reacted swiftly but it took them a year or more to catch up. Their first satellite, Explorer 1, was launched on 31 January 1958, and was followed by Vanguard 1 two months later.

A race had also been started to send a man into space. Here the Soviets also took the lead. In November 1957 Spunik 2 contained the dog Laika, which survived the launch and remained alive for seven days. Then, in August 1960, Sputnik 5 carried the dogs Belka and Strelka, who were safely returned to Earth. The following year, on 12 April, test pilot Yuri Gagarin made history by being the first man in space. He completed a single orbit in Vostok 1 and returned safely to Earth by parachute.

In 1958–59, following the creation of a specialist agency, the National Aeronautics and Space Administration (NASA), the United States regained technical superiority. In that year it launched no less than 19 satellites. Among them was the military satellite Score, highly significant in being the first of many communications satellites. The United States sent its first man into space – Alan Shepard – on 5 May 1961. He made a 15-minute suborbital flight in *Freedom 7*.

▲ When the USSR tested its first nuclear fission (uranium) weapon in August 1949, the USA reacted by deciding to build the "Super", a more powerful weapon designed to release energy in a thermonuclear process of the kind that fuels the Sun. The Sun burns by converting hydrogen to helium; the aim with the hydrogen bomb was to "burn" the heavy form of hydrogen, deuterium. The first test was on 1 November 1952. In the following years other countries, including the USSR, developed the technology to build similar bombs, creating a lethal stockpile, sufficient to destroy the Earth many times over. Here an American H-bomb is tested at Bikini Atoll in the Pacific on 21 May 1956.

The 1955 Atomic Conference

At the end of World War II the United States was the only nation that possessed the atomic bomb. For nearly a decade thereafter, it tried to hang on to the atomic "secret" and, by means of draconian laws and secrecy of a strictness unprecedented in peacetime, to prevent the spread of nuclear knowledge and technology. But in 1949 the Soviet Union exploded its first atomic bomb.

When President Eisenhower took office in 1952 he recognized the inevitability of the spread of nuclear know-how and took two initiatives to internationalize atomic power and ensure that diffusion of the technology was for peaceful and not military purposes. In a now famous "Atoms for Peace" speech to the UN General Assembly in 1953, he proposed the establishment of an International Atomic Energy Agency, to monitor and control the peaceful spread of the technology. The second US initiative led to the International Conference on the Peaceful Uses of Atomic Energy, held at the Palais des Nations in Geneva from 8 to 20 August 1955.

Some 450 full scientific papers were presented at the conference. Seventy-three states and eight specialized agencies of the United Nations sent delegations – 1,428 delegates in all, with a further 1,350 observers mainly from universities and commercial companies. The meeting was a major media event: 905 journalists covered the conference. A galaxy of talent in nuclear science attended. It was an outstanding success, not least because the conference had been designed to be technical rather than political. Scientists from different countries who had previously been working in isolation were able to compare notes and found that they had arrived at broadly the same conclusions.

Science and technology have moved on since 1955, and for many countries what then seemed the boundless promise of peaceful nuclear technology no longer appears so benign. But the exchange of information between East and West, North and South, helped ease international tensions, and helped make successful the much trickier business of drawing up the constitution of the International Atomic Energy Agency, which came into being in 1957 and has been a successful international agency.

Atomic energy

Although it was the culmination of several years intense research and development, the advent of the atomic bomb took the world at large totally by surprise. It was immediately apparent, however, that two major lines of development had to be addressed.

One was the purely military one: every major power had to be able to deploy the new weapon independently if its military credibility was to be maintained. At that time the only form of defense seemed to be a demonstrable capacity to retaliate in kind. The second was the possibility of using this totally new source of energy not catastrophically, as in a bomb, but by the development of wholly new technologies to control its release as a source of industrial power.

These two aspects were in fact closely inter-related as events in the UK illustrated. There the atomic energy program had, up to 1951, been largely directed toward the military aspects which in effect meant the production of plutonium. Thereafter the two began to converge, as atomic piles designed to produce plutonium could also be used to generate electrical power. One such was Pippa, which produced plutonium as a main product and power as a by-product. This achieved fame as the power unit of Calder Hall, the world's first major nuclear power station, which opened in 1956.

Rather surprisingly, in view of the vast amount of effort and money they had devoted to the Manhattan Project, the Americans showed little interest in the development of atomic energy as a controlled power source in the immediate postwar years. Only the US navy showed any serious interest, recognizing the immense potential for keeping ships indefinitely in commission without need for refueling. In 1955 the submarine

◄ The world's first nuclear power station, Calder Hall in Britain, was opened in 1956. In a nuclear power station the chain of fission reactions in uranium is controlled in a reactor, releasing energy in a steady manner to heat water so that the steam produced can drive turbines to generate electricity. The reactor at Calder Hall was the first in a series of "Magnox" reactors built in Britain. These use carbon dioxide gas to transfer heat to the boilers and a magnesium alloy (Magnox) to form the cans to hold the fuel rods of uranium. Stations of this kind provide power for about 40 years but then need a long and expensive period of decommissioning.

▼ In uranium reactors not all the uranium undergoes fission, but some is converted to plutonium. This can be used to fuel a "fast reactor", so called because the neutrons produced in the fission reactions do not need to be slowed down, as plutonium absorbs neutrons at higher energies. The world's first fast reactor came into operation at Dounreay in Scotland in 1959.

Nautilus was launched, the first of a series of atomic-powered underwater vessels. In 1959 the Russians commissioned the nuclear-powered ice-breaker *Lenin*. Not until 1957, however, did the first American nuclear power plant go into service, at Shippingport in Pennsylvania.

In the event it was again the Soviets who had taken the lead, with the opening in 1954, at Obninsk (near Moscow), of a small power station using enriched uranium as fuel and graphite as a moderator. The heat generated in the core of the reactor was extracted first into a closed-circuit water system at high pressure, and then transferred to a separate water system generating steam to drive turbines. A somewhat similar water-cooling system was developed by the USA for their Shippingport reactor and by Canada for the Candu reactors developed in the 1950s.

Water-cooling has the advantages of being simple and cheap, but has its critics. It is argued that in the event of an emergency – too much heat being generated in the core – the water will be turned to steam and so no longer be able to fulfil its vital cooling function. For this reason, France and the UK favored gas-cooled reactors, as gas does not change its state however much it is heated. The first reactor in the UK, at Calder Hall, was gas-cooled.

Meanwhile, in the 1950s, a new type of reactor – the fast breeder – was being developed. This was fueled with a mixture of uranium-238 and plutonium-239. Neutrons generated by the plutonium interact with the uranium to produce ("breed") more plutonium: this extracts several times more energy from a given weight of uranium. The first breeder reactor was commissioned at Dounreay, Scotland, in 1959 and the similar Phénix reactor became operative at Marcoule, France, shortly afterward.

During the late 1940s considerable attention was directed towards the use of yet another form of coolant for reactors. This was the metal sodium, which melts at 98°C, slightly below the boiling point of water and well below the normal operating temperature of reactors. On thermodynamic grounds it is attractive as a heat-transfer medium but it has inherent disadvantages. It is highly reactive chemically in a general way, and is then liable to corrode most materials with which it comes in contact. More particularly, it reacts explosively with water. This eventually ruled it out as a coolant for submarine reactors (in which the US navy was interested), though some studies were done in 1948 at the Argonne Laboratories near Chicago and by General Electric at Schenectady.

▲ While nuclear-powered transportation has remained mainly in the minds of science-fiction writers, one area where it soon came into its own was in the powering of submarines. *Nautilus* was the world's first nuclear-powered submarine, built by the USA and launched in 1955. The great advantage of nuclear-powered engines over diesel engines is that they require no oxygen, so the submarine can remain underwater for months, even years, rather than a day or so.

SCIENCE AND ETHICS

At the opening of the century, the idea that there could be ethical problems of science was nearly a contradiction in terms. Since science claimed to provide objective truths, and its application led to increased human power over nature, what ethical problems could there be? Centuries of struggle against religious obscurantism had fostered a vision of rational science identified with the progress of humanity in contrast to irrational, reactionary theology.

The first qualms about the application of science came in World War I when soldiers suffered lung damage and blindness from poison gas, a weapon which was developed by highly respected physicists and chemists yet seemed "inhumane". In World War II the previously respectable science of "eugenics", or the improvement of the human species, was applied with Nazi logic in an attempt to destroy whole peoples, with Jews as the principal victims.

Many of the anti-Nazi refugee scientists who created nuclear weapons in America were slow to realize that their inventions could be turned against all humanity. This supreme ethical dilemma of science was reflected in the mass movements against nuclear weapons of the 1960s and later. Also in the 1960s came America's war in Vietnam, in which both natural and social sciences were deployed and corrupted.

Ethical problems within the activity of pure research had long been recognized, in connection with the infliction of pain on sentient beings. Paradoxically, the demand for greater safety in consumer products for humans led to an increase in animal tests, some of which were doubtless very painful or degrading. Protests against vivisection, experiments and testing on animals have continued since early in the century, but in the 1980s scientists using animals both for primarily biological or medical research and for commercial chemical work were plagued by new forms of militancy by "animal liberation" activists.

Biologists have proclaimed that their science is no longer purely descriptive, but now includes its own sort of engineering. With these new powers come problems, as a worried and mistrustful public demands assurances that such engineering entails no environmental risks. But it is impossible to prove the impossibility of an unwanted event; and so the question is whether a process ought to be deemed dangerous until proved safe, or vice versa. Also, ever more invasive techniques of ending infertility create opportunities for alleviating the loneliness of childless couples; but the resulting problems of parenthood and identity create new tangles for which psychology, ethics and even theology are all relevant. Some people wish to end all research on the human embryo, arguing that the embryo, with its complete set of human genes, should be protected in law just like a child or adult. Others claim that such research is validated by the increased happiness brought to infertile couples by successful new techniques of fertilization and implantation of the eggs.

▲ The British philosopher Bertrand Russell was a prominent member of the antinuclear movement in Britain in the 1950s, questioning the ethics of the political and scientific decision-making involved in nuclear policy. In 1961 he led a massive sit-down protest in London.

▶ As research became more sophisticated, ethical problems grew beyond those of simple cruelty. Psychology probed human propensities for self-deception. One classic experiment, by US psychologist Stanley Milgram in 1965, involved subjects who (erroneously) believed that they were inflicting pain on others, as instructed by the experimenter in the cause of "science". This was intended to display the ordinariness of evil; but many protested at this abuse of people's trust in science.

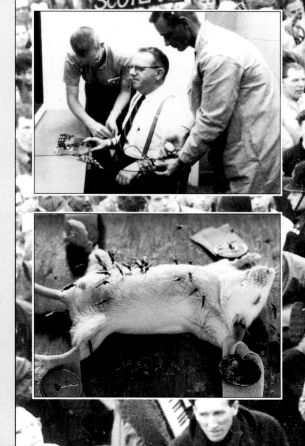

▶ The use of animals for scientific research dates back to the early 20th century, justified by the argument that the resultant medical advances save human suffering. In this image an anesthetized gerbil was exposed to malaria-carrying mosquitoes, in an attempt to learn more about the disease. Opponents of animal experiments claim that similar information could be learned by other means. In the 1980s "animal liberationists" sometimes attacked eminent scientists to gain publicity.

▲ The dilemma of modern science was epitomized in "Agent Orange", a chemical which had originally been discovered by a plant physiologist Arthur Calston in his PhD work. Later it was patented and used as a herbicide, and eventually used to expose or starve out the Vietcong guerrillas.

◄ Reproduction engineering seems to offer the ultimate benefit of consumer technology: total control over the process of making children. In practice, the real benefits are mixed with troubling problems, ethical and practical. Techniques such as laparoscopy, shown here, seem harmless, but they make possible in-vitro fertilization which raises ethical questions. Knowledge of the sex and of the possible defects of a fetus presents parents with new problems of choice. The possibility of a gross gender imbalance in male-oriented societies also looms ahead. The techniques themselves raise additional problems, as in the fate of the extra embryos.

◄ The antinuclear movement of the 1950s focused its opposition on the application of nuclear technology to bombs, while the peaceful use of the same technology to generate electricity seemed benign. By the 1970s this clear distinction was less easy to propose, as civil nuclear power research produced plutonium for use in bombs.

139

Datafile

The end of World War II meant that there could be a vast injection of state funds into scientific and medical research. Both the increased scale of research funding and the ever increasing internationalism of scientific research were impressive. Funding bodies like the Rockefeller Institute in the United States shifted their attention to agricultural research and combating the population explosion: it marked an extension of the range of scientific activity.

Cancer deaths in USA

Poliomyelitis cases in USA

▲ Medical research has only dealt successfully with certain diseases. Although there is an improved understanding of some causes of cancers, there has been limited success in therapy (top). Infectious diseases have been an area of greater success, as is illustrated by the use of the antipolio vaccine (above), developed by J.E. Salk.

Reliability of the Pill

Causes of mortality

USA 1950

Finland 1951

New Zealand 1951

Chile 1950

- Heart disease
- Cancer
- Influenza/pneumonia etc
- Tuberculosis
- Other

▲ With the conquest of major epidemic diseases, heart diseases and cancers have become more prominent causes of mortality in developed countries. This is less evident in Chile. There continues to be a marked contrast between poor countries with high rates of infant mortality and "developed" countries.

German hospital births

▲ Whereas in Imperial Germany, the state attempted to impose domiciliary midwifery services, modern trends have been toward delivery in hospitals.

◀ In the 1960s the contraceptive pill came of age. These statistics on the use of the pill in San Juan, Puerto Rico, derive from "field tests" conducted in 1954.

▶ By the 1950s Latin America was producing innovative scientists, despite the difficulties of research under authoritarian regimes. These statistics showing percentages of students in main subject areas suggest great effort in medical and economic sciences.

Buenos Aires University

1947

1956

- Economics
- Medicine
- Physical sciences
- Agronomy

In the postwar years the biological sciences preserved their identity but increasingly invoked the aid of the physical ones. This generalization is well exemplified by penicillin, for the discovery and development of which the British scientists Alexander Fleming, Howard Florey and Ernst Chain shared the Nobel Prize for Physiology or Medicine in 1945. Popular accounts refer to the people in Oxford who purified penicillin as the "Oxford Team" as though this were some sort of task force of diverse experts assembled by Florey at short notice to tackle the particular problems of penicillin. In fact nothing could be further from the truth.

Long before he had any particular interest in penicillin, Florey had recognized that medical problems could not be solved by medical people alone, and that progress depended on a multidisciplinary approach. For this reason, all the people needed for the penicillin project were already to hand in his laboratory (the Sir William Dunn School of Pathology) when it was launched in 1939. They included an organic chemist, a biochemist and two bacteriologists.

Interestingly, in the following year (1946) the American geneticist H.J. Muller was awarded the same prize for his work on mutations – many of which he created by the use of X-rays, a very particularly physical phenomenon, for the discovery of which Röntgen had been awarded the very first Nobel Prize for Physics.

Increasingly, too, the biological sciences became quantitative rather than qualitative, especially invoking the use of statistics. For example, the connection between smoking and lung cancer and coronary heart disease was established largely on statistical evidence. A very significant report on this was issued by the American Heart Association in 1960.

The postwar years saw increasing popular reference to a new field of research: molecular biology. This is an interdisciplinary study involving structural chemistry, biochemistry and genetics and concerns itself with the vital functions of cells at the molecular level. The term was in fact coined in 1938, but not until 1951 did the National Research Council in America recognize it as one of its six fields for the provision of grants: the foundation of the *Journal of Molecular Biology* in 1959 at last made the term fashionable. By then, however, its definition was somewhat more limited: it was most particularly concerned with the chemical messengers which regulate the function of the genes. The years 1955–70 were ones of great achievement as the nature of the genetic code was unraveled.

Penicillin – or, more strictly, its unique therapeutic properties – was essentially a wartime discovery, developed in great secrecy in the United

THE SECRETS OF HEREDITY

Molecular biology

Salk's antipolio vaccine

Crick and Watson discover the structure of DNA

The development of oral contraceptives

Science in Latin America

States, and reserved almost entirely for military use. Not until the late 1940s did it become generally available for civilian use: in 1950 the United States, by far the largest producer, manufactured nearly 200 tonnes. A similar quantity of streptomycin, another potent antibiotic, was made at the same time. For a time it seemed that antibiotics of one kind or another held the answer to the cure of all infections, but this was not to be. Many bacteria were, or rapidly became, resistant to appropriate antibiotics, some patients are allergic, and – most particularly – viral infections, among the most deadly, are unaffected. Moreover, antibiotics are effective only when infection is established and cannot, therefore, be used to counter epidemics.

Among the greater viral scourges is poliomyelitis (infantile paralysis). Very often this causes no more than an influenza-like fever, but – especially in countries with high health standards and correspondingly low acquired immunity – it can lead on to a severe and permanent paralysis. In the 1950s J.F. Enders, F.C. Robbins, and T.H. Weller in the USA showed that various viruses

▼ Although W.J. Kolff's kidney machine could not support life indefinitely, it could sustain a patient while kidney damage healed or was repaired. This picture shows one of the first machines to be used in the UK, at the Hammersmith Hospital, London (1946).

could be grown on living tissue kept free of bacterial infection by penicillin. This led on to important clinical results, notably J.E. Salk's development in the USA in the early 1950s of an antipolio vaccine, followed, in 1956, by Albert Sabin's oral version of it.

Polio is only one of many viral diseases which afflict mankind. Some cause only minor symptoms, as indeed polio may often do, but others can be severe and even life-threatening. They include the various forms of influenza viruses – epidemics of which periodically sweep the world with devastating results – and those which cause mumps, yellow fever, German measles, hepatitis, sandfly and tick fevers, herpes simplex, and certain kinds of wart. Others attack domestic animals, like the viruses which cause equine encephalitis, rabies in dogs and scrapie in sheep. There is no chemotherapy for viral diseases, as there is for those caused by bacteria, and the best defense lies in prevention by means of viral vaccines. Many such vaccines have been developed since World War II, either live but attenuated, or killed.

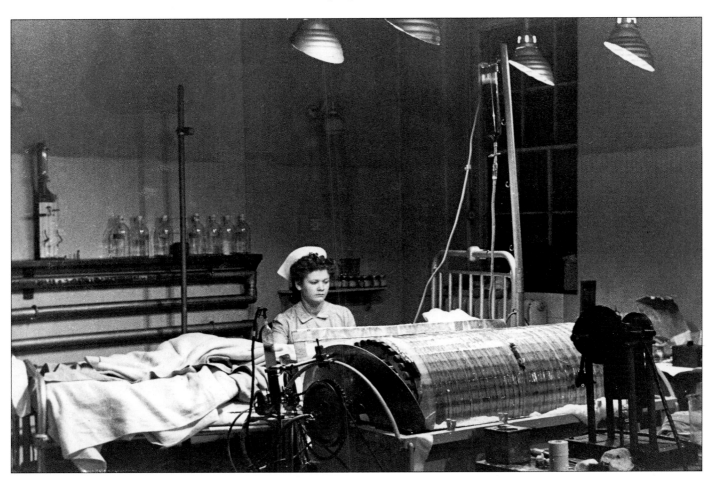

▼ Recollections of the American polio epidemic of 1943, in which 1,200 died and thousands more were crippled, emphasized the importance of J.E. Salk's vaccine, widely introduced in 1954. Here children queue for vaccination in aptly named Protection, Kansas (1957). Sabin's oral vaccine – taken on a lump of sugar – followed in 1956 and was much more convenient and pleasant. It was quickly adopted in America but more slowly in Europe.

Much pioneer work in this field was also done by J.F. Enders, at Harvard Medical School. Initially, he found working with viruses was very much hindered by the limited means of maintaining cultures – for example, in living chick embryos. He found that although living cells were necessary, whole animals were not required if bacterial infection of tissue cultures was prevented with penicillin. Between 1948 and 1951 he successfully cultured the viruses of mumps, polio and measles, typically using media consisting of ox serum and chick embryo cells. On this basis, he developed a vaccine for measles in 1951 and this was widely used by the 1960s. In this research his colleagues were F.C. Robbins and T.H. Weller, and all three shared the Nobel Prize for Medicine or Physiology in 1954.

A promising, but so far not fully realized, advance in virology was made in 1957 when A. Isaacs and J. Lindenmann discovered the group of substances known as interferons. They are released by cells when they are invaded by viruses and they are able to protect noninfected cells. They also have some antitumor activity. However, this limited availability and the incidence of side-effects, often not dissimilar to those of viral infections themselves, have so far limited their clinical use.

Chemically, interferons – which vary slightly from one animal species to another – consist largely of proteins, with molecular weights around 20,000, though some carbohydrate is also present. No specific structure has so far been assigned to any interferon.

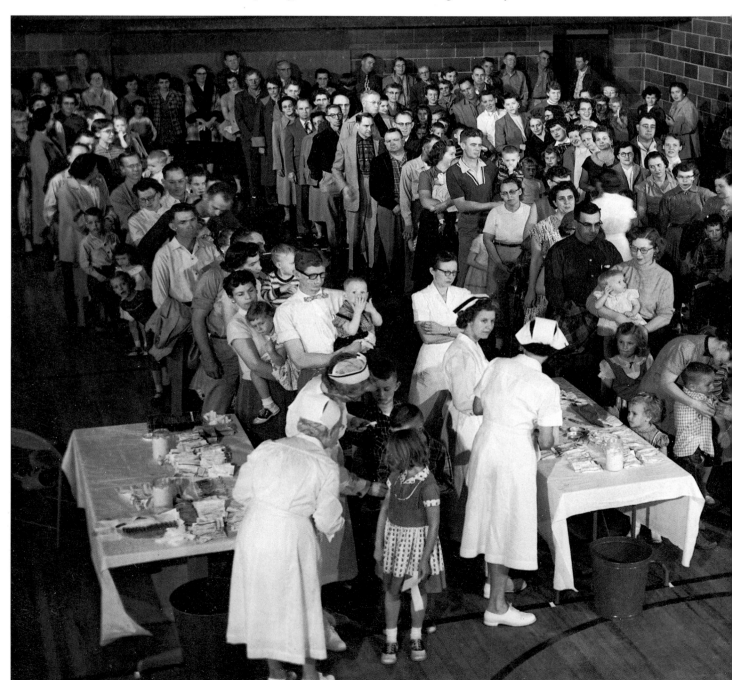

The discovery of DNA structure

That physical characteristics are passed on from parents to offspring in both humans and domestic animals is a matter of common observation. From time immemorial empirical methods have been used to bring out desirable traits and fix them so that they breed true generation after generation. In this field, the great achievement of the 20th century has been to put breeding on a logical basis and unravel the mechanisms which govern inheritance.

Even as late as the middle of the 19th century the role of sperm in reproduction was not clear, for it seemed to do no more than trigger off the development of the ovum. However, by the turn of the century it was becoming clear that the hereditary factors were carried on twin structures, called chromosomes, present in virtually every cell of the living organisms. When the cell divides so do the chromosomes, so that each new cell has the same hereditary makeup as its parent. The only exceptions, but ones of crucial importance, are the cells involved in reproduction. When these are formed, only one chromosome of each pair goes to the ovum or the sperm cell. When these fuse at the moment of fertilization, the new cell then receives its full complement of paired chromosomes. However, as one member of each pair comes from each parent, the new cell combines hereditary features from both. It follows that although they look alike under the microscope the chromosomes of every individual must differ in their fine structure so that some, for example, confer blue eyes rather than brown. In fact, the chromosomes resemble a string of beads, each bead – known as a gene – being different from all the others and responsible for the expression of a particular hereditary characteristic. For a general understanding of the mechanics of inheritance this simple picture is adequate, but in practice most characteristic features result from the combined action of several genes.

Chromosomes were thus identified as the general carriers of inherited characteristics and genes as the means of transmitting special features from one generation to the next. But to delve yet another layer down in the microscopic structure of the cell, what is the fine structure of the gene that permits the great variations observed in nature?

The story of DNA begins in 1869, when the German chemist Friedrich Miescher discovered nucleic acids, a century before their genetic significance became apparent. In 1909 A.E. Garrod – investigating inborn errors of metabolism – reported that genes act by interrupting essential steps in metabolic processes. In the early 1940s G.W. Beadle and E.L. Tatum, at Stanford University, investigated this further and deduced that the synthesis of a particular enzyme is controlled by a particular gene. The serious blood disease sickle-cell anemia proved to be due to a gene altering the sequence of amino acids in the sidechains of hemoglobin. Meanwhile research on bacteriophages – viruses which attack bacteria – proved them to consist of only a protein and a nucleic acid. This greatly simplified the picture. At first it seemed that the genetic activity resided

in the protein moiety but later the activity agent was identified as a nucleic acid, and specifically deoxyribonucleic acid (DNA). This was confirmed by the work of A. Hershey and M. Chase in America in 1952 but already the American biochemist James Watson and the English biophysicist Francis Crick, working at the Cavendish Laboratory in Cambridge, had seized on the key role of DNA as the basic material of heredity. They worked intensively on its structure with the New Zealand X-ray crystallographer M.F.H. Wilkins, who was then working in London. Already in 1952 Linus Pauling in America had proposed a spiral structure for DNA, and this they confirmed unequivocally in April 1953.

The DNA molecule is itself very complex, which is why it can provide the vast array of genes that determine all the wide variation that can occur within a single species. Each DNA molecule consists of four subunits known as a nucleotide, each of which consists of a sugar linked to a phosphate group and one of four nitrogenous bases – adenine, cytosine, guanine and thymine respectively. These details need not concern us, but such complexity obviously provides great scope for variation, just as the relatively few units of a child's set of building blocks can build an almost infinite variety of models, from suspension bridges to windmills. The nucleotides join together at random to form long chains but it was discovered that in fact the chains are double, with a ladder-like structure. The sides of the ladder consist of alternate sugar and phosphate molecules and the rungs of pairs of the nitrogenous bases, either cytosine/guanine or thymine/adenine. Research by Wilkins at King's College London suggested that the ladder was not straight but twisted, like an old-fashioned barley sugar stick. At Cambridge Crick and Watson took this further and proved that the DNA is in the form of two helixes wound round each other – the now famous Double Helix. For this brilliant research Crick, Watson and Wilkins shared the Nobel Prize for Physiology or Medicine in 1962.

▶ The unravelling of the structure of DNA, culminating in its formulation as the Double Helix, was one of the great scientific achievements of the 20th century. This was internationally acknowledged in 1962 with the award of a Nobel prize to the three scientists most closely involved – F.H.C. Crick, J.D. Watson and M.F.H. Wilkins. Watson and Crick (right) are seen here with a model of the Double Helix.

Fertility and contraception

There are few better examples of the yin-yang dichotomy of complementary relationships than fertility and contraception. Throughout history there have been communities to which human fertility was a blessing – closely identified with the fertility of the soil, to be sought through elaborate, sometimes cruel, rites. To others the avoidance of conception has been a no less desirable aim.

In the 20th century rather little has changed at the practical level, though there have been some interesting changes of attitude. The major advances, of course, are the oral contraceptives, containing hormones which suppress ovulation and cause rejection of the sperm.

The pioneers here were the US scientists G.W. Pincus and M.C. Chang who from 1951 investigated the antifertility effect of steroid hormones (especially progesterone) which act by suppressing ovulation. Synthetic hormones physiologically similar to progesterone became available in the 1950s, Pincus and Chang investigated this use to control fertility. Field trials in Haiti and Puerto Rico in 1954 showed them to be very successful oral contraceptives.

Much of the chemical work on synthetic hormones (especially norethistoreone) was done in the 1950s by Syntex in the USA. The chemists particularly concerned there were G. Rosenkranz, C. Djerassi and F. Sondheimer.

Oral contraceptives were first approved for use in the USA in 1959 and soon were in use worldwide. Of all methods of contraception, it is the most widely acceptable and effective – the failure rate is no more than 0.2 percent. It has been estimated that by 1980 some 55 million women around the world were taking the pill, but most in North America and Western Europe. Worldwide, therefore, its impact has not been great. This is not altogether surprising, for the complex chemical constituents are expensive, and the necessity for conscientious daily use makes it unsuitable for use in primitive communities even if it could be afforded. Moreover, its use is no longer growing, and may even be declining, due to concern about possible side effects, particularly thrombosis.

The history of intrauterine devices (IUD), an old device revived in new coil form in the 1960s, has not been good. Thus the generality of the world's population has still to resort to traditional methods, though it must be remembered that for very large numbers of people – such as the estimated 600 million Catholics – the practice of artificial contraception is forbidden.

Chief among the older devices is the condom or contraceptive sheath, supposedly invented by a Monsieur de Condom, a friend of the French writer La Rochefoucauld in the mid-17th century. In the 20th century the introduction of rubber latex, in place of animal tissue, in the 1930s resulted in a much improved product, and has been sold in hundreds of millions. Interestingly, the condom was originally used not only to prevent contraception, but to avoid transmission of the venereal diseases syphilis and gonorrhoea. From 1987 this original use of condoms has reas-

▲ Together with land drainage and insecticide spraying, chemotherapy is the chief weapon against malaria. Until 1926 – with the advent of the synthetic product plasmoquin – the only effective drug was quinine. This was followed by mepacrine (1932) and paludrine (1945). The latter was developed by the British company ICI and introduced after clinical trials in the tropics. As shown here, these involved children as well as adults.

sumed new importance worldwide as a means of preventing the spread of AIDS (Acquired Immune Deficiency Syndrome).

Birth control is not solely a matter for the individual. In many countries, notably those of the Third World, resources – especially food production – have not kept pace with an explosive growth in population. Population control has then become a matter of national policy, but the options are, by the nature of the circumstances, strictly limited on grounds of cost: thus the pill could not even be considered. India, for example, has pursued a policy of voluntary male vasectomy, backed by material inducements. China, with a current population of over one billion, one-quarter of the world's total, has resorted to draconian measures. Except in special circumstances, every family is limited to one child.

▼ The postwar years saw increasing emphasis in many countries on maintaining health rather than curing disease. This involved setting up specialized clinics, devoted to such fields as birth control and infant welfare. This picture shows a typically informal gathering in a maternity center (Paddington, London, 1950).

Science on the Periphery: Latin America

In 1947 Bernardo Houssay (1887–1971), physiologist at the University of Buenos Aires in Argentina, won Latin America's first Nobel prize, in medicine, for his work on the endocrinology of carbohydrate metabolism. The award was all the more dramatic because it came soon after Houssay, a democrat, had been dismissed from his teaching and research posts by the dictator of Argentina, Juan Perón. It provides a vivid demonstration of the precariousness of science in Latin America.

The medical disciplines have been the premier fields of scientific excellence in Latin America, dating to the establishment of modern bacteriological institutes patterned after the French Pasteur Institute in the last years of the 19th century. The most renowned example is Brazil's Institute of Experimental Pathology, founded at Rio de Janeiro in 1902 by Oswaldo Cruz. In its *Memorias* were published significant research findings such as the identification in 1909 of the pathogen of Chagas's disease ("American Sleeping Sickness") by Carlos Chagas (1879–1934). A parallel development was the solution to Carrión's disease, endemic in Peru, by Alberto Barton, an English-trained bacteriologist in Lima in the same year. Also notable was the work of the Cuban Carlos Finlay (1833–1915) in identifying the vector of yellow fever.

Next to bacteriology, a number of important centers of physiological research functioned in Latin America. Those led by Alexander Lipschutz in Chile and Arturo Rosenblueth in Mexico, were part of an international network whose epicenter was Walter Cannon's laboratory at Harvard. A separate center was the Institute of Andean Biology and Pathology, founded in 1934 in Lima by Carlos Monge (1884–1970), where research in high altitude or environmental physiology was performed with significant results and a high level of funding by American agencies.

Research in the physical sciences has been less consistently institutionalized, but certain institutes, such as the Geophysical Laboratory at Chacaltaya, Bolivia, or the Centro Atómico Bariloche in Argentina, have produced high-quality work. At Chacaltaya, for example, a group of physicists led by the Brazilian Cesare Lattes discovered the pi meson in 1947, in coordination with a team of British researchers.

Latin American science did not begin to be integrated into the international scientific system until the 1940s. This happened as the result, first, of the coming together of groups of national scientists in various national associations for the advancement of science (Peru, 1920; Argentina, 1933; Brazil, 1948; Venezuela, 1950) and second, because of the stimulation provided to the whole range of scientific disciplines by refugees from the Spanish Civil War, and Jewish and other refugees from Nazism, both groups arriving in the 1930s. Such persons arrived with international affiliations already formed. In the 1960s and '70s government organization of science was consolidated in the formation of national research councils. In the decade 1973–84, five Latin American countries (Brazil, Argentina, Mexico, Chile and Venezuela) accounted for 89 percent of the region's scientific publications.

▲ An experiment in progress at the National School of Agriculture, La Molina, Peru, in 1952. This research is testing fertilizers for the cultivation of potatoes. Much research in agronomy, medicine and public health in Latin America has been sponsored by the Rockefeller Foundation in the USA. Its participation in Latin American biomedicine dates to its post-World War I assessments of medical education in most countries of South America, and its heavy involvement, contemporaneously in public health projects, notably the eradication of yellow fever in Peru. The Rockefeller Foundation backed talented scientists in several South American countries pursuant to its policy of "making the peaks higher".

THE STRUCTURE OF DNA

The elucidation of the structure of DNA (deoxyribonucleic acid) in 1953 is often regarded as the birth of molecular biology because it marked the key to understanding the chemistry of reproduction – the characteristic which most readily distinguishes the living from the nonliving world. DNA was isolated by the Swiss biochemist Johann Miescher in 1871 from the nuclei of dead cells, but it was another 60 years before evidence emerged that it was the molecule of inheritance. Proof came in 1944 when the Canadian Oswald Avery and colleagues showed that the DNA isolated from one strain of a bacterium could alter the appearance of a second strain by mixing the DNA and cells.

Through chemical analysis it was established that DNA is a polymer (long repeating chain) comprising a sugar phosphate and four bases, termed adenine (A), guanine (G), thymine (T) and cytosine (C). The Austrian-American Erwin Chargaff found that the ratios of A to T and C to G were constant, although A to G varied between DNA isolated from different sources. This finding ruled out the prevailing idea that DNA had a constant repeating motif, but the real significance was revealed in 1953 by the researchers in Britain, James Watson and Francis Crick. X-ray diffraction patterns, obtained by Rosalind Franklin, had shown that DNA was a helical structure of constant width. In order to fit the bases into this structure, Crick and Watson argued that the larger purine bases (A and G) could only pair with the smaller pyrimidine bases (T and C). Furthermore, they proposed that A always pairs with T and G with C, thereby accounting for Chargaff's findings.

This specific pairing suggested how the molecule could replicate – the key question to showing how DNA could operate as the molecule responsible for inheritance. If each strand of the helix was unwound it could direct the synthesis of a complementary strand. Meselson and Stahl proved this in 1958 by growing bacteria on a food source containing a heavy nitrogen isotope until the DNA was "labeled". Such DNA is denser than normal and could therefore be separated from unlabeled DNA. The bacteria were then switched back to an ordinary food source and samples of DNA analyzed. The first generation had DNA of an intermediate density, showing the daughter DNA was made up of one labeled strand from the parent plus one newly synthesized strand.

Crick also suggested that DNA directs the synthesis of RNA, a related molecule implicated in the synthesis of protein. Subsequently the genetic code which defines the relationship between DNA and protein was elucidated. Thus DNA ultimately directs the synthesis of all the components of the cell through control of the synthesis of enyzmes and other proteins. These discoveries allowed a deeper understanding of the mechanisms of inheritance and the growth of an organism. The ability to manipulate an organism's DNA by "gene splicing" made possible the new science of biotechnology.

▼ The double helical structure of DNA revealed a key property of a molecule concerned with inheritance: the potential for self-replication. Each strand is made from a repetitive sugar-phosphate backbone indicated by the helical ribbon. The bases point inward and pair up with bases from the other strand as in the rungs of a ladder. The strands can unravel and form the template for the synthesis of two new identical strands.

► Molecules scatter X-rays to form a kind of image on a photographic film called a diffraction pattern. If the molecule contains a repetitive feature then the scattered X-rays reinforce each other at certain points and give rise to spots on the film. The arrangement and distance between the spots are related to the position and distance between the atoms in the molecule. In the case of DNA, the cross-shaped pattern indicated a helical structure.

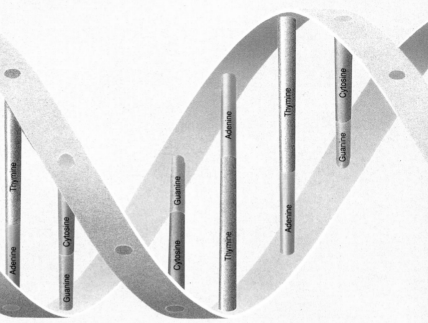

▼ ► Chromosomes in the nuclei of living cells were first implicated as the vehicles of inheritance by observation during cell division using the light microscope. During meiosis (the formation of egg cells), the chromosomes separate so that the resultant sperm or egg contains one of each pair, the full complement being reestablished on fertilization.

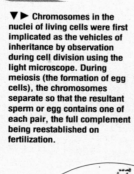

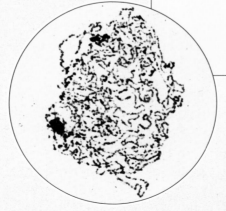

► The original model of the DNA molecule was constructed by Crick and Watson using balls and sticks to represent atoms and bonds respectively. Today such molecular models are usually drawn on a computer screen, which readily allows the observer to modify the structure and also to look inside the molecule at parts which would be obscured in a physical model.

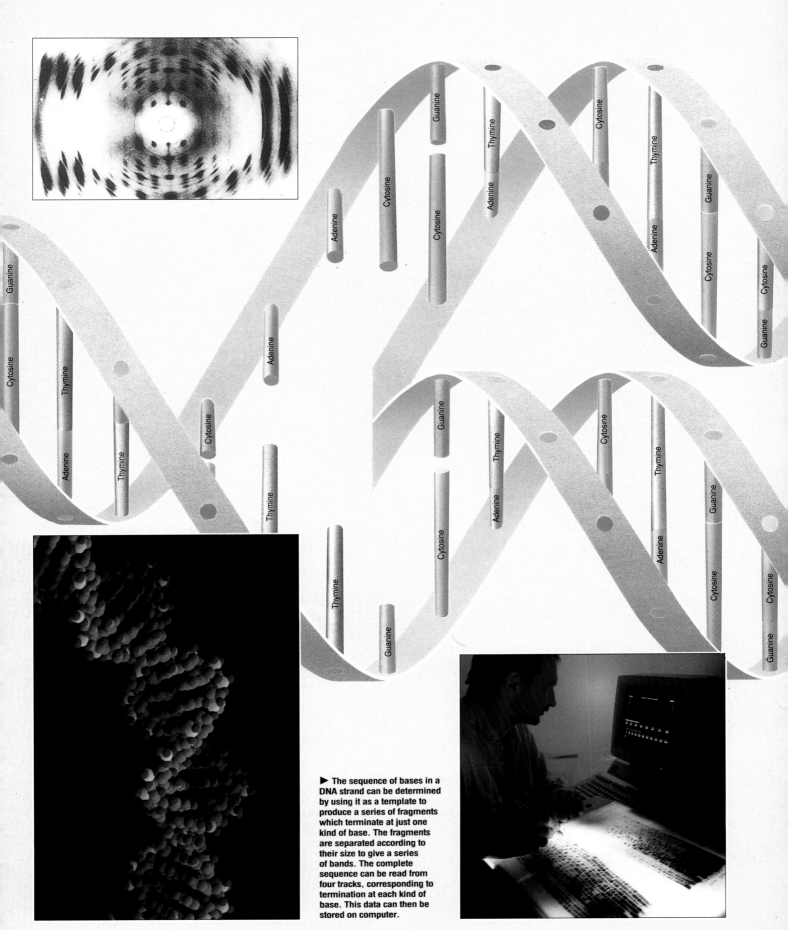

▶ The sequence of bases in a DNA strand can be determined by using it as a template to produce a series of fragments which terminate at just one kind of base. The fragments are separated according to their size to give a series of bands. The complete sequence can be read from four tracks, corresponding to termination at each kind of base. This data can then be stored on computer.

147

Datafile

In the postwar years science, and particularly physics, became an increasingly expensive business. Large-scale equipment was required to probe the atomic nucleus and to observe the Universe; respectively high-energy particle accelerators and giant telescopes. The wealth of the United States, coupled with the effect of the war on Germany, shifted the center of scientific activity across the Atlantic Ocean and away from Europe where it had been since the 17th century.

▼ Scientists from the United States dominated the awards list after the end of World War II. They won for work on viruses, nuclear magnetic resonance, the transistor and the chemical action of genes. This period also saw the first awards to scientists from Japan, China, and the USSR.

▼ The number of doctorate degrees awarded in the United States increased dramatically after 1960, reflecting the increasingly technological employment being offered for which a higher level of training was required. It also reflected the greater availability and attraction of higher education in the 1960s and provision of funds.

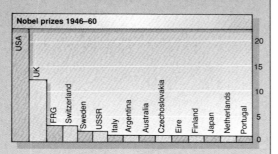

Nobel prizes 1946–60

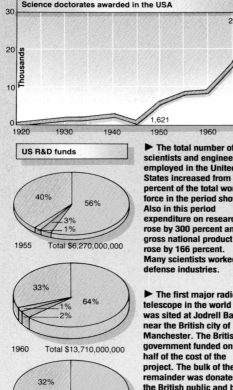

Science doctorates awarded in the USA

▶ The total number of scientists and engineers employed in the United States increased from 1 to 2 percent of the total work force in the period shown. Also in this period expenditure on research rose by 300 percent and the gross national product rose by 166 percent. Many scientists worked in defense industries.

▶ The first major radio telescope in the world was sited at Jodrell Bank near the British city of Manchester. The British government funded only half of the cost of the project. The bulk of the remainder was donated by the British public and by the philanthropist Lord Nuffield, who had made a fortune from automobiles.

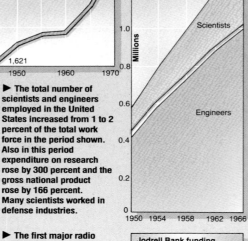

US jobs in science

Jodrell Bank funding

Total £658,900

- Department of Science & Industrial Research
- Nuffield Foundation
- Public appeal
- Lord Nuffield
- US Deep Space Network

US R&D funds

1955   Total $6,270,000,000

1960   Total $13,710,000,000

1965   Total $20,470,000,000

- Federal government
- Industry
- Universities/colleges
- Others

◀ Although the total expenditure on research and development in the United States increased by three times between 1955 and 1965, the distribution of these costs remained similar. An increase in federal government military expenditure was almost matched by an increase in industrial investment in advanced technologies.

In 1963 Derek de Solla Price, then professor of the history of science at Yale University, published a book with the intriguing title *Little Science, Big Science*. Its purpose was to draw attention to an important dichotomy in the sciences which had developed since the end of the war.

Briefly, there were on the one hand the scientists whose research was possible only with the aid of large, complex and – above all – very expensive apparatus. Such were, for example, the new breed of radio astronomers, searching the heavens with their vast steerable bowls to locate a distant source of radio waves. Whereas Lord Rutherford had taken pride in the immensely significant results he and his colleagues at Cambridge in the UK had achieved in the 1920s and '30s with "string and sealing wax", many postwar physicists were calling for vast particle accelerators weighing hundreds of tonnes and costing millions of dollars. These were the representatives of big science. By contrast the practitioners of little science – the chemists, for example – were still capable of making great progress with relatively modest apparatus. True, the traditional array of bunsen burners, test-tubes, flasks, distillation columns and the like no longer sufficed and there was a need for more sophisticated apparatus – such as infrared spectrometers and ultracentrifuges – which could be used by laboratory technicians or even students. By prewar standards these were expensive, but not in the same league as the big science equipment.

This development raised two fundamental economic issues. The first was that the overall demand for funds for scientific research of all kinds had reached a level at which it would no longer be absorbed without major repercussions elsewhere. The second was to find acceptable criteria for determining how available funds should be allocated between the vast array of competing claims. That no such criteria have been established is not surprising, for many imponderable factors are involved and attitudes are subjective. Compromises have to be made.

Two basic questions have to be addressed when considering the allocation of funds for research. Will expenditure lead to the advancement of science by increasing the store of human knowledge? Will it, additionally, confer some benefit on mankind generally? These are questions that cannot be answered with certainty, though collective wisdom and experience can shorten the odds against error. A classic and oft-quoted example of the difficulty of prediction was provided long ago by the British scientist Michael Faraday (1791–1867). When a visitor to his laboratory asked what was the use of all his discoveries in electricity, he replied tersely: "What is the use of a new-born baby?"

BIG SCIENCE

Whether in spite of, or because of, the increasingly objective allocation of research funds the 15 years that followed the war were certainly productive ones for the physical sciences. There was, for example, a fruitful marriage between physics and chemistry. Up to 1939 it had seemed that no more than 92 elements – from hydrogen to uranium – could exist but the Manhattan Project had produced a transuranium element, plutonium, not known in nature. This was followed by a succession of others, though many of these on a scale almost invisibly small.

Physical methods were also unravelling the structure of complex organic molecules by the method of X-ray crystallography developed by W.H. and W.L. Bragg. Here it was the computer that transformed the scene. The principles had been fully worked out before the war but the methods of computation then available were so laborious that months of tedious calculation were necessary to work out even a fairly simple structure. With the aid of the new computers results could be obtained in a small fraction of the time and with much reduced effort on the part of the research workers.

Again the phenomenon of nuclear magnetic resonance (NMR), independently discovered in 1946 by the physicists Felix Bloch and Edward Purcell, was quickly exploited by the chemists. This depends on the fact that many atomic nuclei have a magnetic moment which is "quantized": that is to say, it can adopt only a limited number of orientations. As the magnetic environment of an atomic nucleus is influenced by the configuration of neighboring electrons NMR can be used to give a great deal of useful information about molecular structure.

To this period belong also the invention of the maser (1954) and the laser (1960). The latter gave new life to the technique of holography already invented by D. Gabor in 1948. The essence of the laser is that it produces a nondivergent intense beam of monochromatic light, but in other fields the effects of intense light irradiation had been investigated by other means. At Cambridge in 1945 George Porter and R.G.W. Norrish began a long research program to investigate the photochemical effects of light. They were particularly interested in the very short-lived free radicals that are formed in gases, and for this purpose developed a technique of flash photolysis which made it possible to investigate such radicals which had lives of no more than a picosecond (10^{-10} second). Later this technique was used also to investigate biochemical problems.

The new discoveries in physics fertilized yet another field of science, archaeology, when W.F. Libby in 1947 developed the technique of radiocarbon dating to measure the age of artifacts.

▼ From its infancy in the years immediately after World War II radio astronomy rapidly graduated to the "Big Science" league. This 185m (600ft) steerable dish, weighing 20,000 tonne, was completed in the USA in 1962, at a cost of 80 million dollars.

The International Geophysical Year

The postwar years were ones of great international tension, of mistrust and hostility. The time hardly seemed propitious, therefore, for launching any considerable project involving worldwide collaboration. Nevertheless, the decade 1950–60 saw the realization of one of the greatest collaborative ventures in science ever launched: the International Geophysical Year (IGY).

It began almost casually, in 1950, with a suggestion from the American physicist Lloyd V. Berkner that 1957–58 should be designated a third International Polar Year, on the lines of those of 1882–83 and 1932–33. This proposal was brought before the International Council of Scientific Unions (ICSU) who not only approved it in principle but extended its scope to include geophysical observations in all latitutdes. Eventually 56 nations took part, including the USSR, even though it was not a member of the ICSU. The World Meteorological Organization also cooperated. A steering committee (the *Comité Spécial de l'Année Géophysique Internationale* or CSAGI) decided that the "year" should in fact extend from July 1957 to December 1958. The choice

▲ In the 1940s the French physicist Auguste Piccard turned his attention from exploring the stratosphere in balloons to plumbing the ocean depths in submersibles (bathyscaphes) with no surface link. This shows the bathyscaphe *Trieste* in 1957, under contract to the US navy to make dives in the Mediterranean.

▶ In the 1950s the National Academy of Sciences in the USA began a feasibility study to explore the possibility of "drilling a hole to reach the Mohorovičić discontinuity" in the Earth's crust. This soon became known as the Mohole Project. Here the experimental drill-ship *CUSS 1* is seen in operation off Guadalupe Island in 1961.

of epoch was calculated. The Polar Year 1932–33 coincided with a time of minimum sunspot activity, when the Sun was relatively quiescent. By contrast 1957–58 was expected to be a year of maximum activity. This prediction was amply fulfilled: the IGY coincided with solar activity unprecedented during more than 200 years of scientific observation.

From the beginning the project was planned on a generous scale. The CSAGI met annually – in Rome, Brussels and Barcelona – to plan and establish the observing stations and arrange for close communication between them. Some already existed but many were quite new: a full-time co-ordinator and secretariat was established in 1956. Again, particular attention was focused on the polar regions. The influence of the Antarctic on global weather and on atmospheric and oceanographic dynamics was realized but not fully understood. Additionally, close and prolonged observation of the aurora borealis and the corresponding aurora australis was needed, together with detailed measurements of polar variations in the Earth's magnetic field. All this required much preparation. One observing center was established at the South Pole itself, eleven on the Antarctic continent and ten more on neighboring islands. Equally important – for investigation of the general circulation of the atmosphere, with its important effect on world climate – were three chains of meridian stations. One ran approximately along the meridian 80°W from the Arctic, through North and South America to the Antarctic. The second was along the meridian 10°E and the third along the meridian 140°E. As this last ran partly through Soviet territory – before continuing south through Japan, New Guinea and Australia – the participation of the Soviet Union was particularly welcome.

Finally, there were equatorial stations, located to make observations of the equatorial jet stream

in the upper atmosphere and to carry out studies of cosmic rays and terrestrial magnetism.

These strategically located centers ensured a steady observation throughout the year of the principal geophysical parameters. These included meteorology, geomagnetism, the aurora, the airglow, cosmic rays, ionospheric physics, latitude and longitude determination, glaciology and oceanography. But it was realized that unexpected short-term events might have to be noted. While the general trend of solar activity was predictable there might – and indeed were – sudden outbursts with far-reaching consequences. There was, therefore, provision for special World Intervals, during which observing centers would be on special alert.

While many relevant observations could be made at ground level, the history of plans for making them in the upper atmosphere is very informative. Initially it was expected that sounding balloons and high-flying aircraft could be used for this purpose: balloons might ascend to 30km (18mi). For higher altitudes – up to around 100km (125mi) – rockets might be pressed into service, but had the disadvantage that their flight time was no more than a few minutes. By 1956, however, a new possibility had arisen. Rocketry had by then reached a stage at which it was possible to think in terms of using rockets to launch satellites that would circle the world at a height of a few hundred kilometers, radioing back to Earth the readings of instruments on board. Both the Soviet Union and the United States announced plans to launch such monitoring satellites: the USA hoped to have 12 in orbit early in 1958. In the event, however, it was the Soviet Union which won this race into space, launching the first satellite (Sputnik 1) on 4 October 1957. The USA was not far behind, however, and when the CSAGI held its meeting in August 1958 it was announced that three Soviet and four American satellites were in station. They had already shown that the Earth's atmosphere extends further into

space than had been supposed and that above a few hundred kilometers there is a layer, previously unsuspected, in which there are electrons of considerable energy, though much less than those of cosmic rays.

Technically, the IGY ended on 31 December 1958, but many of the stations remained operative for at least another year.

Nuclear physics after the Manhattan Project

The end of World War II left physics in a curiously unbalanced position. In the USA the successful conclusion of the Manhattan Project had resulted not only in a capability to produce a weapon of unparalleled power but, incidentally, in a wealth of exclusive new knowledge about the atom and its constituent particles. To some extent this new knowledge had been shared with British scientists, but fears that other combatants had come near to achieving an independent atom bomb proved unfounded. While they had certainly considered the possibility, neither Germany nor Japan had mounted any substantial project. The Soviet Union remained an enigma. While the US

government had a touching faith that strict security measures could preserve the secret of the bomb indefinitely, dispassionate observers had no doubt that the Soviet Union had the scientific and technological potential to succeed independently, the more so as they had the assurance – which the Americans did not until 16 July 1945 – that the project was absolutely sound. Disillusionment came in 1949, when the Soviet Union exploded the first of its atomic bombs: Britain followed suit in 1952. By 1964 even China, starting from a low scientific and technological level, was in the atomic race. The USA countered with the even more terrible hydrogen bomb in 1952, but close behind came the Soviet Union (1953) and the UK (1957).

While much skilled scientific manpower was absorbed in these military projects and in concurrent projects (all government sponsored) to develop atomic power for peaceful purposes, many of the scientists involved during the war were resuming their research careers, reinforced by a new government of physicsts who had had no involvement with military affairs. But they did

▲ The International Geophysical Year (IGY) extended from July 1957 to December 1958. It included a 1,600km (1,000mi) traverse of the Antarctic. Here seismic explosions are being used to measure the thickness of the ice cap.

I would like first to point out that a meteorological satellite was one of the most important parts of an original IGY satellite program, and because of the ease with which most people could recognize its potential significance, its inclusion did much to point out the immediate economic significance of space science.

JOSEPH KAPLAN

▲ The first tracks of subatomic particles to be observed in a liquid hydrogen bubble chamber (1954). It was built by John Wood at the Lawrence Berkeley Laboratory, California.

◄ From the late 1930s the English physicist C.F. Powell pioneered the investigation of cosmic rays by monitoring their effect on photographic emulsions. This false-color picture shows a sulfur nucleus producing a fluorine nucleus (green) and other particles (blue).

▼ The computer was applied to the elucidation of the mechanics of atomic fission in the late 1950s. This picture (1949) shows the Selective Sequence Electronic Calculator at Princeton University, used to investigate the fission of Uranium-235 to form plutonium.

not merely resume where they had left off in 1939. Not only was there a wealth of new knowledge but vastly more money to support a field of science which had so convincingly demonstrated its capability. In particular, money was forthcoming for the construction of particle accelerators far more powerful than any previously available. Not surprisingly progress was rapid.

In 1947 C.F. Powell and G.P.S. Occhialini, at Bristol University in England, discovered in the course of their cosmic ray research the pi meson (pion), predicted by H. Yukawa in 1934 as the carrier of the force binding neutrons and protons in the nucleus. This particle decays in flight to form the muon. In 1948 pions were made artificially with the cyclotron at Berkeley in the USA, marking the beginning of the use of increasingly powerful accelerators to study subatomic particles rather than the nucleus. From this time too dates the development of quantum electrodynamics (QED), the blueprint for theories of particle interactions, capable of predicting very tiny effects. QED provides a quantum theory of electromagnetism, and describes the electromagnetic force between subatomic particles in terms of the exchange of photons, the "particles" of light.

Paul Dirac's concept of antimatter was validated by the discovery of the antiproton (1955) and the antineutron (1956), also at Berkeley. W. Pauli's prediction of a very light neutral particle (neutrino) was experimentally confirmed by F. Reines in 1956. Finally, in 1957, C.S. Wu discovered that mirror symmetry (parity) is violated in beta decays, confirming a theoretical suggestion by C.N. Yang and T.D. Lee in 1956. This proved to be an important characteristic of the "weak" force underlying such decays. The invention of the bubble chamber by Donald Glaser in 1953 provided a powerful new means of detecting the new particles. Charged particles leave trails of bubbles in a superheated liquid, which can be photographed to provide a permanent record. It was well suited to detecting particles from the new high-energy accelerators and came into its own in the 1960s.

The optical giants
One of the first areas of "Big Science" was astronomy. Even by the 20th century astronomical telescopes were not only highly sophisticated scientific instruments but also major engineering achievements. These great telescopes are exemplified by the reflecting Hale telescope erected on Mount Palomar in California which has an aperture of 5m (200in). The huge disk of borosilicate glass from which the mirror was machined weighed some 20 tonnes when it was cast in 1934: to avoid setting up strains in it a full year was allowed for its cooling. This was only the beginning of the story, however. Grinding the mirror to a very exact specification and the manufacture and assembly of a wealth of ancillary equipment was unavoidably time-consuming: adding wartime delays, it was 1948 before the Mount Palomar telescope was finally dedicated.

The Hale telescope on Mount Palomar has been used for the never-ending task of mapping precisely the stars, galaxies and other major features of the universe. In the 1950s it helped to produce the massive "star atlas" entitled *National Geogrpahic Society – Palomar Observatory Sky Survey*. More specific and important problems can also be investigated. By using spectroscopic attachments and invoking the Doppler principle (which concerns the relative velocity between bodies), such telescopes can measure the rate of recession of the most distant stars and help to resolve the controversy between the "Big Bang" and "Steady State" theories of the origin of the Universe. Among discoveries made with the telescope at Mount Palomar is that of A.R. Sandage who, in 1960, equated a radio source with a star-like optically visible object with an unusual spectrum. This was the first example of a quasar (from quasi-stellar radio source), an extremely bright object that appears from its spectrum to be a great distance away.

The Hale telescope remained the world's largest optical telescope until 1974 when the 6m (236in) telescope on Mt Pastukhov in the USSR was inaugurated. With conventional design this is probably about the limit in size that can be achieved. But in the early 1950s proposals were well advanced for ground-based optical telescopes with an effective aperture of nearly 25m (1000in), giving 20 times the light-gathering capacity of Mount Palomar. This can be achieved by dividing the total aperture into segments.

Much can still be learned about the Universe with the aid of these optical giants, as evidenced by the fact that their observational time is always booked solid for years ahead. But they give us only a narrow glimpse of the Universe, limited by the range of the spectrum of visible light for which glass is transparent.

Radio astronomy
Until the beginning of the 20th century virtually all our information about the nature of the Universe was gained by studying the light emitted by the various heavenly bodies. In the main it was obtained by the use of optical telescopes, but these could be adapted in various ways. For example, the British scientist J.N. Lockyer in 1868

Pluralism, Profits and Practicality: Science in the USA

◄ In the mid-20th century the USA became generally recognized as the world's leading scientific power. Symbolic of the country's ascendancy was its success in the Nobel prizes in 1946, when Americans won all the science prizes. The US laureates were, from left to right, P.W. Bridgman (physics), J.B. Sumner, J.H. Northrop and W.M. Stanley (chemistry – followed by Otto Hahn, the German winner of the chemistry prize for 1944), and H.J. Muller (medicine or physiology).

► Only wealthy countries can afford the largest optical telescopes. The Hale telescope at Mount Wilson, California, was the world's largest from 1948 to 1974.

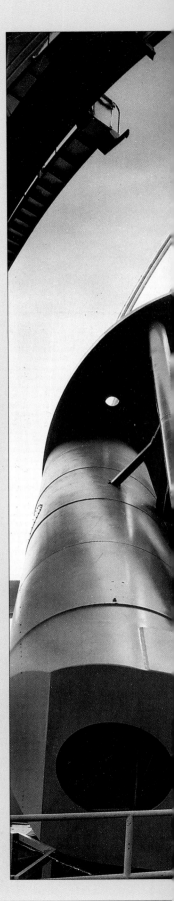

In 1941, *Life* publisher Henry R. Luce proclaimed "The American Century", grandiosely declaring the United States to be "the intellectual, scientific and artistic capital of the world". Although nine American scientists had won Nobel prizes during the 1930s, his verdict about American science was a bit premature, but it was certainly prophetic. Four years later the making of the atomic bomb in the Manhattan Project in the United States heralded an age of American scientific hegemony, topped by a double sweep of the Nobel science prizes in 1946 and 1983.

Credit for postwar American scientific supremacy is commonly given to four sources. Firstly, before 1940 philanthropic foundations had taken deliberate steps to bolster research. The Rockefeller Foundation had funded National Research Fellowships; the General Education Board had constructed academic laboratory facilities; the Chemical Foundation had supported projects ranging from establishment of the American Institute of Physics to construction of Ernest Lawrence's cyclotrons at Berkeley; the Carnegie Institution of Washington had paid for some of the world's best physics; and the Bamberger fortune had underwritten the Institute for Advanced Study at Princeton.

Secondly, scientists themselves had helped to build national institutions for the coordination of research. Most notable was the National Research Council, which was the brainchild of astronomer George Ellery Hale.

Thirdly, economic hardship and Nazi persecution had driven dozens of preeminent European scientists to the USA, where they suffused indigenous research with talent, especially in theoretical areas like atomic physics, physical chemistry and mathematics.

Finally, the wartime emergency unleashed massive direct federal patronage, via research contracts and grants. Shrewd planners, most notably Vannevar Bush, devised entitles like the National Science Foundation that maximized support for top-ranked science at elite institutions by somewhat insulating distribution of public research monies from political accountability. Although dwarfed by presence of the military-industrial complex, such institutions crucially shaped fundamental research.

These factors were an addition, however, to an established national system of research. Where 20th century patronage tended to be centralized, the nation's 19th-century tradition was pluralistic. A variegated and geographically dispersed university infrastructure had been nurtured by religious enthusiasm, regional boosterism, dreams of upward social mobility and multiple federal props, including the Morrill Land Grant College Act (1862), the Hatch Act (1887); and the Adams Act (1906). The success of industrial research complemented that structure, reinforcing its practical orientation and justifying a large population of graduate students. The result has been a complex, interconnected, profit-oriented blend of fundamental and applied work, best understood by the single term "technoscience".

A quarter century of scientific breakthroughs attests to that system's viability. After World War II nuclear physics took the spotlight, with the discovery of subatomic processes and particles sharing center stage with discovery of transuranium elements. There followed many related developments connected with quantum physics, most notably the invention of the transistor, subsequent work in solid state physics, fuller understanding of the chemical bond, invention of radiocarbon dating and a veritable revolution in thinking about gravity and cosmology.

Alongside that work came equally impressive achievements in the chemistry of life and of synthetic macromolecules, including award-winning work on enzymes, vitamins, and viruses, elucidation of the structure of DNA and the mechanisms of inheritance, determination of metabolic pathways, and the burgeoning of polymer science. The race for space combined with wartime development of radar to boost radio astronomy, lunar and planetary geology and the atmospheric sciences. Computer development spawned discrete mathematics, while submarine warfare, among other things, stimulated both oceanography and the plate tectonics revolution in geology. Finally, concern over pollution has boosted population biology and ecology.

The United States remains dominant in science today, albeit less successful than some countries at translating that position into technical innovations.

used a spectroscope to detect the presence of helium in the Sun nearly 30 years before the British chemist W. Ramsay identified it as one of the rare gases of the atmosphere. The overriding importance of light arose from two causes. First, the eye is sensitive to it and not to other kinds of radiation: second, glass is transparent to it, so that it can be investigated by a variety of optical systems. But light is only a narrow part of a broad spectrum of electromagnetic radiation, which includes radio waves and X-rays.

One of the problems of radio reception in its early days was the crackling sound known as static. In 1932 an American radio engineer, Karl Jansky, was investigating static to try to reduce its interfering effect and accidentally discovered a powerful source of radio emission located in the direction of the Milky Way. Curiously, this unexpected discovery attracted little attention at the time and not until after World War II was serious attention directed to the nature and source of the radio waves falling on the Earth from outer space. The lead was taken by Australia and the UK.

A major practical problem at once arose. The ability of a telescope to distinguish between close objects – depends on ratio of the telescope's diameter to the wavelength of the radiation involved. As light waves are about 10,000 times shorter than radio waves, it follows that to obtain anything like comparable results radio telescopes must have apertures far bigger than those of optical telescopes. The first important instrument of this kind was that built for the British astronomer Bernard Lovell at Jodrell Bank near Manchester, completed in 1957. Its bowl had a diameter of 76m (250ft); as it had to be very precisely steerable, without distortion, its construction was an engineering triumph. In the event, one of its first major uses was to track the series of satellites that appeared in space following the launch of Sputnik 1 in 1957. The mechanical problems of constructing very large "lenses" (dishes) for radio telescopes were overcome in two ways. One was to combine a series of smaller dishes to give the effect of one very large single one. An alternative is to have a fixed dish that scans the sky as the Earth rotates. This means that only a limited area of the sky can be scanned, but it avoids the mechanical problems of steerage.

Radio telescopes have identified various objects in space not revealed by optical telescopes. Quasars, first observed in 1960, are relatively small bodies at the limits of the known Universe; they emit a hundred or more times as much energy as a single galaxy. (Not all quasars emit radio waves, however.) Pulsars, detected in 1967, are apparently dead stars that rotate at enormous speed and are so called because they emit intense pulses of radio energy. Radio telescopes have also been employed to scan the Universe for radio sources. In 1959 Martin Ryle, the director of the Mullard Radio Astronomy observatory in the University of Cambridge, UK, published a catalog of 500 radio sources in the Universe – increased to 5,000 by 1965. He overcame problems of resolving power by interferometry, that is by linking a series of small dishes and mixing the waves from each receiver.

1960 · 1973

THE GIANT LEAP

Time Chart

	1961	1962	1963	1964	1965	1966	1967
Nobel Prizes	• *Chem*: M. Calvin (USA) • *Phys*: R. Hofstadter (USA), R.L. Mössbauer (FRG) • *Med*: G. von Bekesy (USA)	• *Chem*: M.F. Perutz, J.C. Kendrew (UK) • *Phys*: L.D. Landau (USSR) • *Med*: J.D. Watson (USA), F.H.C. Crick, M.H.F. Wilkins (UK)	• *Chem*: G. Natta (It), K. Ziegler (FRG) • *Phys*: E.P. Wigner, M.G. Mayer (USA), J. Hans, D. Jensen (FRG) • *Med*: J.C. Eccles (Aus), A. Hodgkin, A. Huxley (UK)	• *Chem*: D.C. Hodgkin (UK) • *Phys*: C.H. Townes (USA), N.G. Basov, A.M. Prokhorov (USSR) • *Med*: K.E. Bloch (USA), F. Lynen (FRG)	• *Chem*: R.B. Woodward (USA) • *Phys*: R.P. Feynman (USA), S. Tomonaga (Jap), J.S. Schwinger (USA) • *Med*: F. Jacob, A. Lwoff, J. Monod (Fr)	• *Chem*: R.S. Mulliken (USA) • *Phys*: A. Kastler (Fr) • *Med*: F.P. Rous, C.B. Huggins (USA)	• *Chem*: M. Eigen (FRG), R.G.W. Norrish, G. Porter (UK) • *Phys*: H.A. Bethe (USA) • *Med*: R. Granit (Swe), H.K. Hartline, G. Wald (USA)
Technology	• 12 Apr: Y. Gagarin is the first man in space in *Vostok I* (USSR) • World's largest computer, *Atlas*, is installed at Harwell to aid atomic research and weather forecasting (UK) • Patenting of the silicon chip by Texas Instruments (USA)	• US astronauts J. Glenn (February) and M.S. Carpenter (May) are put into orbit (USA) • 10 Jul: Launch of TV satellite Telstar from Cape Canaveral (USA) • Solar furnace completed under the direction of F. Trombe in the French Pyrenees	• Invention of friction welding (USSR) • Carbon fiber developed (UK) • Semiconductor diodes become commercially available • First satellite in geosynchronous orbit launched (USA) • Introduction of the first car to feature a Wankel rotary engine, the Japanese NSU Spyder	• 31 Jul: *Ranger VII* succeeds in obtaining close-up photographs of the Moon's surface (USA) • *Mariner IV* (USA) and *Zond II* (USSR) are launched with equipment for photographing Mars • First word-processor introduced by IBM in the USA and Europe	• 18 Mar: Cosmonaut A. Leonov leaves spacecraft *Voskhod II* and floats in space for 20 minutes (USSR) • 15 Jul: *Mariner IV* transmits close-up photographs of Mars (USA) • Dec: First space rendezvous between two Gemini spacecraft (USA) • 26 Nov: Launch of the first French satellite	• Lunar Orbiter I photographs two million square miles (5.18 million square kilometers) of the Moon's surface (USA) • Fuel injection for automobile engines developed (UK) • Unmanned Soviet spacecraft *Luna IX* makes the first successful soft landing on the Moon	• Invention of Dolby noise-reduction system (USA) • China explodes its first hydrogen bomb • World's largest hydroelectric power station, the Krasnoyarsk Dam on the Yenisei River in Siberia, is finished
Medicine	• Invention of the Barnet Ventilation electric lung pump • F.L. Horsfall announces that all forms of cancer are caused by mutations in the DNA of cells • J. Lippes develops an inert plastic intra-uterine device (IUD) as a contraceptive	• T.H. Weller develops a vaccine against rubella • First use of lasers in eye surgery • L. Harrington develops an operation for the correction of scoliosis (curvature of the spine) (USA)	• F.D. Moore and T.E. Starzl perform first liver transplant • M. De Bakey uses an artificial heart during cardiac surgery (USA) • J.D. Hardy performs the first human lung transplant • A.M. Cormack and G.N. Hounsfield develop X-ray tomography (CAT scanning) (UK)	• Home kidney dialysis introduced in the UK and USA • B.S. Blumberg discovers the "Australian antigen", key to the development of a hepatitis B vaccine (USA) • US Surgeon-General's report *Smoking and Health* links lung cancer with smoking	• Introduction of a measles vaccine • Invention of soft contact lenses • J. Ochsner uses Marlex, a plastic mesh, to bridge large defects in body tissue without risk of rejection (USA)	• H.M. Meyer and P.D. Parman develop a live-virus vaccine against rubella • France adopts brain inactivity as definition of death • The first radio immunoassay kit, to test for minute quantities of hormones, is devised (UK)	• 3 Dec: First heart transplant performed by C. Barnard (SA) • I.S. Cooper introduces cryosurgery as a treatment for Parkinson's disease (USA) • First successful growth in the laboratory of the leprosy bacillus
Biology	• M.W. Nirenberg discovers the first known "letter" of the genetic code when he demonstrates that three uridylic acid bases in a row in RNA form the code for phenylalanine, an amino acid (USA) • F. Crick and S. Brenner demonstrate that the genetic code of DNA consists of a string of non-overlapping base triplets (UK)	• Publication of R. Carson's *Silent Spring*, alerting the general public to the introduction of chemicals into the ecosystem • M.F. Perutz makes his research into the structure of hemoglobin and other globular proteins (UK)	• C. Sagan detects ATP in a mixture of chemicals reflecting what is thought to be the early conditions on Earth (USA) • A. Hodgkin and J. Eccles make discoveries in the transmission of nerve impulses	• Beginning of the "Green Revolution" by the International Rice Research Institute with the introduction of improved strains of rice • Publication of W.D. Hamilton's *The Genetical Evolution of Social Behavior*, a study of the sociobiology of bees and ants	• H. Harlow demonstrates the negative emotional effect of rearing monkeys in total isolation • R. Holley discovers the structure of a molecule of transfer-RNA, the protein-building molecule	• Publication of K. Lorenz's *On Aggression* (Aut) • S. Spiegelman and I. Haruna discover an enzyme allowing RNA molecules to duplicate themselves	• A. Kornberg announces his synthesis of biologically-active DNA • H. Green forms hybrid cells containing both mouse and human chromosomes to study the possibility of genetic engineering (USA)
Physics	• R. Hofstadter discovers the structure of protons and neutrons: a central positive core and two shells of mesons (USA) • M. Gell-Mann and Y. Ne'eman develop a scheme for classifying elementary particles (USA)	• G. Daley and colleagues establish that there are two types of neutrino. Later comes the belief that there is a third neutrino, the tauon neutrino (USA) • B.D. Josephson discovers tunnelling between superconductors (UK)	• N. Cabibbo develops a theory of weak interactions leading indirectly to the electroweak theory • Discovery of anti-xi-zero, a fundamental atomic particle of antimatter	• Discovery of the fundamental particle omega-minus using the *Nimrod* cyclotron • M. Gell-Mann and G. Zweig independently propose the existence of quarks as the building blocks of protons and neutrons (USA)	• Opening of the Stanford Linear Accelerator Center (SLAC) (UK) • Moo-Young Han and Yoichiro Nambu introduce the quark concept later known as color	• Announcement by the US Atomic Energy Commission of its intention to build a 200 GeV particle accelerator near Chicago	• S. Weinberg, A. Salam and S.L. Glashow propose electroweak unification theory (USA) • B. Matthias and colleagues discover an alloy of niobium, aluminum and germanium that sets a new record for high temperature superconductivity (FRG)
Chem.	• Creation of lawrencium by A. Ghiorso and colleagues (USA)	• N. Bartlett makes first noble gas compounds by preparing xenon platinum hexafluoride (USA)			• Rare earth complexes first separated by gas chromatography		
Other	• M. Ryle concludes from radio astronomical observation that the Universe changes with time, challenging the "steady state" theory		• F.J. Vine and D.H. Matthews show the history of seafloor spreading using changing polarities of rocks (UK) • M. Schmidt discovers the large red shift of the object 3C273, the first recognition of a quasar	• A.A. Penzias and R.W. Wilson detect cosmic background radiation, providing decisive evidence for the "Big Bang" theory (USA)	• F. Reines and J.P.F. Sellshop detect neutrinos from cosmic rays deep in a South African gold mine	• H. Friedman, E.T. Byram and T.A. Chubb discover a powerful source of X-rays in the constellation Cygnus, coming from the galaxy Cygnus-A	• First pulsar discovered by A. Hewish and J. Bell • Theory of plate tectonics introduced by D.P. McKenzie, R.L. Parker (UK) and W.J. Morgan (USA)

1968	1969	1970	1971	1972	1973
• *Chem*: L. Onsager (USA) • *Phys*: L.W. Alvarez (USA) • *Med*: R.W. Holley, H.G. Khorana, M.W. Nirenberg (USA)	• *Chem*: D.H.R. Barton (UK), O. Hassel (Nor) • *Phys*: M. Gell-Mann (USA) • *Med*: M. Delbrück, A.D. Hershey, S.E. Luria (USA)	• *Chem*: L.F. Leloir (Arg) • *Phys*: L.E. Néel (Fr), H.O. Alfvén (Swe) • *Med*: J. Axelrod, B. Katz (USA), U. von Euler (Swe)	• *Chem*: G. Herzberg (Can) • *Phys*: D. Gabor (UK) • *Med*: E.W. Sutherland (USA)	• *Chem*: S. Moore, W.H. Stein, C.B. Anfinsen (USA) • *Phys*: J. Bardeen, L.N. Cooper, J.R. Schrieffer (USA) • *Med*: G.M. Edelman (USA), R.R. Porter (UK)	• *Chem*: E.O. Fischer (FRG), G. Wilkinson (UK) • *Phys*: L. Esaki (Jap), I. Giaever (USA), B.D. Josephson (UK) • *Med*: K. Lorenz (Aut), N. Tinbergen (UK), K. von Frisch (FRG)
• Jan: *Surveyor VII* lands softly on the Moon (USA) • Tidal power station opens in France • Soviet Tupolev Tu-144 is the first supersonic airliner • 21 Dec: Astronauts J.A. Lovell, W. Anders and F. Borman complete the first flight around the Moon (USA) • Completion of the new Aswan Dam (Egy)	• 16 Jul: N. Armstrong and E. Aldrin land on the Moon and on 20 July, Armstrong is the first to set foot on its surface (USA) • Invention of "bubble memory" devices for computers, which retain information even when the computer is turned off • Concorde supersonic airliner makes its maiden flight from Toulouse (UK/Fr)	• Development of the "floppy disk" computer memory by IBM • TV lasers are installed to monitor air pollution in Duisberg (FRG) • *Venera VII*, an unmanned spacecraft, lands on Venus (USSR) • Boeing 747 jumbo jets enter transatlantic service • First demonstration of the AEG Telefunken Decca Teledec, a forerunner of the video disk (FRG)	• Soviet space station *Salyut* launched • 12 May: Opening of the world's largest radio telescope (FRG) • 13 Dec: *Mariner IX* sends pictures from the moons of Mars (USA)	• Launch of space probe *Pioneer X* (USA) • Launch of *Landsat I*, the first of a series of Earth resources technology satellites (USA) • Experimental coal-fired power station built, in an attempt to increase the efficiency of generating electricity by gasifying coal (FRG)	• First demonstration of Oracle and Ceefax information to home TV receivers (UK) • 25 May: Launch of space station *Skylab I* (USA) • 1 Oct: Opening of a natural gas pipeline from the Ukraine to West Germany • Development of the image intensifier, a TV camera tube for use in low light conditions (FRG)
• A mother gives birth to sextuplets as a result of new fertility drugs (UK) • M. Arnstein develops a vaccine against meningitis (USA) • US Government move back the people of Bikini Atoll after claiming that the radioactive contamination from the 1956 hydrogen bomb explosion has diminished to a tolerable level • Discovery of the negative effects on children of the thalidomide drug	• M. Perutz discovers the composition of hemoglobin (UK) • US government removes cyclamates from the market and limits the use of monosodium glutamate, after experiments linking food additives to cancer • D. Cooley and D. Liotta make the first implant of an artificial heart, however the patient, H. Karp, lives for less than three days (USA)	• Introduction of mass vaccination for children in Western countries • First successful nerve transplant (FRG) • Nuclear-powered heart pacemakers are implanted in three patients to correct "heart block" (Fr/UK) • G. Cotzias pioneers L-dopa therapy as a relief to patients with Parkinson's disease	• Development of the diamond bladed scalpel (UK) • US Food and Drug Administrations bans the prescription of diethylstilbestrol (DES) to control morning sickness in pregnant women due to evidence that the drug renders their daughters susceptible to cancers of the reproductive tract • Isolation of the herpes virus from the lymph-cell cancer, Burkitt's lymphoma (USA)	• Introduction of computerized axial tomography (CAT or CT scanning) to give cross-sectional X-rays of the human body (UK) • J. Prineas confirms that multiple sclerosis is caused by a virus (Aus) • I. Cooper develops a "brain pacemaker" for epileptics. Platinum electrodes are implanted in the cerebellum to receive radio signals which relieve spasticity, paralysis or seizures (USA)	• Radio immunoassay used to test for spina bifida in the unborn fetus (Fin) • Introduction of the nuclear magnetic resonator (NMR) to form images of soft body tissues (UK) • Discovery made that communications through the corpus callosum – linking the two hemispheres of the brain – are defective in schizophrenics (UK)
• Publication of J.D. Watson's *The Double Helix* (USA) • M. Ptashne and W. Gilbert independently identify the first repressor genes • New Mexican wheat strains introduced to India increase that country's production by 50%	• G.M. Edelman finds the amino acid sequence of immunoglobin G (USA) • Successful isolation by J. Beckwith and colleagues of a single gene (a bacterial gene for a step in the metabolism of sugar) at Harvard Medical School (USA)	• H.M. Temin and D. Baltimore discover reverse transcriptase , an enzyme causing RNA to be transcribed on to DNA, in viruses • H.G. Khorana produces the first manmade gene (analine-transfer RNA) assembled directly from chemical components (USA)	• D. Nathans and H. Smith develop various enzymes that break DNA at specific sites (USA) • R.B. Woodward synthesizes Vitamin B12	• Use of DDT is restricted in the USA due to its adverse effect on the environment • M. Calvin and M. Tributsch obtain minute quantities of electricity from a solar cell containing chlorophyll, the principle agent of photosynthesis (USA)	• Calf produced from a frozen embryo for the first time • S.H. Cohen and H.W. Boyer demonstrate that molecules of DNA can be cut with restriction enzymes, joined with other enzymes, and reproduced by inserting them into the bacterium *Escherichia coli*
• Soviet physicist A.D. Sakharov campaigns in favor of nuclear arms reductions, thus coming into conflict with the Soviet government • J. Weber reports the detection of gravitational waves, first postulated by Einstein in 1916, but his experiment is not considered valid	• R. Wilson founds the Fermi National Accelerator Laboratory near Chicago			• The large particle accelerator at Fermi National Accelerator Laboratory begins operation (USA) • M. Gell-Mann initiates quantum chromodynamics, the theory linking quarks and color forces in particle physics (USA)	• P. Musset and colleagues at CERN discover neutral currents in neutrino reactions, a partial confirmation of the electroweak theory (Fr)
	• Scanning electron micrograph perfected for practical use	• Choh Hao Li synthesizes the growth hormone somatotropin (USA)			• H.W. Boyer uses recombinant RNA to produce chimera (USA)
• Smallest ever time measurement is made at Bell Laboratories – pulses from a laser measure the picosecond (USA) • Using radar, astronomers at Cornell University map a third of the surface of Venus (USA)	• Cocke, Taylor and Disney are first to identify a visible star associated with a pulsar, the pulsar in the Crab Nebula • A new species of swallow, the white-eyed river martin, is discovered in Thailand	• First Chinese and Japanese satellites launched • Completion of the 100m radio "dish" at Bonn (FRG) • Completion of the reflecting telescope at Kitt Peak, and the first large telescope erected on Mauna Kea, Hawaii (USA)	• I. Shapiro discovers superluminal sources by using very long baseline interferometry (USA) • US astronomers discover two "new" galaxies adjacent to the Milky Way	• R.N. Manchester calculates the galactic magnetic field strength by measuring the Faraday rotation for different wavelengths of the polarization plane of radio waves emitted by pulsars	• Complex molecules, including CH, HCN and water, are discovered in Comet Kohoutek • Gamma ray bursts originating in outer space discovered by the Vela satellite

Datafile

The 1960s proved the transistor to be one of the major technological developments of all time, comparable with the wheel in its effect on the most advanced contemporary societies. Initially its main impact was in the control systems for satellites and missiles which led to both the space race and arms race. Simultaneous with these high-tech advances great improvements were achieved in the productivity of the world's agricultural land.

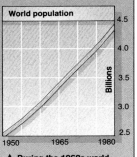
World population

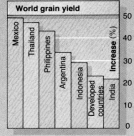
World grain yield

◀ Between the early 1960s and the early 1970s world agricultural production increased by an extraordinary 20 percent. This was achieved by using new strains of crops and more mechanized farming methods. Unfortunately the populations of the developing countries increased even more rapidly.

▲ During the 1960s world population increased markedly, to the point that United Nations agencies warned of serious consequences. It now doubles every 35 years. Birth rate exceeds death rate by 250,000 per day; three every second. It is doubtful whether the planet can support this rate of growth.

Metro systems

▶ During the 1960s the computer market increased rapidly in the United States. Most of the applications for computers at this time were associated with military requirements, for missile-guidance systems, and for the space program. The USA has remained the largest market for scientific, business and home computers.

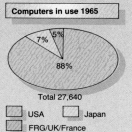
Computers in use 1965

7% 5% 88%

Total 27,640

USA ☐ Japan
FRG/UK/France

▲ Industrialization usually encouraged migration from country to city, creating huge urban areas. These can only operate efficiently if they are served by internal transport systems. In the 1960s and 1970s many new metro systems were opened, mainly in the wealthy cities of Europe, Japan and the United States.

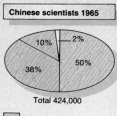
Chinese scientists 1965

10% 2%
38% 50%

Total 424,000

☐ Supporting staff
☐ Lecturers/assistants/technicians
☐ Industrial engineers/scientists
☐ Professors/Academy researchers

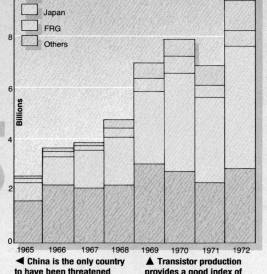
Transistor production

USA
Japan
FRG
Others

Billions

1965 1966 1967 1968 1969 1970 1971 1972

◀ China is the only country to have been threatened with the use of nuclear weapons by both the USA (1954, 1958) and the USSR (1969). Even in the period of intellectual repression following the Chinese Cultural Revolution of 1966 much scientific progress was made. In 1964 they became the fifth nuclear power.

▲ Transistor production provides a good index of national adaptation to electronics-based industry. During the 1960s Japan surged ahead, its economy growing 11 percent each year. By the 1980s it dominated the world electrical goods market, having put out of business electronics companies in many countries.

Perhaps the most significant factor in the relationship between technology, science and society in these years was the growing size of the society concerned. In 1960 world population was a little over 3 billion; by 1973 it had risen to almost 4 billion, with the expectation that by the end of the 20th century it would exceed 6 billion.

Its relationship with science and technology was twofold. On the one had, new medical discoveries (notably penicillin) and improved public health measures (purer water supplies, improved sanitation, better housing) had all contrived to increase the expectation of life. At the same time, however, the increased population put an increased demand on the available food supplies, especially in the poorer countries where the increase was most rapid. This was offset, but only partially so, by improved agricultural productivity through the use of fertilizers and agrochemicals and by improved techniques for preserving foodstuffs from times of plenty to those of shortage, and transporting them from high-production areas to world markets. The gloomy prophecy of eventual world overpopulation, made in 1798 by the British economic theorist Thomas Malthus in his *Essay on the Principle of Population*, had not been fulfilled, but to many it seemed that it had still been only postponed. This unease was increased by the fact that some of the means used to increase food production presented hazards, which often took long to be recognized. The pesticide DDT, for example, could not only destroy beneficial insects but could reach levels dangerous to humans and other mammals through accumulation in food chains: the USA effectively banned its use in 1969. Also in 1969, 39 nations met at a conference in Rome to discuss growing concern about marine pollution, an early expression of alarm which was to emerge as the "Green" movements of the 1980s.

Technological progress and concerns

Concern about adequate supplies of food were matched by others about a shortage of energy. While atomic power was making an increasing contribution, the world was still largely dependent on fossil fuels, coal and oil. While estimates of world resources of these varied greatly, what was certain was that they were finite and must eventually be exhausted: as a measure of the inroads being made, coal production in 1960 was round 2 million tonnes and oil around 1 million tonnes. Two major events strikingly demonstrated the dependence of the Western world on fossil fuels, and on all the high technology involved in their extraction and refining. In August 1959 a joint Shell/Esso survey team struck natural gas in enormous quantities at Slochteren, in The Netherlands. This led to the discovery of huge

THE SPACE AGE

Technological responses
to population increase

Science in China

Developments in
computer technology

High-speed rail travel

The USA reaches for
the Moon

The first space stations

deposits of both gas and oil beneath the North Sea. This had far-reaching effects on the economies of the countries concerned, especially Britain which through geographical chance and international law had an entitlement to the lion's share. The second event took place in 1973, when the Arab oil-producing countries began an embargo on oil shipments to Western Europe, Japan and the United States, ostensibly as a protest against their support of Israel. This precipitated a world energy crisis which took some 15 years to resolve. The social importance of technology in assuring adequate power supplies was further demonstrated in 1965, when the failure of a simple relay switch in Ontario deprived 30 million people of electricity for a considerable time in the northeast USA and parts of Canada.

Other events of the 1960s encouraged the belief that mankind was gradually establishing a mastery over nature. In Egypt the Aswan Dam, completed in 1968, at last controlled the vagaries of the

▼ Personal mobility is a prized result of 20th-century technology. But as this Los Angeles flyover illustrates, a high price has to be paid in terms of land usage and urban congestion.

Nile, upon whose annual flooding the economy of the country had depended for thousands of years. In 1969, as the culmination of by far the biggest research and development program ever launched, the first man set foot on the Moon. Three years later the Anglo-French Concorde made its debut as a supersonic airliner, while 1971 saw the launch of the biggest ship ever built, the 372,000 tonne Japanese tanker *Nisseki Maru*.

Biology and the Green Revolution
The "revolutions" beloved of historians are generally more leisurely events than their names imply, and indeed few can be precisely defined. The Scientific Revolution, in which the Greco-Islamic view of Nature was transformed into modern science, lasted roughly from 1550 to 1700 and the Industrial Revolution from about 1760 to 1830. The Green Revolution is no exception. It represents a phase in the history of world agriculture, in which productivity was greatly increased

by extensive mechanization, intensive use of fertilizers, pesticides, and herbicides; and systematic improvement of crop plants. Although it is largely a post-World War II phenomenon, its beginnings can be traced back to the start of the synthetic fertilizer industry in the early 20th century.

By 1960 world consumption of synthetic nitrogenous fertilizers was around 7 million tonnes annually, with comparable quantities of phosphates and potash. But the pattern of distribution was very significant: 46 percent went to Europe, 53 percent to North America and the Far East, and the remainder to the whole of Africa, the Near East and South America. In some of the latter countries, of course, heavy use of fertilizers is precluded by lack of sufficient rainfall to wash them into the ground. Even among the big users, there were significant differences: Belgium and the Netherlands, for example, applied four times as much per hectare or acre as the UK.

While judicious use of fertilizers, whether natural or artificial, can increase yields, these can be eroded by pests and diseases. Some pests, like locusts, are omnivorous and will destroy vegetation of every kind, while others, like the cotton boll weevil, are quite specific. Once they have swarmed, locusts are virtually uncontainable because of the sheer weight of numbers. With them, the secret of success is to monitor their home breeding areas and to spray these with insecticide at the first sign of population growth. With most pests, however, the solution is to spray the crops regularly, preferably to prevent infection or at least to control it. Ideally, spraying needs to be done several times a season, timed to keep in step with the growth stages of the pest.

The successful use of insecticides depends in part on the availability of a large battery of alternatives, for no one insecticide can be used universally. Insect pests vary enormously in their sensitivity; some may be virtually unaffected by an insecticide which is deadly to others. More seriously, sensitive pests can quite quickly develop an immunity. Equally, crops vary in their response to the insecticide itself: some may be so damaged that the cure is worse than the disease.

Pests are not the only natural enemies of plants: a whole range of fungi can be equally destructive. The Irish famine in the 1840s, caused by repeated blighting of the potato crop by Phytophthora, caused many thousands of deaths and led to mass emigration to America. Early fungicides were based on sulfur or salts of heavy metals, such as copper, but after World War II a range of synthetic products was developed. As with pesticides, a wide choice is necessary to take account of varying circumstances and, particularly, the spontaneous development of resistance.

Finally, in the field there is the problem of weeds which, if allowed to grow unchecked, can smother a crop and diminish the yield. In countries where labor is still cheap and abundant, traditional methods of hand-weeding and hoeing may suffice, but for plantation crops and in countries where labor is dear and scarce, chemical herbicides are essential. With these the emphasis is on selectivity: the herbicide must destroy the

Progress and Reversals: Science in China

Although China had a highly advanced scientific culture in antiquity and the middle ages, no indigenous scientific revolution occurred. Modern Western science was introduced to the country only in the mid-19th century. Probably due to their conservatism and resistance to the imperialist influence under the so-called "unequal treaties" most Chinese were then reluctant to accept anything "foreign", including science, until after the defeat by Japan in the war of 1895. Modern science only started to be implanted in China early in the 20th century, especially after the founding of the Republic of China in 1912.

In the 1920s and early 1930s scientific endeavor was gradually built up under the leadership of a number of enthusiastic scholars who had studied abroad. High-quality research was done in biology, chemistry, geology, mathematics, physics and physiology. A notable Chinese success was a modification of the Solvay process for the manufacture of soda, developed by Te-Pang Hou (Debong Hu).

The Japanese invasion and conquest of eastern China (1937–45) and the civil war after World War II caused a serious stagnation in science. Many scientists fled abroad. Nonetheless, some elegant work continues to be done, such as the theory of geotectonics developed by J.S. Lee (Siguang Li), the intrinsic proof of Gauss-Bonnet theorem and the development of Chern characteristic classes by S.S. Chern (Xingshen Chen), and the structure of polyatomic molecules by Ta-You Wu (Dayou Wu).

Since 1949 the reconstruction of China has been independently undertaken by two governments, one on the mainland and the other on Taiwan. Many scientists returned from abroad and played indispensable roles in the development of science and technology. However, the policies of the two governments have been very different. On the mainland the Communist-ruled People's Republic followed the Soviet pattern. Heavy industries, advanced military techniques, material science, traditional medicine, agriculture, mining and metallurgy, as well as basic sciences, were all emphasized. Science education was strengthened and scientists were strongly supported. Unfortunately links with the Western scientific world were prevented until about 1978, which hindered the progress of Chinese science. In addition, an intensive repression of science resulted from the "Cultural Revolution" of 1966–76.

In spite of these difficulties mainland China managed to develop the areas of advanced science that could give it the weapons of a superpower. In 1964 it exploded its first atomic bomb, in 1967 its first hydrogen bomb. In 1970 it launched its first satellite and in 1980 its first intercontinental ballistic missile. It has also been able to provide highly sophisticated research facilities, such as the synchrotron radiation facilities and the electron-positron collider (the latter used for research in particle physics). Its scientists have made notable contributions to basic science, such as the best solution of the Goldback problem (one of the most significant achievements in postwar mathematical research), pioneer work in superconductivity and the total synthesis of insulin. Studies in more traditional fields continue to yield interesting results; for example, work in herbal medicine has produced a new antimalarial drug, arteannuin. But for all

► The progress of technology in China has been uneven. On the one hand, it is the only country in the world still building steam locomotives. On the other, it quickly mastered the technology of atomic bombs and rockets capable of delivering them or launching satellites. The first (conventional) atom bomb was exploded in 1964; an H-bomb in 1967. This illustration shows the explosion of an H-bomb.

心臓

◀ In China traditional medicine, based on acupuncture and herbalism, is practised in parallel with Western methods. This class of Communist "young pioneers" is having a lecture on the anatomy of the heart, displayed in the poster on the wall. The teacher is demonstrating with a plastic model which can be taken apart. Later, in the classroom, they will dissect an animal's heart to reveal the valves and blood vessels. In the more squeamish West, such laboratory work is being discontinued, or at least made optional.

these advances, scientific achievements and technical breakthroughs have contributed little to economic development.

By contrast, in Taiwan state investment is generally directed toward economic development. The import of foreign industrial technology has been encouraged and agriculture, medicine and public health have had priority over pure science. The research and development in the former subjects have therefore been especially fruitful. Pharmacological studies on alpha-bungarotoxins and the use of polyamide thin-layer chromatography for biochemicals have won international admiration. Sugarcane genetic breeding and fish culture studies are examples of agriculture achievements, which in turn stimulated economic growth from the 1970s.

In the 1980s there were signs that promised a good future for science. Mainland China was gradually recovering from the "Cultural Revolution", and in Taiwan government support for research was becoming more generous. The number of publications registered for the *Science Citation Index* nearly quadrupled from 1981 to 1989; mainland China ranked No. 15 (6,428 publications) and Taiwan No. 29 (2,168).

▲ For many years the equipping of industry and the teaching of science were handicapped by lack of appropriate instruments and one of the main objectives of the Sixth Five-Year Plan (1981–85) was to remedy this defect. Instrument manufacturing works were established at various centers. Here a student is carrying out laboratory work with the aid of a microscope made in China.

weeds without doing more than marginal damage to the crop itself. The first such compounds appeared in the 1930s. More were developed in the UK during World War II (particularly 2,4-D and Methoxone), in response to an urgent need to produce more home-grown crops. Similar research was done in the USA, and by 1950 American production alone of 2,4-D had reached 10,000 tonnes. Subsequently many new types were developed.

In 1954 there appeared in the laboratories of ICI a new herbicide, paraquat, which was destined to revolutionize age-old agricultural practice. This was found to have a unique combination of properties: it rapidly killed green tissue but was almost instantly inactivated in the soil. At first this seemed to be a disadvantage rather than otherwise. It was then realized, however, that such a herbicide could be used to clear ground of weeds which could then be immediately planted. By direct drilling of seeds or seedlings after paraquat treatment the need for plowing could be obviated. This idea of minimum tillage, or chemical plowing, was quickly developed in particularly favorable circumstances. Plowing is expensive in terms of both labor and equipment: so, too, is fuel, especially after the OPEC crisis of 1973. By the late 1980s minimum tillage with paraquat was practiced for numerous crops in many parts of the world.

Finally, an important role in the Green Revolution was played by the development of improved varieties of crop plants. Improvement was sought along various lines. These include higher yields; greater resistance to pests and diseases; wider climatic adaptability; increased tolerance of drought; and reduced liability to storm damage. While many new varieties were developed by traditional plant-breeding methods, increasing use was made after the 1970s of tissue-culture techniques which produce results much more rapidly. Much of the development was done by commercial breeders, but a major part was played by CGIAR (Consultative Group on International Agricultural Research) set up in 1960 as an informally organized group of some 35 donor organizations composed of governments, international research organizations and research foundations.

Computer technology

The first generation of computers were massive machines, using thermionic tubes (valves), mostly custom-built for a particular user. The public at large knew little about them and were only mildly interested. In 1952 the success of UNIVAC in predicting the election of Dwight D. Eisenhower as president of the USA did make people aware of new possibilities but the turning point was the advent of the transistor in the early 1950s. This radically changed the situation, enormously reducing the size of computers and their power requirements. A few years later, in 1958, the first integrated circuit was demonstrated: a small structure containing about 10 individual components. This idea was developed in the 1960s. By the early 1970s thousands of integrated circuits could be incorporated on a silicon wafer or "chip" smaller than a postage stamp. Chips eliminated the kilometers of wiring that had been necessary in the original machines. An enormous market quickly developed: in 1962 the US government placed an order for more than quarter of a million integrated circuits for use in its Minuteman missiles. Then, in 1971, Intel of California introduced the microprocessor, essentially a computer on a silicon chip.

As a result of all these developments an enormous new industry mushroomed, soon to have a bigger turnover than the steel industry, which had been dominant in the 19th century. It was an industry with two main divisions: hardware and software. The former comprised the tangible apparatus, including magnetic tapes and floppy disks; the latter all the programing material which tell the computer step by step what to do. There was, of course, an international business but there were problems in making the industry homogenous. With development so rapid and competitive, different manufacturers developed hardware systems which were incompatible:

◀ Technological advances have interesting and sometimes unpredictable consequences for employment patterns. In the 1960s computers were taken up by businesses and administrative organizations. Old, slow and often tedious methods of calculation quickly became redundant but large numbers of employees were required to prepare data for feeding into the new machines. In the 1960s, dominated by "second generation" computers incorporating transistors, information was usually prepared in the form of punched cards. Here employees of the Royal Dutch-Shell Company in London are transcribing data onto cards using mechanical equipment (1962).

▲ The Soviet space control center at Kaliningrad near Moscow. The space programs would have been unthinkable without computers. As computing power grew, computers not only provided information for crews and controllers but increasingly controlled the space vehicles themselves.

▼ In the 1960s the advent of the integrated circuit made it possible to incorporate mainframe computer capability into a few tiny chips of silicon. Continuing miniaturization then produced the microprocessor, as seen here; in effect a miniature computer.

programs devised for one would not necessarily be accepted by another. Similarly with software: some form of internationally acceptable language had to be developed for encoding the tens of thousands of instructions that comprise a program. Various computer languages were developed, according to the complexity of instructions to be given. Early examples were COBOL (Common Business Oriented Language) for commercial programs and FORTRAN (FORmula TRANslation) and ALGOL (ALGOrithmic Language) for scientific and mathematical programs. Within these languages complex systems of abbreviated coding instructions had to be created: SQRT, for example, to give an instruction for introducing a square root. PL/1 (Programing Language 1), sought to combine commercial and scientific computer languages within a single language.

Originally, in the 19th century, the computer was conceived, as its name implies, for performing mathematical calculations. The electronic computers of the 1940s continued in this tradition, but it was soon realized that they had immense possibilities for storing and retrieving information because this is capable of being converted into a digitized form. This a computer can digest and process like any other mathematical instruction, and then reconvert into words. Such stores of information, data bases, can be sampled on demand. For example, a data base comprising the populations of the world's leading cities could provide, virtually instantaneously, the names of all cities with a population over a million, or under half a million, and so on. Moreover, data

bases can be consulted internationally through the ordinary telephone network. An inquirer in Paris can, through the ordinary dialling system, link up with a database in Tokyo. Such information can be displayed on a television screen and later printed out on paper for permanent record.

High-speed rail travel

The social impact of the railroads in the 19th century was enormous. Whereas horse-drawn coaches averaged only 9–11km/hr (6–7 mi/hr) the Great Western Railway in the UK had some services timetabled at 96km/hr (60mi/hr) as early as 1847, and the same company's *City of Truro* exceeded 160km/hr (100mi/hr) in 1904. Thereafter progress was slow so far as speed was concerned, though there were great improvements in passenger comfort and amenities. In 1936 the UK locomotive *Mallard* set the highest speed ever achieved by a steam train, 202km/hr (126mi/hr), but by then other forms of traction – diesel and electric – were assuming increasing importance. In France in 1955 two electric locomotives (CC107 and BB9004) achieved a spectacular 331km/hr (206mi/hr), but only when pulling three specially designed lightweight coaches. But speed in special record-breaking runs is much less significant than high speeds consistently achieved on regular schedules, and from the 1960s this received increasing attention.

A major factor here was the growth in domestic airlines. Railroads could never compete on speed alone, but they had the great advantage of running direct from city center to city center: airline

◀ By the 1960s several countries had trains regularly timetabled for speeds of 200km/hr (125mi/hr) or more, but excessive wear and tear on the track was sometimes a limiting factor. Germany's *Blauer Enzian*, a high-speed train introduced in 1966, had to be withdrawn and running was not resumed at such speeds until 1978. This picture shows the Japanese Shinkansen "bullet train" (which first ran in 1964).

▶ From the first days of the aircraft industry in the early 20th century wind tunnels were extensively used as a basis of design. This picture shows a wind tunnel experiment conducted in the design of Concorde. By the 1980s, however, designers began to make increasing use of mathematical modeling to solve aerodynamic problems.

▼ The idea of the hovercraft, a vehicle riding on a cushions of compressed air, was conceived by the British inventor Christopher Cockerell in 1955 but development took time. On 25 July 1959 – the 50th anniversary of Blériot's historic flight – the first hovercraft crossing of the English Channel was achieved. This picture shows the inventor (center) with an early model.

passengers had to make their way to out-of-town airports and then face a considerable wait before takeoff. Economic and technical considerations led to the conclusion that top speeds of abouy 200km/hr (125mi/hr) would be a reasonable target for main-line express services. There were, however, many considerations other than purely mechanical ones to be observed: such speeds would be disastrous, for example, on bends beyond a certain radius and only minor gradients were acceptable. Thus certain routes had a notable advantage. In France, for example, the Paris–Bordeaux line was one of the most nearly ideal in Europe. The USA with its emphasis on freight and poor standard of track, had to be content with a target of 180km/hr (110mi/hr).

One of the most remarkable achievements, however, was in Japan, where the new Tokaido–Osaka line was opened in 1964 – the first of the Shinkansen ("New Trunk Line"). It was a bold conception, involving the building of 515km (320mi) of new electrified track, cutting through urban and country districts alike. The "bullet trains" that were built to run on the line, so called from their frontal appearance, were originally designed for a top speed of 260km/hr (160mi/hr) but this was finally reduced to 210km/hr (130mi/hr): the scheduled time was 3 hours 10 minutes.

In several countries – including France, Italy, the UK and Canada – a novel approach to the problem of safely maintaining high speeds round bends was made in the 1960s and 1970s. This took the form of an automatic tilt mechanism to compensate for the centrifugal force generated on bends. However, it had limited success, for two main reasons. The first was purely mechanical: a satisfactory system proved difficult to design. The second was that the tilt caused the rolling stock to overhang the inter-track area – a hazard to oncoming traffic. This could be avoided by careful timetabling, but only at a price. Equally, much lineside equipment, such as signal and telegraph posts, had to be shifted.

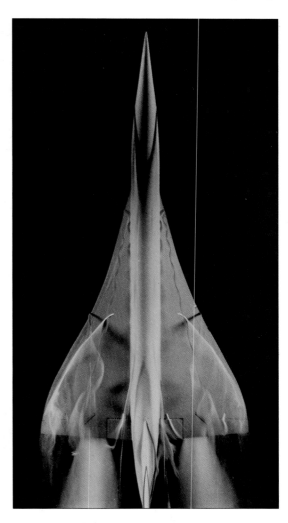

New kinds of track-guided vehicles

In general these developments in high-speed rail travel were sophisticated developments of existing forms of traction, rolling stock, track, and signaling but the 1960s saw the appearance of new forms of track-guided vehicles. Several countries (including France, Italy, Germany, the UK and USA) experimented with jet-propelled track-guided hovercraft during the years 1965–75. Typical was the French Aérotrain, which in 1974 topped 426km/hr (226mi/hr) on a special track from Paris to Orleans. For economic and technical reasons, however, this has not been pursued.

Two other forms of levitation have also been explored, again primarily with the object of virtually eliminating friction between vehicle and track. In electrodynamic systems use is made of repulsion forces between electric conductor plates in the track and magnets in the vehicle. Although studies were made in many countries, it has not so far been adopted, though the availability of high-temperature superconducting magnets may change the economies. In electromagnetic levitation – pursued particularly in Germany and Japan since 1970 – the repulsion is between electromagnets in the vehicle and ferromagnets in the track. Extensive trials have been made but rather few installations: one in the UK links Birmingham Airport, the National Exhibition Centre, and the international railroad station.

► The 1960s were a period of intense military rivalry between the United States and the Soviet Union, especially in the field of rocketry. The goal was to develop Intercontinental Ballistic Missiles (ICBMs) capable of carrying a nuclear warhead from one continent to another. This picture shows a range of American missiles, the largest being the 3,000 tonne Saturn V rocket, used from 1967 to 1973 for the Moon-landing program. The Saturn V is the most powerful rocket ever built. A three-stage vehicle, it stood 110m (363ft) high and on takeoff generated 4 million kg (9 million lb) of thrust. The first stage was jettisoned at 66km (41mi) above the Earth, the second at 190km (120mi), the third at 300km (190mi).

Now it is time to take longer strides – time for a great new American enterprise – time for this nation to take a clearly leading role in space achievements, which in many ways may hold the key to our future on Earth. I believe that this nation should commit itself to achieving the goal before this decade is out, of landing a man on the moon and returning him safely to Earth.

JOHN F. KENNEDY
25 MAY 1961

Space exploration

A conspicuous feature of modern science is the speed with which the impossible becomes an everyday occurrence and the pundits are confounded. When the newly appointed British "astronomer royal" arrived in England from South Africa in 1956 and the press invited his opinion on space travel, he described it as "bunkum". Yet this was only five years before the Soviets put Yuri Gagarin into orbit in Sputnik 1 and only 13 before Neil Armstrong and Edwin Aldrin set foot on the Moon, watched by an estimated 600 million viewers on television. The latter was itself a service then barely 30 years old, yet already there were 200 million television sets in the world. By the early 1980s more than a hundred people had ventured into space.

The manned Moon landing was the culmination of a carefully planned series of increasingly sophisticated trials. The first was a dramatic failure. Faced with the Soviet Sputnik in 1957, the United States had only their navy's Vanguard rocket with which to put their own satellite into orbit. In December 1957 this blew up on the launching pad. A second attempt in the following month had to be called off at the last moment. The United States then embarked on the Explorer series of satellites, launched by the military ICBM Jupiter C: Explorer 1 went into orbit in January 1958, and was followed within the next two years by nearly a score of other satellites, carrying a range of experimental apparatus.

But these were no more than nursery slope excursions. In October 1958 the National Aeronautics and Space Administration or NASA – set up to coordinate all civilian space projects – launched Pioneer 1, intended to orbit the Moon and relay information about its surface back to base: unfortunately its booster failed and it fell back to Earth.

Meanwhile the Soviets had some spectacular successes with their Luna probes. Luna 1 passed the Moon at a distance of 7,000km (4,300mi) and then went on into orbit round the Sun – the first artificial planet. Luna 3 caused a sensation in October 1959 when it passed over the far hemisphere of the Moon – which is never turned toward the Earth – and on reappearing beamed back a series of pictures of the hitherto invisible surface.

Soviet luck then turned, however, and it was six years before they had another considerable success, but again it was spectacular. In January 1966 Luna 9 made the first soft-landing on the Moon and televised a long series of pictures back

to Earth. What these revealed was a great relief to the Americans, who were still unsure whether the surface might not be covered in dust so deep as to swallow up a spaceship and its crew. In fact, the Soviet camera showed a barren rocky crater-strewn landscape – totally inhospitable, but solid.

Meanwhile, the United States had been accumulating essential information about the lunar geography, primarily as a guide toward selecting a suitable landing site. Their Ranger series of space probes got off to a bad start, the first six ending in failure. But Rangers 7, 8 and 9 successfully relayed thousands of pictures before plunging to destruction: with their flair for publicity the United States televised the photographs taken by Ranger 9 to the world, labeling them "Live from the Moon". Then, in May 1966, the United States, too, achieved a soft landing with the first of the Surveyor series, which collectively sent back tens of thousands more pictures from their touchdown points.

Simultaneously, lunar orbiters were taking numerous pictures of large areas of the Moon from orbits high above its surface. By February 1967 all that could be learned by way of preparation from unmanned spacecraft had been learned and the next stage – the development of technology for a Moon landing – could begin.

The method finally chosen by NASA for the landing was the Lunar Orbit Rendezvous. This involved putting a duplex spacecraft, with a crew of three, in orbit round the Moon. At the appropriate moment the lunar module (LM), with two of the astronauts aboard, would detach itself from the command module (CM) and, controlled by its rockets, descend to the Moon's surface. After the two astronauts had disembarked and completed their allotted tasks they would return to the LM, fire rockets to put their craft back into orbit, and rejoin their companion in the CM. All three would then make the long journey back to Earth, and finally descend by parachute to splash down in the Pacific.

As a first step a series of two-man flights was mounted (the Gemini program), with the main object of perfecting a technique for docking two satellites in space. After several nerve-racking missions had all been safely accomplished the way was clear for the final stage, the Apollo missions, in which astronauts were to be launched on their way to the Moon by the enormous Saturn 5 rocket. (At nearly 3,000 tonnes it was the biggest rocket ever built and represented 98 percent of the total equipment launched.) Several trial orbits of the Moon were to be carried out before the first landing was attempted.

The program began tragically, when three astronauts died in a fire in the CM during a practice run on the ground. This delayed the whole operation for more than a year, but finally – on 21 December 1968 – three men were set en route for the Moon in Apollo 8. They orbited it and returned safely. In the next mission (March 1969) the LM was tested in Earth orbit and then, in the Apollo 10 Mission (May 1969), in Moon orbit. The Apollo 10 LM astronauts took their craft to within 14km (9mi) of the surface before returning to the CM.

▼ The Telstar I Communications Satellite, launched from Cape Canaveral on 10 July 1962. It was covered with 3,600 solar cells and topped by a helical antenna transmitting a beacon signal for tracking by ground stations and transmitting data from scientific instruments on board. The antenna was also used to receive signals to maneuver the satellite and to activate the transmission circuits. Total weight 77kg (170lb).

◀ The Gemini series of spacecraft (1964–67) – so-called because of their two-man crew – signaled that the USA was back again in the space race. Weighing 4 tonnes, it needed a powerful rocket, the Titan II, to propel it into orbit around the Earth. This picture of Gemini 8 shows astronauts D.R. Scott and N.A. Armstrong in their capsule after splashdown. Launched on 16 March 1966 it made the first docking in space, with an Agena target vessel. However, malfunction in a thruster rocket then caused the spacecraft to yaw dangerously and an emergency landing was made after only seven orbits. Rescue personnel are seen in attendance.

▲ On 21 July 1969 Neil Armstrong emerged from the lunar module Eagle to be the first man to walk on the Moon. He and Buzz Aldrin (seen here) wore $100,000 moonsuits – designed like vacuum bottles to provide insulation – and a backpack to provide oxygen for four hours. He carried a TV camera which beamed pictures of the lunar landscape to 600 million viewers.

Finally, on 20 July 1969, came the triumph of Apollo 11 when the astronaut Neil Armstrong became the first man to walk on the moon: "One small step for man, one giant leap for mankind." Of six more Apollo missions, one (Apollo 13) almost ended in disaster without a moon landing at all but the other five went smoothly. The last landing was in December 1972, since when the Moon has reverted to its age-old solitude.

The ultimate cost of the whole program is said to have been around 25 billion US dollars, but what dividend did it pay? In the way of tangible assets, it yielded some 400kg (880lb) of rock samples of great geological interest, but their significance remains to be seen. The Soviets convincingly demonstrated that the same result could be achieved more cheaply and with no hazard to life with unmanned lunar landings. The same may be said of the scientific experiments that have been planted on the surface by the American astronauts. In terms of exploration, a minuscule area of the Moon was actually covered by them and with modern techniques the whole of it can as well be surveyed from circling orbiters. As a means of making scientific observations, the project was literally astronomically expensive.

The aim of the Soviet space program is unclear, for while the United States worked in a blaze of publicity the Soviets were extremely reticent. That they sent four unmanned Zond spacecraft round the Moon in 1968-70 suggests that a manned landing might have been planned, but if so a catastrophic explosion in 1969 with their G1 launcher rocket – bigger even than the American Saturn – seems to have been a blow from which they never recovered.

Instead they embarked on the program known as Soyuz (Union) in which permanently orbiting satellite laboratories (Salyut) were periodically remanned and restocked by Soyuz spacecraft which docked with it. Salyut 1 was successfully launched in April 1971 and a three-man crew spent 23 days aboard it in June. Tragically this crew died on the homeward journey due to a simple valve fault in their Soyuz spacecraft. Three later Salyut launchings were unsuccessful, including COSMOS 557 on 11 May 1973. Three days later the USA launched its first massive Skylab. This ran into serious difficulties as one solar panel was ripped off and another failed to extend. Only some extremely ingenious and hazardous mid-space plumbing eventually saved the day, and it became operational a fortnight later. Not until the end of 1977 did the Soviet Salyut program achieve a comparable success.

The main dividends of these pioneering activities in space were twofold. First, they generated within an extraordinarily short space of time a great deal of know-how about maneuvering vehicles in space. This laid the foundations for the present complex of satellites which now provide greatly improved telecommunications systems, worldwide weather surveillance, and information about the world's mineral resources and changing pattern of vegetation. And, of course, the consequent advances in rocketry were of profound military significance. Second, the Moon-landing program gave a much needed boost to American morale after what they regarded as the humiliation of the Sputniks – though it has to be said that by 1969 the superiority of American technology had been convincingly demonstrated in countless other ways.

The last landing of men on the Moon took place in 1972 but the Soviet Union demonstrated that unmanned landings can produce comparable results. In November 1970 Luna 17 safely delivered a remotely controlled Moon rover, the Lunokhod (followed by Lunokhod 2 in 1971). They covered 50km (31mi); transmitted thousands of picture; collected and analyzed rock samples; and made cosmic ray measurements.

◄ On 16 January 1969 the USSR established the first experimental space station by locking together two space modules, Soyuz 4 and 5. Each unit weighed 6 tonnes and there was cramped accommodation, about 9cu m (320cu ft), for the four-man crew. Solar panels on two large wings provided power and recharged batteries.

► Between October 1968 and December 1972 11 Apollo missions were launched, of which seven made Moon landings. This picture was taken during mission 15 (1971) and shows the command and service modules as seen from the lunar lander. During their three-day stay on the Moon the Apollo 15 astronauts extended the range of their exploration with a four-wheeled lunar rover, traveling 28km (17mi) in all.

SEEING BEYOND THE VISIBLE

Our eyes can see only visible light, a very small part of the electromagnetic spectrum, but radiation beyond the visible range can be "seen" using specialized instruments, revealing details hidden to the naked eye. Although 19th-century scientists could detect the X-ray, infrared, and ultraviolet parts of the spectrum by photography, during the 20th century a whole new range of instruments were developed to overcome the limitations of the naked eye, making exciting new developments possible. Photography relied on the use of sensitive film to make an image, but the development of the cathode ray tube and television technology made it possible to build detectors which gave television pictures showing how things looked in different parts of the spectrum, without having to wait for film to be developed. Looking at things in other parts of the spectrum made it possible to see through opaque materials using X-rays, or, with radar, to spot enemy aircraft at great distances, day or night, cloudy or clear, by looking at them with radio waves.

Microscopy was transformed by the development of electron microscopes from the 1930s. Optical microscopes could give magnifications of 20,000 times, but when electron microscopes were eventually made they could magnify 1,000,000 times, revealing viruses and the interior of cells. By 1960 a further development, the field-ion microscope, gave the first images of atoms, magnifying 10,000,000 times.

When radar and infrared detectors were used to look at the Earth itself, the new science of remote sensing was born, and found immediate applications in weather forecasting, mapmaking, monitoring crop damage and detecting natural resources. Astronomers too found uses for such sensors, and new branches of astronomy were born. Space technology made it possible to place remote sensing satellites in orbit above the Earth, to monitor the weather and map natural resources, and send space probes to the farthest reaches of the Solar System.

A key development was the introduction – originally for military use – of computer image processing from the 1960s. When applied to remote sensing, greatly improved pictures could be created, revealing even more detail of the Earth and planets. Applied to X-ray technology, "body scanners" could be built which revealed hidden details of the human body.

Improved techniques also made it possible to see things that happened too fast for the eye. By 1908 photographs were being taken of fast-moving splashes of liquids, and by the 1970s it was possible to "freeze" waves moving at the speed of light.

"Seeing" was no longer restricted to using the electromagnetic spectrum. Sonar, developed during World War II to detect submarines by sound waves, was developed into a way of making pictures of the seabed, and, in the 1950s "ultrasound" which detected objects in a watery environment such as the womb, was first used to see inside the human body.

► Electron diffraction microscopy was developed in the 1930s. An image is formed by putting a beam of electrons through the specimen and studying the ways in which the beam is diffracted. This image is formed by the crystallic structure of the atoms of pure silicon. In the 1980s several other techniques of microscopy were developed to reveal detail at the atomic level.

▼ The scanning electron microscope gives a three-dimensional image of the specimen (such as *Staphylococcus* bacteria, shown magnified 5,000 times).

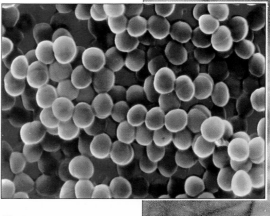

▼ A false-color image of the planet Neptune, (eighth planet from the Sun) photographed by the space probe Voyager 2 in August 1989.

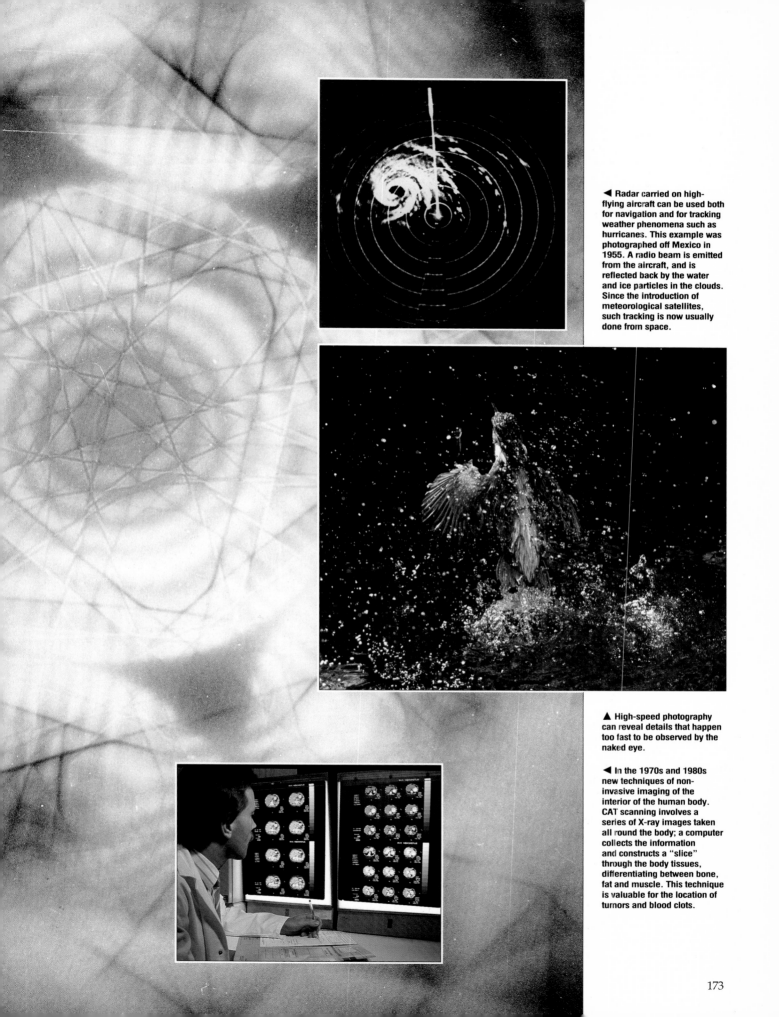

◄ Radar carried on high-flying aircraft can be used both for navigation and for tracking weather phenomena such as hurricanes. This example was photographed off Mexico in 1955. A radio beam is emitted from the aircraft, and is reflected back by the water and ice particles in the clouds. Since the introduction of meteorological satellites, such tracking is now usually done from space.

▲ High-speed photography can reveal details that happen too fast to be observed by the naked eye.

◄ In the 1970s and 1980s new techniques of non-invasive imaging of the interior of the human body. CAT scanning involves a series of X-ray images taken all round the body; a computer collects the information and constructs a "slice" through the body tissues, differentiating between bone, fat and muscle. This technique is valuable for the location of tumors and blood clots.

Datafile

During the 1960s the pace of research quickened in the biomedical sciences. Medical technology became increasingly sophisticated (and costly) as screening procedures and surgical techniques were improved. New drugs and vaccines were developed. Yet despite numerous achievements, there was also new awareness of environmental implications of scientific advances, and of sometimes tragic side effects.

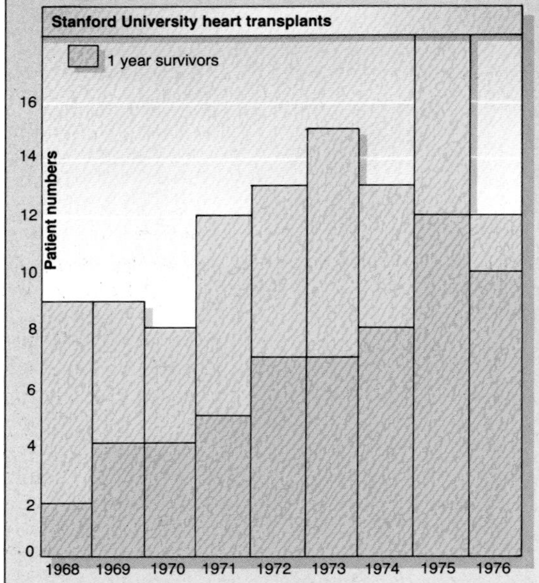

Causes of mortality

USA 1970 — 52%, 28%, 3%, 17%

Hungary 1964 — 55%, 23%, 2%, 3%, 17%

Guatemala 1964 — 72%, 7%, 4%, 15%, 2%

Taiwan 1964 — 30%, 44%, 9%, 11%, 6%

- Heart disease
- Cancer
- Influenza/pneumonia etc
- Tuberculosis
- Other

Stanford University heart transplants

1 year survivors

Patient numbers (y-axis 0–16); years 1968–1976

▲ Heart transplants were an innovation of the 1960s. They became possible because of the development of immunosuppressant drugs, able to control rejection. The graph shows the transition in one hospital from an experimental phase to one when the operation did improve chances of extending life. Other transplant operations became increasingly routine, as for kidneys. Cost–benefit ratios have increasingly determined whether transplants can be made generally available.

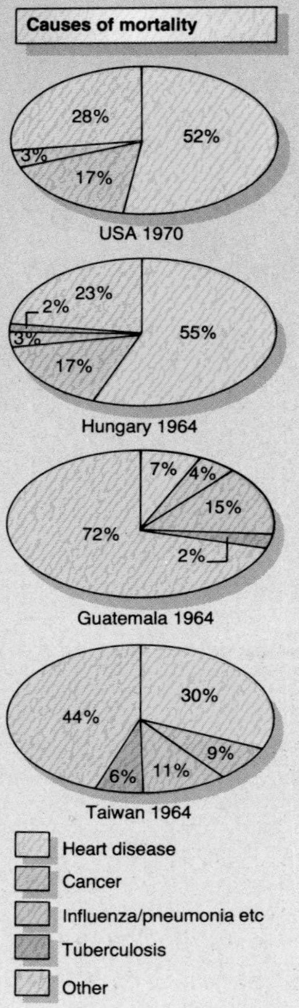

Drug industries 1967

France, FRG, UK, Spain, Netherlands, Belgium, Sweden, Denmark, Austria, Finland

US dollars (billions) 0–1.0

▲ While European and North American mortality rates showed similar trends, contrasts persisted with Third World countries. This is exemplified by the marked differentiation between the United States and Guatemala in Central America.

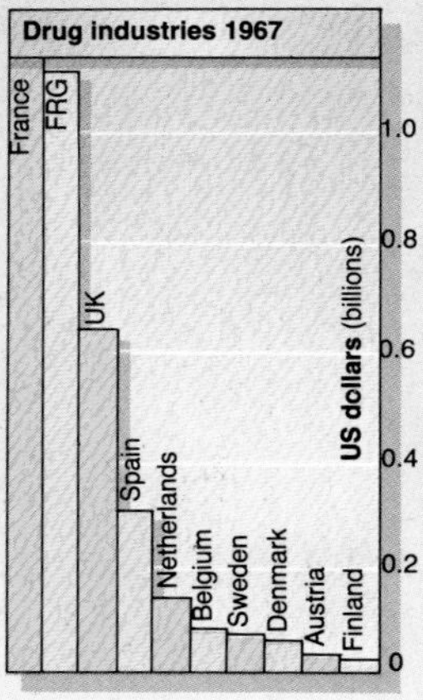

◀ Despite the high cost of research, drug manufacture has become a highly profitable business. The rising costs of developing new drugs were outstripped by the demands of doctors and patients for convenient drug-based therapies.

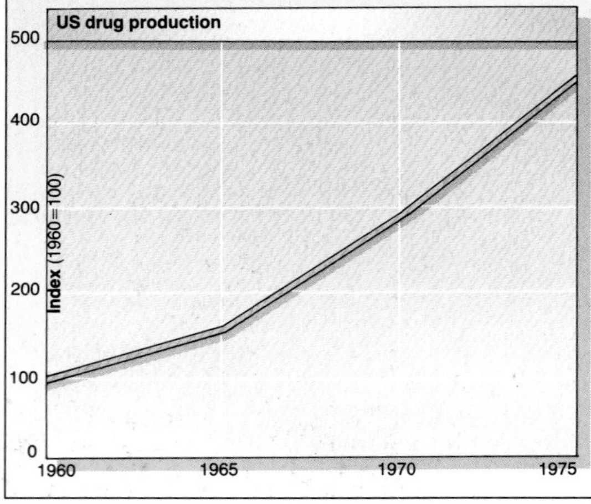

US drug production

Index (1960=100); years 1960–1975

◀ Between 1960 and 1980 the production of drugs and pharmaceuticals increased at a rate of 8–9 percent per year. Among the factors contributing to this growth have been rising prosperity, increasing life expectancy and innovations in drug therapy which have replaced hospitalization and surgery. The pharmaceutical industry exemplifies how highly profitable science-based enterprises can be, but the development of new drugs is a highly complex and costly process involving long research periods.

In science it is very rare indeed for a discovery to be made which suddenly opens up new prospects. This is not to say, however, that there are not occasional changes of course over relatively short periods of time. One such change can be clearly seen in the biological sciences around 1960, when, following the discovery of the structure of DNA, molecular biology came decidedly into the ascendant. The change can indeed be identified with a particular event, namely the launching of the *Journal of Molecular Biology* in 1959. This did not, of course, mean that research into life at the molecular level began then, but simply that the kind and volume of research being done made existing journals insufficient. The new breed of molecular biologists needed a platform of their own, rather than relying on the chemical and biological journals already in existence.

The 1960s saw many major advances in this new field. In 1960 John Kendrew and Max Perutz – who shared the Nobel Prize for Chemistry in 1962 – used X-ray diffraction to determine the molecular structures of the very important oxygen-transporting and oxygen-storing products hemoglobin and myoglobin. In 1961 Sydney Brenner and Francis Crick firmly established the nature of the genetic code which determines the inheritance of natural characteristics. They showed that the double helix structure of DNA, proposed in 1953, consists of a sequence of amino acids and nitrogenous bases and that it is the precise nature of this sequence – the genetic code – that controls the synthesis of proteins (acting via a DNA-like molecule called ribonucleic acid or RNA). In 1968 J.D. Watson interpreted their new discoveries to a general readership in his book *The Double Helix*.

At this time it was assumed that the sequence DNA-RNA-protein was a one-way system. Then, in 1970, the American virologist David Baltimore showed that "reverse transcription" is possible – RNA can be transcribed into DNA – and that this phenomenon occurs in some tumor viruses. The enzyme responsible is known as "reverse transcriptase": it was discovered independently by H. Temin, another American virologist, who shared a Nobel prize with Baltimore in 1975.

Finally, in 1973, Herbert Boyer developed the technique of recombinant DNA. He discovered that an active DNA can be created by splicing together fragments of different DNAs (plasmids). This new variety of DNA could then be introduced into a microorganism known as *Escherichia coli*, which then acquired synthetic powers derived from both plasmids. In this way the *E. coli* could be made to synthesize valuable biological products; eg insulin. This was the foundation of the field that became known as biotechnology.

HIGH-TECH MEDICINE

Medical developments

But while these exciting new fields were being explored there was no lack of progress on other biological and medical fronts. In 1963 initially successful operations were performed to transplant a heart or a lung, and the basic problem of rejection showed promise of solution with the introduction (by the pharmaceutical company Wellcome) of the immunosuppressant drug Imuran. Four years later there was the first successful heart transplant, though the patient succumbed 18 days later from postoperative complications.

In preventative medicine the battery of vaccines available for immunization was considerably strengthened by the development in 1962–63 of an antirubella (German measles) vaccine. This was particularly important for women, as infection in pregnancy may cause injury to the fetus. A vaccine against ordinary measles had been introduced in 1958, but it was not until 1981 that a reliable vaccine against serum hepatitis emerged.

The development of rubella vaccine removed one of the hazards of pregnancy, and in 1961 the cause of another was identified as a blood problem. One of the most important of the 30 or so blood group systems is the Rh (Rhesus), so called because it was first identified in Rhesus monkeys. Some 85 percent of the human population are Rh-positive: the remainder are Rh-negative. If an Rh-negative woman is carrying an Rh-positive child she may slowly develop Rh-antibodies (defensive proteins produced by white blood cells). This can have two serious consequences. First, the woman may show a violent reaction to transfusion with Rh-positive blood; secondly, the children of subsequent pregnancies may show a blood disease known as hemolytic disease of the newborn. This can be prevented by immunizing the mother to prevent the production of the maternal antibodies.

While important information about human blood groups was being discovered, other medical researchers were revealing the existence of a new category of virus. Working in Papua New Guinea after the war, D.C. Gajdusek investigated a disease called kuru, which afflicts the Fore people. Apparently spread by ritual cannibalism, associated with the eating of brains, kuru was the commonest cause of death among these people, especially women. The disease is unusual among viral infections in that it develops very slowly, causing total incapacity and death in the course of about a year. For this reason the virus is categorized as a slow virus. Scrapie, a disease found in sheep, is caused by a similar agent, but the only other certainly identified incidence in humans is Creutzfeldt-Jakob disease, somewhat similar in its symptoms to Alzheimer's disease. Slow viruses are difficult to identify as

they produce no antibody response and do not show up in electron micrographs: they are detectable only by the symptoms they cause. For his highly original research and description of kuru, Gajdusek was awarded a Nobel prize in 1976.

The development of organ transplantation

The human body is a curious mixture of resilience and vulnerability. Some parts can be lost without fatal consequence but the loss or serious impairment of others causes almost instant death. The same is true of a complex machine, such as an automobile, but there a defective part can be replaced at leisure and all is well again. In theory human organs should be equally interchangeable given the requisite surgical skill.

Such skill did in fact begin to become available in the second half of the 19th century with the advent of anesthesia and antisepsis. Nevertheless, it was the 1950s before organ transplantation was developed as a recognized surgical procedure. In 1953 John Merrill in the USA successfully performed the operation of kidney transplant on a human patient and since then tens of thousands of such operations have been performed successfully. In 1967 the South African surgeon Christiaan Barnard carried out the first heart transplant, but the record of success here has been much less satisfactory, with many patients dying soon after the operation and others surviving for relatively

▼ The 1960s saw mounting concern about the threatened extinction of many species of wildlife, both plants and animals. Among them were some kinds of whales, the American bald eagle and the gorilla. Sometimes the cause was ruthless commercial killing – as with elephants for the sake of ivory – or more often the shrinking of the natural habitat necessary for survival. When the World Wide Fund for Nature was set up it chose as its symbol the panda – perhaps the most endearing of all threatened species.

short periods. Liver and lung transplants have also been carried out and even combined transplants of all three organs. Transplants of bone marrow, in which blood cells are created, have assisted patients suffering from severe leukemia.

With modern surgical techniques, and in particular with great advances in anesthesia, the diffiulty lies less in the operation itself than in the fact that the body tends to reject the donor organ. This hazard can be reduced by typing the tissues of patient and donor to ensure that they are as nearly as possible compatible, and by the development of immunosuppressant drugs. The relatively low rate of success in the case of heart, lung and liver transplants is in part due to the fact that these are single, vital organs and can therefore be obtained only from recently dead donors. Kidneys, however, are paired organs and it is quite possible to survive with only one. Thus transplant kidneys can if necessary be obtained also from living donors – often close relatives, among whom there is a higher than average chance of finding tissue compatibility.

Very often potential heart transplant patients die before a suitable new heart can be found. For such situations artificial hearts have been developed, which enable the patient to survive for a few days even if his own heart has failed completely. The first such device (the Orthotopic Cardiac Prosthesis) made from Dacron and silicone rubber, appeared in 1969. It kept a middle-aged

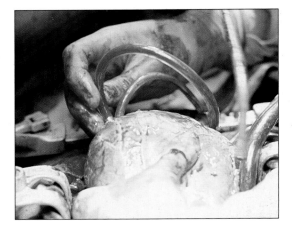

patient alive for three days. Possibly the way forward in this field will not be through transplants but with artificial hearts capable of working within the body for long periods. At a less ambitious level, since 1959 the pacemaker has been a conspicuous success for regulating heartbeat.

Currently the two major obstacles to widespread, successful transplantation of organs are shortage of donors and – despite steady improvements – the limitations of currently available immunosuppressant drugs to prevent rejection of the graft. In a few countries, such as India, the sale of a kidney by a donor is acceptable, but generally speaking this is regarded as unethical.

◄ The heart, with its key role of maintaining the circulation of oxygen-carrying blood, has long been recognized as the most vital of all organs: even brief interruption of its function causes death. For this reason heart surgery presents peculiar difficulties and little success was achieved until after World War II.

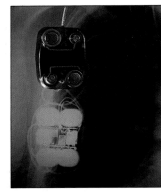

▲ In 1910 the English physician Thomas Lewis identified the sinoatrial node of the heart as the regulator of the cardiac rhythm. In 1959, thanks to the availability of microelectronic devices, use was made of this to introduce pacemakers providing a regular stimulus to maintain a sufficient heart rate in patients whose natural rhythm is deficient. In the early 1980s a computerized pacemaker appeared which would automatically adjust to a change in heartbeat. In the late 1980s half a million pacemakers were implanted annually.

► Cardiac surgery was transformed with the advent of the heart-lung machine in the early 1950s. This is a device for oxygenating and pumping the blood outside the patients body, bypassing the heart and lungs. This enabled the surgeon to operate on a dry nonpulsating heart. It was invented by the American surgeon John Gibbon and first used for a hole-in-the-heart operation at Jefferson Medical College, Philadelphia, in 1953.

The International Biological Program

On a train between Cambridge and London in March 1959, during an after-dinner conservation between the Italian biologist Giuseppe Montalenti and the new president of the International Council of Scientific Unions (ICSU), the British biochemist Sir Rudolph Peters, a seed was sown that eventually grew into the first major international research program in ecology. Their initial idea was for a biological counterpart to International Geophysical Year (1957–58) which had stimulated extensive transnational collaboration in the Earth sciences and had greatly increased research funding. By the time the project was inaugurated in Paris in July 1964, however, it had escalated into an eight-year International Biological Program (IBP).

IBP was ambitious. Working with the theme "Man and Ecology", it was intended to be a worldwide stocktaking and analysis of the biosphere. It encompassed not only biological productivity in land, freshwater and marine environments but also boldly stimulated the study of the human species. At the time this broad compass did not reflect the widely accepted view of ecology: the British geneticist C.H. Waddington, who provided the major drive behind the IBP in its formative years, noted that then "The general idea among biologists at large seemed to be that ecology dealt with the blow-by-blow account of a day in the life of a cockroach, woodlouse or sparrow." For this reason, many scientists, particularly in the UK and USA, were initially reluctant to give IBP wholehearted support although later all the scientifically developed countries joined in.

IBP could not support an independent research program. Rather, it brought together groups of

potential collaborators, a process for which national dues totaling 50–100,000 US dollars per annum were collected from participating countries. Loans from the ICSU and grants from UNESCO significantly bolstered IBP's finances. The IBP process was to define a problem, bring together a small number of specialists to formulate a research plan that would be supported, for the most part, by national funding bodies. IBP often brought an air of internationalism to preexisting research programs which fell, or could be made to fall, within its schemes. In other cases IBP stimulated funding and activities that were clearly an adjunct to national efforts. For instance, the US Congress voted around 5 million dollars for IBP work; in the USSR funds were found to support over 100 expeditions in the Human Adaptability segment.

In the developing countries, however, lack of international monies hampered research despite the establishment of a number of binational partnerships involving "sponsorship" from richer countries. External politics, too, dogged IBP: the crushing of the 1968 "Prague Spring", for instance, isolated Czech researchers within or outside the country. Nevertheless, the internationalism engendered by IBP is one of its most precious achievements.

Scientifically IBP was a brave enterprise – a project whose time had, perhaps, not come even by 1974 when IBP was wound up. IBP helped pioneer the idea of whole Earth ecology. It led directly to the "Man and Biosphere Program" project organized by UNESCO and, more importantly perhaps, shaped the way scientists (and ultimately the general public) thought about Earth and its occupants.

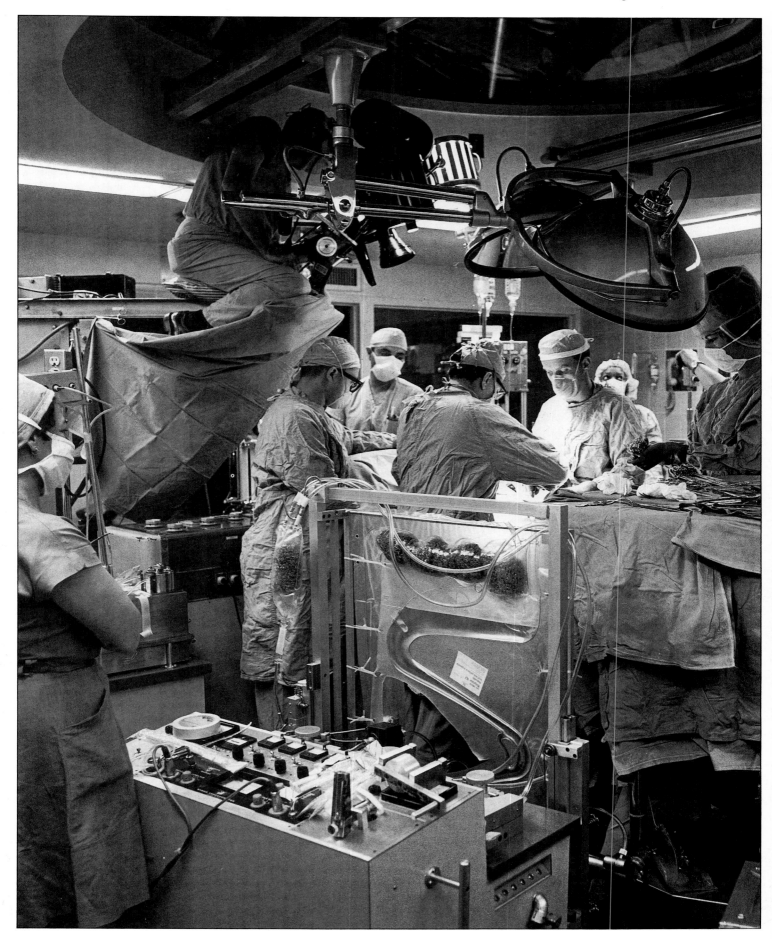

Body scanning

In some respects the problems of the physician are similar to those of the geologist: he has to deduce from surface indications what lies underneath. Over the years various instruments were developed to assist diagnosis, some of them quite simple. The stethoscope, a simple device for listening to the heart and lungs invented by T.R.H. Laennec in 1819 became commonplace. Later a variety of optical instruments were invented, such as ophthalmoscopes and otoscopes, and made it possible to examine the inner structure of organs such as the eyes and ears. Undoubtedly, however, the greatest single advance in diagnostic techniques was the discovery of X-rays, by W.K. Röntgen in 1895. This made it possible for the first time literally to see inside the body. There were, of course, limitations. Although bone was largely opaque, so that fractures and other defects were sharply delineated, soft tissues were penetrated almost equally by X-rays, making differentiation difficult. However, a variety of techniques were developed to overcome this limitation to some extent. The intestinal tract, for example, would be delineated by feeding the patient a meal containing bismuth, which is opaque to X-rays. While specialized focusing techniques made possible some sectional imaging, X-ray pictures were still essentially shadowgraphs. The 1960s and early 1970s saw the advent of four important new techniques in diagnosis. The first was thermography, introduced in 1962. In this, temperature differences over an area of the body are recorded as a color-coded picture, usually obtained by recording infrared radiation. It is particularly useful for investigating surface blood vessels, but may also reveal deeper-lying disorders, such as tumors.

In 1973, G.N. Hounsfield, of the Medical Research Division of the British electronic company EMI, devised the system known as computerized axial tomography (CAT). This was based on an instrument which could detect the very small differences in X-ray transparency between different soft tissues and then accentuate these by means of a computer to create a sharp image. There were, however, still technical problems to be solved. As the technique was based on a scanning process – rather as a television camera scans the picture to be transmitted – it took some considerable time to build up the image, and the quality of this was much diminished if the patient moved during the examination. Originally, therefore, the method was restricted to the

The biological revolution of the past three decades has placed at the disposal of biomedical science an array of research techniques possessing a power previously unimaginable. It should be possible, henceforth, to ask questions about the normal and pathological functions of cells and tissues at a very profound level, questions that could not even have been thought up as short a time ago as ten years.

LEWIS THOMAS

The Thalidomide Tragedy

By the late 1920s it was generally accepted that congenital bodily malfunctions could be induced in animals, especially fish and birds, by environmental factors. It was believed, however, that mammals, and particularly humans, were immune to change from such factors because the fetus in mammals is completely encased within the mother's body.

This belief was shattered between 1959 and 1962 with the discovery that the German drug thalidomide (widely used as a sedative by pregnant women) could produce severe congenital malformations, especially reduction in length, or total absence, of the long bones of the arms and legs (phocomelia). The total number of children affected is not known, but estimated to be at least 10,000 worldwide. Mercifully, the Food and Drug Administration (FDA) in the United States did not approve the drug, so it was never distributed there.

The episode had two major consequences. The first was to instigate much research on the nature of teratogenic (monster-producing) drugs – the effects produced by thalidomide had never been encountered before. Secondly, there was a general tightening up of the already severe criteria that had to be satisfied before any new drug could be marketed.

The thalidomide story had an interesting sequel. In 1989 it was discovered that – like many other organic compounds – the thalidomide molecule exists in two different forms, one the mirror image of the other, like a left-hand and a right-hand glove. By well established chemical techniques it is possible to separate these stereo-isomers, as they are called. It was then found that one form is responsible for the sedative effect and the other for the teratogenic effect.

▶ **Three children with malformations caused by thalidomide.**

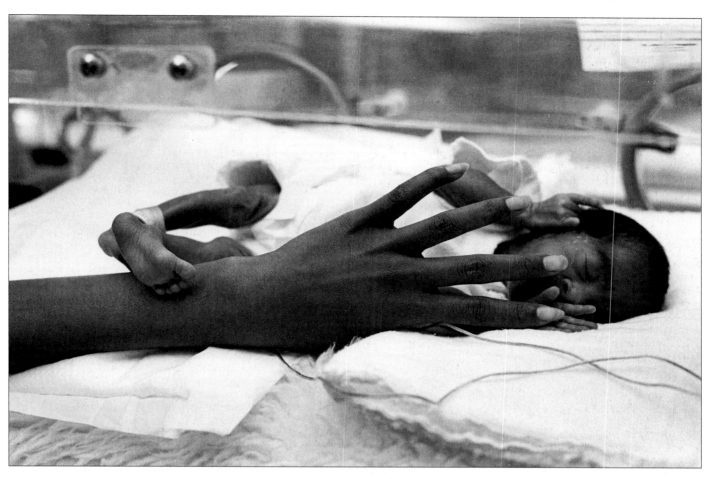

▲ One of the major advances in clinical medicine from the late 1960s onward was the growing success achieved in securing the survival of children born prematurely; that is, after a gestation of less than 38 weeks. This involves strict environmental control in an incubator and constant monitoring of the principal bodily functions. Sadly, such facilities are still very scarce in developing countries, where the need for them is great.

► The application of thermography to medical diagnosis from 1962 was one of the most important advances in diagnosis since the introduction of X-rays at the end of the 19th century. Heat emitted by the body is translated into a color thermal image. Tumors show up particularly well, so the technique has been valuable for breast screening, as here, so that breast cancer can be detected before it becomes fatal.

brain simply because the skull can quite easily be immobilized for the necessary length of time. This made it possible to diagnose and locate abnormalities such as tumors and blood clots without any trauma at all for the patient.

This success encouraged research to achieve scanning of the whole body. This was a major step forward, making it possible to identify disease in many organs at a very early stage, including secondary tumors.

From an early stage after their discovery X-rays were used not only for diagnosis but for therapy, the latter being based on the fact that cancerous cells, in particular, can be differentially destroyed by the rays. There is a problem, however, in focusing on the cells to be destroyed without damaging the surrounding tissue. Here CAT is extremely valuable, making it possible not only to locate a tumor precisely but to establish its relationship to other organs.

Scanners based on ultrasonics were also developed. Basically, these depend on the same principle as SONAR devices for detecting submarines. A pulsed ultrasonic beam is directed at the target, and the timelag of the echo indicates its position. The first prototype instruments in the early 1960s displayed a rather crude black-and-white picture on a cathode-ray tube, but the promise of the method for obstetrics and gynecology encouraged further research and development. By the early 1970s a range of automatic scanning devices had been developed and the speed of scanning had been much increased.

Finally, there was nuclear magnetic resonance (NMR). This depends on the fact that specific atomic nuclei respond to magnetic fields by emitting electromagnetic radiation, which can be detected. The phenomenon was well known in the 1940s, but its medical use as an exploratory technique was pioneered in Sweden from the 1950s by Erich Odeblad.

In 1973, in the UK, Paul Lauterbur succeeded in obtaining the first NMR image, built up from the density of hydrogen atoms in the various tissues of the body. Images of human tissue were achieved in 1977, followed quickly by NMR images of the brain.

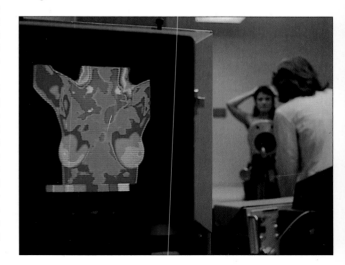

Datafile

In the 1960s and 1970s research in the physical sciences was extended into the unimaginably small world of the atomic nucleus and also out into the unimaginably vast expanses of the Universe. Particle accelerators were constructed – increasingly more powerful, larger and more expensive – which could smash nuclei apart so that a study could be made of the fragments. Images of the debris became more sophisticated. Theories were advanced to explain the increasing number of particles observed. The concept of quarks as the building blocks of nuclear particles was proposed. Many new objects were also discovered in the distant Universe. The quasar and pulsar were observed for the first time by the new field of deep-space radio-astronomy. A low background of radio waves was found to be approaching the Earth uniformly from all points of the sky. This was identified as the remnants of the radiation from the very beginning of the Universe and was powerful evidence in favor of the "big bang" hypothesis.

Radio telescopes		
Date	Place	Aperture (m)
1957	Jodrell Bank, UK	76
1961	Parkes, Australia	64
1962	Green Bank, USA	91
1963	Arecibo, Puerto Rico	305
1964	Haystack, USA	37
1965	Green Bank, USA	43
1966	Sugar Grove, USA	46
1968	Owens Valley, USA	40
1971	Vermilion River, USA	37
1976	Effelsberg, FRG	100
1979	New Mexico, USA	25

▲ Many large radio telescopes have been constructed since the first one was built at Jodrell Bank near Manchester, England. The largest now is the huge single dish at Arecibo in Puerto Rico, in the crater of a dormant volcano. It is fixed, pointing vertically upward, and scans the sky only as the Earth rotates.

Nobel prizes 1961–73

(USA 25, UK, FRG, France, Sweden, Japan, USSR, Argentina, Australia, Austria, Canada, Italy, Norway)

▲ Nobel prizes were awarded during this period for the invention of the laser (1964) and the process of holography (1971). The discovery of the double helix structure of DNA won the physiology prize in 1962. In 1972 and 1973 the physics prize went for research into superconductivity to Brian Josephson (UK).

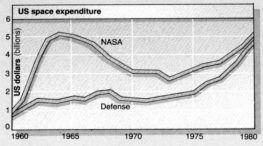
US space expenditure (US dollars, billions; NASA, Defense; 1960–1980)

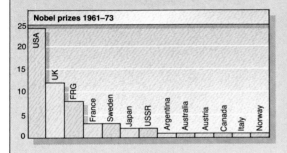
Indian scientists 1972

7%, 26%, 67%
Total 1,607,000

☐ Support staff
☐ Technicians
☐ Scientists/engineers in research/development

▲ In 1961 the first manned space flight by Yuri Gagarin shocked President Kennedy into asserting that the USA would put the first man on the moon. Expenditure by NASA immediately rocketed. Later with Skylab (1973) and particularly with the Space Shuttle (1977), military interest in surveillance satellites dominated funding.

◀▶ In 1972 India was home to 500 million people, 15 percent of the world's population. Of these only 0.3 percent worked in science, a tenth of the proportion in the USA. However, science and technical industries such as chemicals, engineering and energy were expanding. Much effort was put into military research.

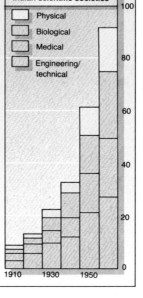
Indian scientific societies

☐ Physical
☐ Biological
☐ Medical
☐ Engineering/technical

(1910, 1930, 1950)

In the 1960s popular interest in the physical sciences continued to be centered on the very large and the very small. The remote parts of the Universe were being explored not only with very large optical telescopes, but with new instruments detecting X-rays and radio waves. These investigations were much enlarged by the possibility of locating instruments in satellites high above the screening and disruptive effects of the Earth's atmosphere. The physical exploration of what may be called near space, culminating in the Moon landings, captured the popular imagination and stimulated interest in the immensely complex science and technology on which it all depended. At the other end of the scale, increasingly powerful and expensive particle accelerators were producing a bewildering array of new atomic particles which theoretical physicists – with only limited success – sought to bring into some systematic order.

One area in which notable discoveries were made in the 1960s was that of Earth sciences. The theory of continental drift put forward by the German meteorologist Alfred Wegener in 1912 gave a plausible explanation of the present pattern of the world's land masses. It left many questions unanswered, however; in particular, it provided no satisfactory explanation of how the motion was maintained. By the 1950s growing knowledge about the ocean floor at great depths revealed that parts of it were much younger than would be expected on general geological grounds and that in mid-ocean there were great ridges and complementary rift valleys. In 1962 H.H. Hess, of Princeton University, put forward his hypothesis of sea-floor spreading. According to this, the mid-ocean ridges were continuously being renewed from the depths of the Earth's crust, and then spreading out sideways to form a new ocean floor. When this flow reached the continents it was thrust under their relatively light rocks and forced back again into the depths. The whole system could be viewed as a sort of vast conveyer belt. A year later Hess's hypothesis gained considerable support from the discovery by D.H. Matthews and F.J. Vine that the magnetism of ocean crust in bands parallel to the ridges alternates in polarity, sometimes running north–south and sometimes south–north. This is consistent with the well-established fact that in geological time the direction of the Earth's magnetic field has reversed very frequently.

In 1967, a new theory was advanced which involved elements of those of both Wegener and Hess. Known as plate tectonics, this postulates that the Earth's crust to a depth of 50–100km (30–60mi) consists of a number of rigid plates which are in motion relative to each other. This relative movement is the source of earthquakes,

QUARKS, LASERS AND SATELLITES

and it follows that mapping the location of major earthquakes provides a means of identifying the boundaries between the plates. By this means eight principal plates have been identified, together with many smaller ones. Interaction at the boundaries of the plates is of three kinds: first, sideways slippage; second, separation and growth, as new material wells up to fill the gap; third, convergence and destruction.

Plate tectonics still fails to explain how the energy requirements of these vast movements are found, but the explanation may lie in convection currents within the Earth's mantle, the region lying between depths of around 40–3,000km (25–1,860mi).

In the field of chemistry a particularly interesting development occurred during the 1960s. The group of six so-called "noble gases" had been thought for decades to be totally inert and unable to combine with other elements to form compounds. In the early 1960s, however, it was discovered that the three heaviest (krypton, xenon and radon) can form compounds with fluorine.

▼ **In 1969 these detectors at the Stanford Linear Accelerator Center in the USA began to provide the first direct evidence for structure within protons and neutrons, so helping to establish the concept of quarks as basic constituents of matter.**

Nuclear research

In the 1950s nuclear research was carried on by the momentum generated by the dramatic success of the atomic bomb in World War II. It is difficult to generalize about so dramatic a decade but two major and related characteristics are apparent. Firstly, the invention of the strong focusing principle made it possible for accelerators to move from the cumbersome cyclotron, with an intrinsic upper energy limit of 25 MeV (million electronvolts), to the typical modern ring synchrotron. As early as 1952 the cosmotron at Brookhaven generated a proton beam above 1 GeV (giga-electronvolts). Secondly, the experimental emphasis moved from the nucleus to subnuclear particles and it was for the first time possible to demonstrate the existence of some of the particles – such as pions and neutrinos – which had been postulated on theoretical grounds.

It is also perhaps fair to say that a feature of the 1960s was to bring some order into interpreting the significance of the plethora of subatomic particles being discovered. By the end of the decade

there was good evidence that the many observed particles were different clusterings of only a few, more basic particles – the quarks.

So far as the nucleus itself was concerned, Maria Goeppert-Mayer and J.H.D. Jensen had worked out a shell structure analogous to the shell structure of Niels Bohr's atomic model and the importance of this was acknowledged in 1963 when they were jointly awarded a Nobel prize. In an alternative theory, however, the nucleus was analogous to a liquid drop. In the event, it transpired that the two concepts were not incompatible, and a collective model of the nucleus, embodying both, was advanced in 1953 by Aage Bohr (son of Niels), L.J. Rainwater, and B.R. Mottelson.

The year 1954 saw the birth of the concept of "strangeness" by M. Gell-Mann, a new quantum number which has to be conserved in all "strong" nuclear intractions. In the early 1960s, with Y. Ne'eman, he used this to classify all the known subatomic particles, on the basis of symmetry, in multiples of 1, 8, 10 or 27 members. One fruit of this was the prediction of an omega-minus particle, demonstrated experimentally in 1964. Gell-Mann (and independently G. Zweig) used the symmetry concept in 1963 to introduce the quark – a purely fanciful name taken from a passage in James Joyce's *Finnegans Wake*: "Three quarks for Muster Mark!" It was a revolutionary move, in that it required quarks to have fractional charges – one-third and two-thirds – compared to that on the electron. The quark was greeted with some skepticism but the concept gained experimental support (1969–72) in investigations at Stanford on the scattering of electrons by protons and at CERN by the scattering of neutrinos by protons.

For the detection of particles the cloud chamber had been superseded by the bubble chamber, but 1969 saw the advent at CERN of multiwire detectors. They have the great merit of being able to be linked directly to a computer, facilitating the automatic monitoring of collisions.

The early 1970s saw the appearance of a great breakthrough in the theoretical interpretation of the forces that act on subatomic particles. The new theory linked the "weak" forces responsible for beta decay with the long-familiar electromagnetic force. A predicted consequence was the existence of heavy particles analogous to photons: Z-particles, which are neutral carriers of the weak force, and W-particles which are charged carriers. The independent architects of this new theory were A. Salam, S.L. Glashow, and S. Weinberg. The neutral carrier of the weak force should give rise to neutral currents, and these were in fact detected at CERN in 1973.

This still left the strong intranuclear force required to bind quarks together. Again the problem of verbalizing highly abstract concepts led to a new word, chromodynamics, in which "color" described a charge-like property of quarks. Yet another word, gluon, was coined to denote the force carriers which unite quarks.

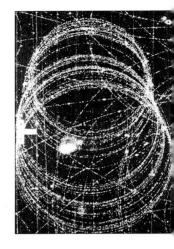

▲ Though otherwise invisible, charged subatomic particles leave a trail of ionized atoms as they pass through matter, and this can be made visible. Here an electron has curled round many times in a magnetic field, to leave a spiral track in a cloud chamber – a detector containing a supersaturated vapor which condenses out on the ionized trail.

▼ In such spark chambers a high voltage induce a gas to spark along the ionized trails of particles.

The Varied Strategies of Science in India

The history of science in India in the 20th century is basically a response to the failure of the colonial imagination. In 1900 little scientific research was sponsored by the British colonial government – not even in the universities. The only research institution under Indian control was the Indian Association for the Cultivation of Science (founded in 1876).

The encounter between colonial and nationalist science can be divided into four overlapping phases. The first spans the *Swadesi* movement (1904) and World War I. Swadesism sought indigenous control of science, technology and education. Three scientists demonstrate well the diversity of activity in the period: P.C. Ray, who established the Bengal Chemical and Pharmaceutical Works; J.C. Bose, botanist and physicist; and the mathematical genius Srinivasa Ramanujan. Gandhi's *Hind Swaraj* (1909) remains the period's most important science policy document. In this pluralist phase, the geologist A.K. Coomaraswamy sought to sustain craft technologies; Patrick Geddes, the British town planner, proposed a post-Germanic science university. Meanwhile Tagore's *Visva Bharati* and the work of A. Howard (farming), A. Chatterton (lift irrigation, windmills) and F. Nicholson (fishing) anticipated the Intermediate Technology Movement. The Institute of Science was established by J.N. Tata in 1911.

The second phase extends from the war years to independence in 1947. In 1916 the British established the Indian Industrial Commission. But Indian scientists, motivated by the failure of the Commission and later inspired by the Bolshevik revolution in Russia, saw a need for planned science. By 1916 Asutosh Mukherjee had established the University College of Science with C. V. Raman, P. C. Ray, M. Saha (FRS – Fellow of the Royal Society of London) as teachers. Raman (awarded the Nobel Prize for Physics in 1930) had already begun to create a nursery of scientists who were to man major institutions after 1947. Saha and M. Visvesvaraya were instrumental in establishing the National Planning Committee (1938) of which Jawaharlal Nehru was the first chairman. The most remarkable institution of this period was the Indian Statistical Institute founded by P. C. Mahalanobis. The British response to this great wave of Indian intitative was initially restricted to the establishment of the Indian Council of Agricultural Research (1929) and the Industrial Intelligence Bureau (1935), but it eventually culminated in the A.V. Hill Report of 1944 which proposed a centralized organization for research.

After the granting of Independence to India the third phase (1947–1964) was marked by both euphoria and the slow bureaucratization of science. S.S. Bhatnagar (FRS) established a chain of industrial research laboratories (CSIR), which tragically displaced talent from the universities. By the 1960s Nehru (prime minister since Independence), disappointed with the CSIR, transferred his hopes to H.J. Bhabha's Tata Institute of Fundamental Research and the Atomic Energy Commission. Stunned by public criticism of science after the Chinese attack (in October 1962), Nehru invited the British physicist P.M.S. Blackett to evaluate the most celebrated CSIR laboratory, the NPL. The Blackett Report initiated questions about the validity of high investment in pure science and emphasized the vital need to integrate science and technology.

◀▲ As early as 1943 the
Indian physicist Homi Bhabha
had a dream that atomic
energy could help to solve
India's poverty. Under premier
Nehru he later became the
first chairman of the Indian
Atomic Energy Commission,
and his memory is now
honored in the name of the
Bhabha Atomic Research
Center at Trombay, shown
here. In 1974 the country
joined the list of nations
capable of building nuclear
weapons, with the successful
explosion of a "deterrent"
based on plutonium produced
by a research reactor at the
center.

In the post Nehru phase (from 1964) efforts were made to rationalize the organization of science and technology. Many developments were partially successful, but often became tinged with irony. The Green revolution under the leadership of B.P. Pal (FRS) and M.S. Swaminathan (FRS) is now seen as an ecological disaster. The Atomic Energy Commission became the epitome of secret and unaccountable science. The gas leak in Bhopal (1984) killed about 2,500 people but produced little response from the scientific community. The innovative waves were restricted to few sectors like biophysics, radioastronomy, statistics and liquid crystals.

Lasers

The laser was once described as a solution in search of a problem. Although its potential was only slowly realised it eventually found a very wide range of applications and has became an important part of the electronics industry. In the early 1990s sales reached 10 billion US dollars.

The laser was invented in 1958 but it is descended from an earlier device known as a maser – derived from Microwave Amplification by Stimulated Emission of Radiation – developed independently in the USA and the USSR in the early 1950s. Its principle is very simple, and depends on the fact that the energy of atoms is quantized: that is to say, it can have only certain fixed values and not intermediate ones. If an atom is irradiated it goes from one energy level to another. Usually atoms reemit this energy at random; but in a maser (or laser) they are made to reemit the energy in unison. However, the technical problems of finding the right sort of atoms to correspond to a desired wavelength, and then deve-

loping a workable device, were considerable. In 1954 C.H. Townes and J. Weber in the USA described a maser in which ammonia gas was the irradiated material: this was followed by a maser incorporating a crystal of ruby. Such devices could be used to amplify short-wave radio signals, such as are used in radar and are received from space by radio telescopes.

In theory, the same principle could be applied to amplifying light signals, since these are simply another form of electromagnetic radiation. The wavelength of light is, however, much shorter than the shortest radio waves and this posed new technical problems. Nevertheless, in 1960 the US physicist T.H. Maiman devised the prototype laser (Light Amplification by Stimulated Emission of Radiation). This, too, was based on a ruby crystal and emitted a pulsed beam of light, but was soon followed by gas lasers giving a continuous beam.

The light emitted by a laser has three important characteristics. First, it is coherent; that is, the

▲ The laser stimulated the popular imagination in the 1960s with its image of producing an invincible "death ray" that could burn through anything. In truth the laser's remarkable properties lie not so much in its basic power, as in its ability to concentrate energy efficiently in the laser beam. Here a low-power laser produces the brightest light on the skyline near the Oakland Bay Bridge in San Francisco, demonstrating the laser's inherent efficiency.

▲ Because their energy is highly concentrated, laser beams are dangerous if the beam enters the eye. However, this very effect can be put to use surgically to remedy a detached retina by "spot-welding" it back in place.

▼ In 1966 František Hoff (left), and Rudolf Konvalinka (right) in Prague demonstrated the successful transmission of a laser beam across a distance of 5km (3mi). Since then lasers have been used to cross far greater distances, for example measuring accurately the distance to the Moon. Light from a laser on Earth is sent back by reflectors left on the Moon by American astronauts and the time for the return journey measured.

waves are in step. This makes it possible to concentrate it into an intense beam. Second, it is monochromatic; that is to say, it consists of light of a single wavelength rather than a mixture. Third, it emerges as an extremely narrow pencil that scarcely diverges, even over enormous distances.

Many applications of lasers depend not so much on the fact that they generate a beam of light as because they represent an exceedingly high concentration of energy. There are, however, three very important applications in the optical field. One is in fiber optic telecommunications, in which electrical signals flowing through metal conductors are replaced by optical pulses transmitted through hair-thin glass fibers. As a strong source of light the laser gives such fibers a very high carrying capacity.

A second major optical application is in holography, a novel form of three-dimensional image-forming invented in 1947 by the Hungarian electrical engineer Dennis Gabor, who was awarded a Nobel prize in 1971. It depends on the interference between two beams of light. One is reflected from the object being depicted onto a photographic film. The other goes straight to the film as a reference beam. When the film is developed and the image viewed in light of the same wavelength, a 3-D representation of the original is created. An essential feature of the process is, however, that coherent light is used and at the time of its invention only relatively weak sources of such light were available. The advent of the laser transformed the situation, for the generation of a strong source of strictly coherent light is of its very essence. In due course holography became familiar in a variety of forms, such as the security logos on credit cards and in advertising. A third major application of the laser is in computer typesetting, where a controlled

laser beam traces out the words to be printed on a photographic film.

Other applications of lasers depend on their ability to concentrate a very high level of energy – around a million watts per square centimeter – on a very small target for a very short space of time. Two major industrial applications are in piercing and cutting. In the first case, it is necessary to heat a small area for just long enough to vaporize the material to be removed. For this purpose what are known as YAG (Yttrium-Aluminum-Garnet) lasers are used: these emit very short pulses near the infrared region of the spectrum, each lasting less than a hundred millionth of a second. Thus a hole – even as small as 0.3mm in diameter – can be drilled with minimum damage to the surrounding material. Again, because of the high energy that can be concentrated on a very small area, lasers have important applications in spot welding sensitive components. Lasers can also be used for precision cutting, but for this purpose gas (carbon-dioxide) lasers are used because these generate a continuous rather than a pulsed beam. A 500 watt CO_2 laser, for example, can cut 1mm steel sheet at around 10cm per second.

Yet another industrial use of lasers is in the "doping" of silicon chips by implanting local concentrations of various atomic species. A tiny spot on the chip is melted, the alien atom introduced, and the silicon allowed to solidify again, trapping the desired species.

Because the laser beam is narrow and virtually non-divergent, it can be used to measure long distances with great precision. The technique – similar to that of radar – is to reflect the beam from the distant object and measure the time lapse between dispatch of the signal and receipt of its reflection. Knowing the speed of light, the total distance is easily calculated. In the 1970s this method was used to determine the distance of the Moon with great precision, using reflectors planted there by American astronauts.

Two other uses of lasers come much more within everyday experience. They are used for scanning in compact disk players – so avoiding any mechanical contact with the disk itself – and in reading at the checkouts of supermarkets and other stores the bar-codes printed on wrapping which store price and other information.

Lasers have also found limited, but nevertheless very significant, application in medicine. They can be used, for example, to repair detached retinas by a kind of spot-welding technique, and to make bloodless incisions, the heat of the laser cauterizing the wound. In certain circumstances, such as operations on hemophiliacs – this can be of crucial importance.

In the 1970s lasers found a new use in chemical research, when Sir George Porter used them his work on photochemistry. This demanded intense flashes of light lasting no more than a picosecond (10^{-12} second). Previously, this had been provided by a bank of capacitors but the laser was simpler and more efficient.

An intriguing use of lasers is in espionage. In 1968 it was discovered that a laser beam can pick up perfectly from the outside the vibrations of window panes caused by conversation indoors.

The Literature of Science

An important feature of science is that it is a cumulative process, each research worker gaining inspiration from the work of his or her predecessors and contemporaries and in turn pointing the way forward for others. To this extent it is subjective, but the interpretation of facts also demands imagination. The great German chemist F.A. Kekulé (1829–96) summarized this succinctly: "Learn to dream, then perhaps we shall find the truth." It follows that the progress of science as a whole, and of the individual worker, depends on the swift and free dissemination of new knowledge.

The 20th century has seen great changes in the way in which this is effected. At its start there had been little change for some 200 years. The number of journals available for the publication of original results was relatively small: they were mostly published by learned societies, for whom they were an important source of revenue, and they were not highly specialized. For students, and for research workers needing an authoritative review of a particular field, books provided the necessary information, and these were normally produced by commercial publishers. As the century advanced, however, science became increasingly specialized and specialized journals were required.

After World War II commercial publishers recognized this as a lucrative new field and new journals proliferated, so much so that it became a major problem for research workers to find the time to read even those of direct interest. This problem was partly met by journals publishing abstracts of papers, but much more effectively in creating computerized databases making it possible to identify almost instantly what was relevant and available.

Meteorology by satellite

In the broadest sense, meteorology comprises all atmospheric phenomena – lightning, auroras, rainbows, and, of course, meteors. For many years, however, it has assumed a more restricted meaning: namely the study of the atmosphere in relation to climate and weather. Only since around the mid-19th century, however, has this been based on interpretation of systematic scientific observations – especially of pressure, temperature and wind speed. Before this, reliance had to be based on experience translated into folklore: the migration of birds, the pattern of clouds, the color of the sunset and so on.

Scientific meteorology presents many problems, and despite much progress these are still not fully solved. The first is that a local pattern of weather may be determined by conditions in distant regions of the atmosphere and the forecaster thus needs data from a large number of observation points, not only from the ground but at different atmospheric levels. Secondly, as weather is so variable, this data must be available almost as soon as it is recorded. Thirdly, there must be a means of interpreting the mass of data quickly enough to be helpful. A forecast for tomorrow is useless if its computation takes a week.

The first step toward fulfilling these conditions was the electric telegraph, which made it possible quickly to assemble information from distant observatories at a central meteorological office. The advent of radio widened the catchment area, making it possible, for example, to get regular reports from ships at sea. Sounding balloons and high-flying aircraft – and after World War II, rockets – introduced a three-dimensional element. Then, from the 1960s, the availability of powerful computers made it increasingly feasible to interpret the growing flood of data quickly enough to produce a useful forecast.

► Hurricanes are severe storms that develop in the tropics and are characterized by high wind speeds. The storms develop over the ocean, but their tracks can bring them across coastal regions where they can have a devastating effect on highly populated areas. They can also cause immense indirect damage through floods. Weather satellites provide vital information about the development and movement of such storms, although accurately predicting the precise path of a hurricane remains a difficult task involving data from many sources.

▼ Between 1960 and 1965 the USA launched 10 weather satellites in the TIROS (Television and Infra-Red Observation Satellites) series. This mosaic was the first complete view of the Earth's weather systems, provided on 13 February 1965 by Tiros 9, which followed a low polar orbit around the Earth. In the 1970s satellites such as Meteosat were launched into geostationary orbit at 35,000km (23,000mi), from where they can view a single large portion of the Earth's surface.

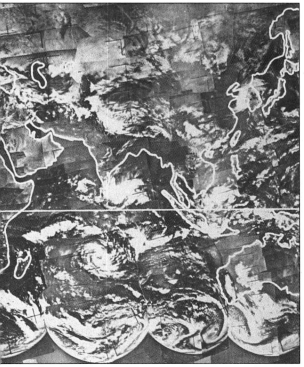

All these advances significantly improved the speed and accuracy of weather forecasting, but the most important development was the introduction of observational satellites which could steadily transmit pictures of the Earth's cloud cover from a viewpoint hundreds of miles above the surface. The earliest ones were the ten American TIROS satellites launched between 1960 and 1965, followed by ESSA in 1966. These in turn were followed in 1970 by ITOS1, the first weather satellite able to relay meteorological data by day or night. In the same year, the UK Meteorological Office ordered the largest computer in Europe to improve its national forecasting service. Then, in 1977, METEOSAT – designed and built by ESRO – was launched in the USA. Since then increasingly sophisticated satellites and instrumentation have made it possible to monitor factors other than cloud pattern, such as snow cover and temperatures. Apart from providing data for ordinary day-to-day forecasts weather satellites can also provide advance warning of catastrophic events, such as hurricanes.

But the biggest single factor in the accuracy of modern forecasting is the computer, which alone makes it possible to digest – within a necessarily brief span – the mass of data collected day by day.

One can visualize an ideal weather service in which satellites automatically fired their global reports into electronic digital computers that process the data and produce weather forecasts without human intervention. One can look much further in time and imagine computer-controlled devices capable of modifying the predicted weather – if it is unfavorable – before it has a chance to develop.

"SCIENTIFIC AMERICAN"
1961

Astronomy and space exploration

In astronomical research the discovery of new kinds of stars is analogous to the discovery of new kinds of atomic particles in physics. Radio astronomy revealed the existence in the Universe of discrete emitters of radio waves, and in 1960 A.R. Sandage, in the USA, made an important advance by identifying one of these with an object faintly visible with an optical telescope. This was the first of the quasars, a contraction of "quasi-stellar objects". These are small but energetic sources, distinguished by a peculiarity in their spectra which shows that they are moving with velocities close to that of light.

In 1967 Anthony Hewish and Jocelyn Bell discovered the first pulsar, a cosmic radio source which fluctuates regularly. Briefly, it was thought that it might be evidence of some distant form of intelligent life trying to communicate, but this was discounted when other examples were found. The frequency of bursts of radio energy from pulsars varies from a few hundredths of a second to about four seconds. Pulsars are apparently collapsed neutron stars – originally massive stars which have collapsed to form matter so dense that their electrons and positrons have combined to form neutrons. They are perhaps no more than a few kilometers in diameter. The pulsation seems to be due to their very rapid rotation, much as the rotating lens of a lighthouse gives a regular flash in any given direction.

In 1962 the astronomical spectrum was furthered widened with the pioneer work of B.B. Rossi on the detection of X-rays from space. More specifically, he identified a discrete source of X-rays in Scorpio X-1. Two years later radio emissions provided yet another insight into the working of the Universe. Using a large telescope designed for communication with satellites, A.A. Penzias and R.W. Wilson, of Bell Telephone Laboratories in the United States, recorded a pervasive background of radio "noise" corresponding with that emitted by a black body at a temperature of 3.5°K. At Princeton University R.H. Dicke and P.J. Peebles were able to provide a theoretical explanation.

According to the "big bang" theory of the origin of the Universe, the radiation then released should have distributed itself and progressively cooled: calculation indicated that the present temperature of the Universe should be about 5°K allowing for a number of uncertainties. It was reasonable to suppose that the radiation observed by Penzias and Wilson was the natural background radiation of a cooled Universe. This is regarded by cosmologists as one of the most convincing pieces of evidence available in favor of the "big bang" theory.

Rossi's experiments were made with the assistance of a rocket-borne probes, a reminder that by 1960 considerable progress had been made in the use of unmanned observational satellites. These had been adumbrated with some confidence as early as 1955, when both the Soviet Union and the United States had provisionally included them as part of their contribution to the International Geophysical Year (IGY). Their advantage over sounding rockets, such as those used by Rossi, was that they can record and transmit scientific information over long periods instead of for just a few minutes. The United States had hoped to have had no less than 12 satellites in orbit at the start of the IGY, but in the event the first success went to the Soviet Union with the launch of Sputnik 1 on 4 October 1957.

Mortified, the USA advanced its plans and promised a launch within 90 days. The first attempt failed but America's first satellite, Explorer 1, was in orbit on 1 January 1958. It had limited capacity, notably a Geiger-Müller counter to register cosmic rays, but it succeeded in locating the two Van Allen radiation belts that encircle the Earth.

Thereafter progress was rapid, not least because the Soviet Union and the United States vied with each other to demonstrate their technological superiority to the world. Several different categories of satellite emerged in the 1960s. The earliest, used for purely scientific purposes, were joined by more sophisticated ones designed for military observation, global weather monitoring, telecommunications, remote sensing and other specialist purposes.

Finally, there were the space probes which paved the way to landing men in the Moon. The Soviet probe Luna 2 (1959) was the first Earth-originating body to land on a celestial body. In 1966 Luna 9 made a soft lunar landing – relieving American fears that the surface might be covered in deep dust – and transmitted thousands of pictures back to Earth. Luna 16 (1970) brought back samples of lunar soil. By the end of the 1970s Soviet and American probes had had close encounters with, or had landed on, several planets, including Jupiter, Mars and Venus.

◀ The world's largest radio telescope dish was built into a natural crater near the city of Arecibo in northern Puerto Rico. Completed in 1963, it has a diameter of 305m (1,000ft), but cannot be steered.

▼ Flights to Venus by the USSR's Venera probes and the USA's Mariner probes established a detailed picture of the planet's atmosphere, revealing its composition to be almost entirely carbon dioxide.

▼ During the 1960s unmanned spacecraft provided scientists with such detailed photographs of the lunar surface, leading to the production of atlases of the Moon. They also landed probes which could analyze the soil and sometimes return to Earth with small samples.

PLATE TECTONICS

During the 1960s the scientific view of the surface of the Earth changed dramatically, credence being given to vague ideas that had been developing for three centuries.

Since the English philosopher Francis Bacon had noticed, in 1620, that the continents of Africa and South America could once have fitted into one another, many have elaborated the idea. The most influential was the German meteorologist Alfred Wegener who, in 1915, proposed the theory of "continental drift": that all the continents were once joined together and that they have since drifted apart. The idea was championed by the British geologist Arthur Holmes and the South African geologist Alexander du Toit in the 1920s and 1930s, but the notion was generally disregarded by the vast majority of scientists.

Acceptance began in 1960 when American geophysicist Harry Hess found that certain findings made by oceanographers in the previous decade fitted well with the idea of moving continents. These findings included the fact that the oceanic ridge up the middle of the Atlantic was part of a ridge system that ran through all the oceans, and the fact that the crust under the ocean was remarkably thin. Hess suggested that the ocean ridges were situated above rising convection currents in the Earth's mantle, the material thus brought up solidified on the surface to form new crust, and the new solid crust moved sideways away from the seat of activity. All this suggested that the crust near the ridges was very new, and became older with increasing distance from the ridge. He called this notion "seafloor spreading".

Proof came with the discovery in 1963 by British geologists Fred J. Vine and Drummond H. Matthews that the ocean crust at each side of the Atlantic ridge was magnetized in parallel stripes, each stripe having an opposite polarity to its neighbors. By 1966 it had been established that the polarity of the Earth's magnetic field had reversed several times in the recent past. The inference was that when each part of the new crust was formed it took on the magnetic polarity of the contemporary magnetic field.

In 1967 the American geophysicist Hugo Benioff observed that earthquake foci in an earthquake-prone area seemed to be located on a tilted plane that dipped beneath the edge of a continent. The Japanese seismologist Kiyoo Wadati made the same observation but it is Benioff whose name is commemorated. The "Benioff zone" represented old crust sinking into the Earth's mantle and being destroyed. Molten crust material finds it way to the surface here, forming volcanoes.

All these phenomena were combined into a single concept in the late 1960s. The surface of the Earth consists of several plates, each continuously created along an oceanic ridge and continuously destroyed in a Benioff zone. The term "plate" was coined by the American geologist W. Jason Morgan, and now the whole concept is known as "plate tectonics".

▲ Pillow lavas, produced as lava erupts on to the ocean floor, lie on the ocean ridges. Away from the ridge crest they are covered by successively deeper sediment layers, indicating their increasing age with distance from the crest. This was confirmed by the JOIDES (Joint Oceanographic Institutions Deep Earth Sampling) expedition in 1968.

▶ According to plate tectonic theory, the Earth's surface consists of a number of "plates", like the panels of a football. Each is being generated in an ocean ridge, moving away from the ridge and being swallowed up beneath another plate and destroyed. Continents are carried about by the movement. The plates consist of the crust and the topmost layer of the mantle, together called the lithosphere, and they move over a soft layer of the mantle called the asthenosphere. The mechanism is probably convection currents within the mantle itself.

◀ "Smokers" – hot springs on the seafloor – are produced by the volcanic activity in oceanic ridges where new crust is being generated. The sudden chilling of the hot water precipitates minerals which form opaque clouds.

Constructive plate margin

Convection currents

▲ The Atlantic ridge southwest of Iceland was studied in detail by Vine and Matthews in 1966, and this showed a pattern of magnetic stripes in the ocean crust. The pattern on one side of the ridge was a mirror image of that on the other, showing that the seafloor was spreading from the ridge axis.

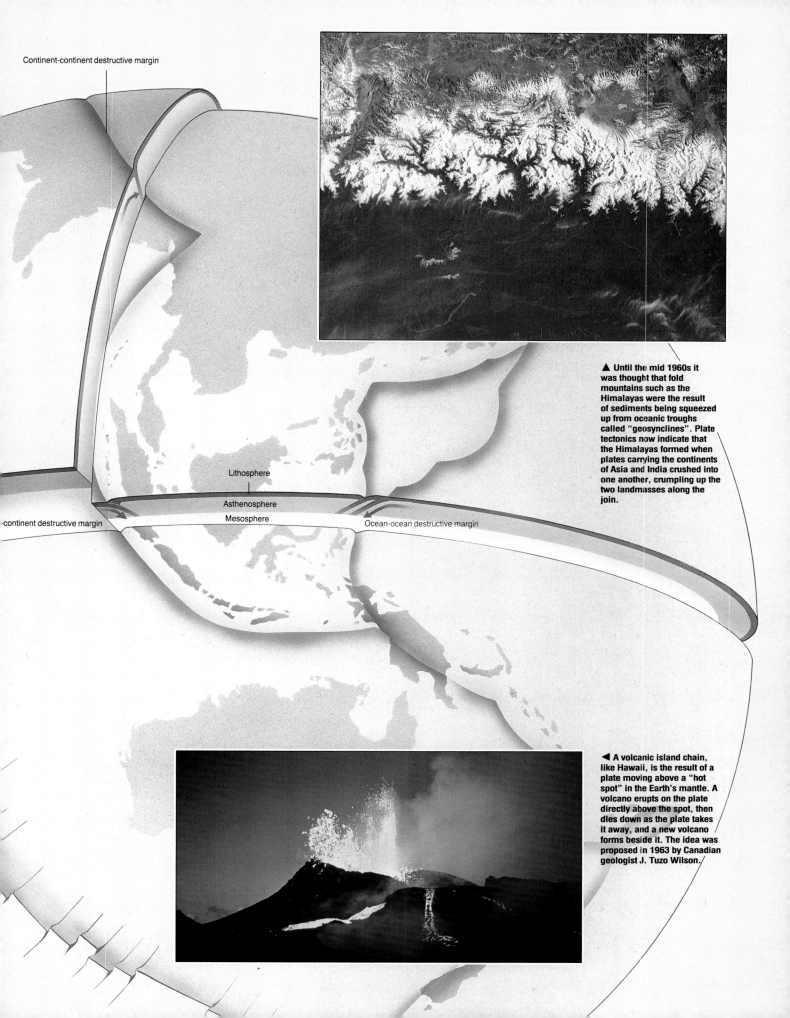

Continent-continent destructive margin

Lithosphere

Asthenosphere

Mesosphere

continent destructive margin

Ocean-ocean destructive margin

▲ Until the mid 1960s it was thought that fold mountains such as the Himalayas were the result of sediments being squeezed up from oceanic troughs called "geosynclines". Plate tectonics now indicate that the Himalayas formed when plates carrying the continents of Asia and India crushed into one another, crumpling up the two landmasses along the join.

◄ A volcanic island chain, like Hawaii, is the result of a plate moving above a "hot spot" in the Earth's mantle. A volcano erupts on the plate directly above the spot, then dies down as the plate takes it away, and a new volcano forms beside it. The idea was proposed in 1963 by Canadian geologist J. Tuzo Wilson.

1973 · 1989

THE
HIDDEN
COSTS

Time Chart

	1974	1975	1976	1977	1978	1979	1980	1981
Nobel Prizes	• *Chem*: P.J. Florey (USA) • *Phys*: M. Ryle, A. Hewish (UK) • *Med*: A. Claude (Bel), G.E. Palade (USA), C. de Duve (Bel)	• *Chem*: J.W. Cornforth (Aus/UK), V. Prelog (Swi) • *Phys*: A. Bohr, B. Mottelson (Den), J. Rainwater (USA) • *Med*: D. Baltimore, R. Dulbecco, H.M. Temin (USA)	• *Chem*: W.N. Lipscomb (USA) • *Phys*: B. Richter (USA), S.C.C. Ting (USA) • *Med*: B.S. Blumberg, D.C. Gajdusek (USA)	• *Chem*: I. Prigogine (Bel) • *Phys*: P.W. Anderson (USA), N.F. Mott (UK), J.H. van Vleck (USA) • *Med*: R. Guillemin, A.V. Schally, R. Yalow (USA)	• *Chem*: P.D. Mitchell (UK) • *Phys*: P.L. Kapitsa (USSR), A.A. Penzias, R.W. Wilson (USA) • *Med*: W. Arber (Swi), D. Nathans, H.O. Smith (USA)	• *Chem*: H.C. Brown (USA), G. Wittig (FRG) • *Phys*: S.L. Glashow (USA), A. Salam (Pak), S. Weinberg (USA) • *Med*: A.M. Cormack (USA), G.N. Hounsfield (UK)	• *Chem*: P. Berg, W. Gilbert (USA), F. Sanger (UK) • *Phys*: J.W. Cronin, V.L. Fitch (USA) • *Med*: B. Benacerraf (USA), J. Dausset (Fr), G.D. Snell (USA)	• *Chem*: K. Fukui (Jap), R. Hoffmann (USA) • *Phys*: N. Bloembergen, A.L. Schawlow (USA), K.M. Siegbahn (Swe) • *Med*: R.W. Sperry, D.H. Hubel (USA), T.N. Wiesel (Swe)
Technology	• Franco–German satellite *Symphony* comes into operation • Development of the holographic electron microscope (USA) • Launch of the ATS-6, a direct broadcast TV satellite capable of linking remote communities (USA)	• *Apollo XVIII* (USA) and *Soyuz XIX* (USSR) couple in space • A thermonuclear power station, Tokamak 10, begins operation to see if power generated from nuclear fusion could become commercially viable (USSR)	• *Viking I* sends back photographs of the Martian landscape (USA) • *Palapa I*, the first of two geostationary satellites, provides the inhabitants of Indonesia's islands with TV, radio and telegraph communications	• Development of the neutron bomb (USA) • Bell Telephone Company use optical fibers to transmit TV signals • First gliding test flight of the Space Shuttle (USA) • Launch of space-probes *Voyager I* and *Voyager II* (USA)	• First geothermal power station in Tibet begins operation near Lhasa • *Seasat I* satellite is launched to measure sea surface temperatures, wind and wave movement, icebergs and ocean currents (USA)	• *Voyager I* and *Voyager II* explore the moons of Jupiter • Canada is the first country to operate a satellite TV broadcasting service • 28 Mar: Accident at the Three Mile Island nuclear power station caused by a defect in the cooling system (USA)	• First compact disks are produced by Philips (Neth) and Sony (Jap) • The first of the *Intelsat V* series of communications satellites is launched • H. Rohrer and G. Binnig develop the scanning tunneling microscope	• First flight of the Space Shuttle (USA) • Launch of IBM personal computer (PC) (USA) • SODAR (sonic detection and ranging) is installed at Frankfurt and other large West German airports to measure air currents
Medicine	• Discovery that certain "anti-antibodies" can be purified to prevent reactions against skin grafts (UK) • Discovery that X-ray irradiation of children can lead to the development of tumours (Isr)	• Successful treatment of infertile women with Bromocriptine, a fertility drug (UK)	• Emergence of a new drug, Cimetidine, for the treatment of peptic ulcers • Discovery that the drug Nootropyl enhances memory and learning performance by improving communications between the two hemispheres of the brain	• Praziquantal is developed to treat the debilitating parasitic disease bilharzia (FRG) • Last recorded case of wild smallpox found in Somalia • Discovery of the virus infection known as Lassa fever, after it sweeps through parts of Zaïre and Sudan	• 28 Jul: Birth of the first test tube baby. P. Steptoe and R. Edwardes fertilize the egg using sperm from the father and keep it in the test tube for 60 hours before implanting it (UK) • Discovery that "Legionnaires' disease" is caused by the bacterium *Legionella pneumophila*	• D. Rees develops leprosy vaccine (UK) • Smallpox is declared "eradicated" by the World Health Organization (WHO) • First observation of Acquired Immune Deficiency Syndrome (AIDS) • The hepatitis virus is cultured by P. Provost and M. Hilleman	• Dornier Medical Systems, Munich, develops the lithotripter, using sound waves to break up kidney stones • M. Cline attempts to treat thalassemia, an hereditary blood disease, by inserting a corrected version of the defective gene into the bone marrow, but this procedure fails to cure the disease (USA)	• AIDS recognized for the first time (USA) • Biotechnologically-grown insulin marketed (USA) • Sleeping sickness is combated by attaching the antibiotic daunorubicin to albumin or ferritin, two human proteins, which easily absorb it and increases its potency
Biology	• National Academy of Sciences calls a halt to research in genetic engineering until safer techniques are developed (USA)	• C. Milstein announces the results of his work in genetic engineering to create identical micro-organisms, or monoclonal antibodies (UK) • J. Hughes discovers morphine-like chemicals, named endorphins, in the brain	• Genentech, the first commercial company for developing products via genetic engineering, is established in San Francisco (USA) • Discovery of the blood-clot preventing characteristics of prostaglandin (UK)	• L.F. Sanger describes full sequence of bases in a viral DNA (UK) • After post-mortem studies of the brains of schizophrenics reveal chemical differences to normal brains, a disorder in dopamine is discovered (UK)	• Completion of the research into finding the genetic structure of virus SV40, enabling scientists also to find the structure of the virus's protein coat • S.C. Harrison announces the first high-resolution structure of an intact virus, the tomato bushy stunt virus		• US Supreme Court rules that a microbe developed for oil cleanup by General Electric can be patented • P. Faulk discovers the protein ferritin which surrounds the fetus and protects it from its mother's immune system (UK)	• First successful cloning of a fish, a golden carp (China) • First transfer of genes from one animal to another achieved by scientists at Ohio University (USA) • Essential blood proteins are genetically engineered in the laboratory (UK)
Physics	• J/psi particle first observed • H.M. Georgi and S.L. Glashow develop the first of the grand unified theories (GUTs)	• Discovery of the tau lepton, or tauon	• Introduction into physics of theories employing supergravity	• K. von Klitzing discovers the quantum Hall effect (FRG) • L. Lederman discovers the upsilon particle, confirming the quark theory of baryons	• Laser cooling of trapped positive ions demonstrated for the first time • W. Pauli succeeds in measuring the life of a neutron, about 15 minutes (USA)	• Observation of the gluon at the Deutsches Elektronen Synchotron (DESY) in Hamburg (FRG)	• Several research groups announce that neutrinos may possess a tiny mass, and this mass may represent the "missing mass" unaccounted for in the context of the Big Bang theory	• Chain of eleven carbon atoms detected in molecules in a star 600 light years distant, the biggest space molecule
Chemistry	• Microbes are used to synthesize hormones for contraceptive pills (Jap)	• *Rhizobium* – is adapted to "fix" nitrogen in the laboratory to the end of increasing the protein content of crops (Aus/Can)	• Givaudan-La Roche Icmesa pesticide plant near Seveso accidentally releases a cloud of poisonous gas spreading dioxins over 730 hectares	• A.J. Heeger and A.G. MacDiarmid discover that iodine doping makes polyacetylene an electrical conductor		• An Antarctic meteorite is found to contain traces of amino acids	• Micro-organisms are developed to ferment organic waste to produce alcohol	• A. Heller, B. Miller and F.A. Thiel announce a liquid junction cell that converts 11.5 percent of solar energy into electricity
Other	• F.S. Rowland and M. Molina warn of the effects of chlorofluorocarbons (CFCs) on the ozone layer	• V.C. Rubin and W. Kent discover the proper motion of the Milky Way, about 500 km per second (USA)	• T. Kibble proposes the idea of "cosmic strings", threads of energy surviving from the Big Bang	• Recognition of the Vela pulsar with a visible star (Aus)	• Pluto discovered to have a satellite, Charon • Development of a form of natural rubber that can be molded and recycled (UK)		• A.H. Guth proposes a modification to the Big Bang, known as the inflationary universe theory • *Voyager I* passes Saturn (USA)	• J.P. Cassinelli and colleagues discover the most massive star known, R136a, with a mass 2500 times greater than the Sun (USA)

1982	1983	1984	1985	1986	1987	1988	1989
• *Chem*: A. Klug (UK) • *Phys*: K.G. Wilson (USA) • *Med*: S.K. Bergström, B.I. Samuelsson (Swe), J.R. Vane (UK)	• *Chem*: H. Taube (USA) • *Phys*: W.A. Fowler, S. Chandrasekhar (USA) • *Med*: B. McClintock (USA)	• *Chem*: R.B. Merrifield (USA) • *Phys*: C. Rubbia (It), S. van der Meer (Neth) • *Med*: N.K. Jerne (Den), G.J.F. Köhler (FRG), C. Milstein (UK/Arg)	• *Chem*: H.A. Hauptman, J. Karle (USA) • *Phys*: K. von Klitzing (FRG) • *Med*: M.S. Brown, J.L. Goldstein (USA)	• *Chem*: D.R. Herschbach, Y.T. Lee (USA), J.C. Polanyi (Can) • *Phys*: E. Ruska, G. Binnig (FRG), H. Rohrer (Swi) • *Med*: S. Cohen (USA), R. Levi-Montalcini (It/USA)	• *Chem*: C. Pedersen, D. Cram (USA), J.-M. Lehn (Fr) • *Phys*: G. Bednorz (Swi), A. Müller (FRG) • *Med*: S. Tonegawa (Jap)	• *Chem*: J. Deisenhofer, R. Huber, H. Michel (FRG) • *Phys*: L.M. Lederman, M. Schwartz, J. Steinberger (USA) • *Med*: J.W. Black (UK), G. Elion, G. Hitchings (USA)	• *Chem*: T. Cech, S. Altmann (USA) • *Phys*: N. Ramsay, H. Dehmett (USA), Wolfgang Paul (FRG) • *Med*: H. Varmus, M. Bishop (USA)
• Introduction of compact disk players • Introduction of the supercomputers *Cray I* and *CYBER 205*, which can perform 100 million arithmetical operations per second (USA)	• Beginning of the Strategic Defence Initiative (SDI) program (USA) • First flight of *Spacelab*, an orbiting laboratory (USA) • First flight of European *Ariane*, a satellite-launching rocket	• Introduction of optical disks for data storage • IBM introduce a megabit RAM memory chip with four times the memory of previous chips	• AT&T Bell Laboratories success in sending the equivalent of 300,000 simultaneous phone conversations over a single optical fiber (USA)	• 28 Jan: Space Shuttle *Challenger* explodes shortly after takeoff, calling a temporary halt to Shuttle launches (USA) • European satellite *Giotto* passes close to Halley's comet • Nuclear reactor explosion at Chernobyl near Kiev (USSR)	• 9 Mar: The Numerical Aerodynamic Simulation Facility, a supercomputer capable of a top speed of 1,720,000,000 calculations per sec, begins operation • Work begins on the Channel Tunnel (UK/Fr) • Laying of a transatlantic glass fiber cable (TAT 8)	• The human-powered aircraft *Daedalus 88* sets a new record for human-powered flight (Gr) • F.C. Moon and R. Raj build an almost frictionless high-speed bearing using a superconductor • Stealth bomber developed by US Air Force	• West Germany hosts the "Solarmobile 1989", an exhibition of solar-powered cars • *Voyager II* reaches Neptune, sending back pictures to Earth (USA) • Dec: First use of the non-nuclear model of the Stealth bomber during the US invasion of Panama
• M. Epstein identifies the first virus implicated in human cancer, the Epstein–Barr (UK) • Successful treatment of severe infectious hepatitis with interferon (Isr) • First computerized heart pacemaker is fitted to a 49-year old man (UK)	• Identification of HIV retrovirus from which AIDS can result • J. Buster and M. Bustillo perform the first successful human embryo transfers (USA) • A. Weinberg discovers that cancer results from the combined effect of two mutated genes (USA)	• American Heart Association lists smoking as a risk factor for strokes for the first time • First successful surgery on a fetus before birth by W.H. Clewall (USA) • World Congress of In Vitro Fertilization and Embryo Transfer in Helsinki (Fin)	• B.L. Vallee and colleagues discover the tumor angiogenesis, renamed angiogenin • Gene marker for polycystic kidney disease found on chromosome 16, and a gene marker for cystic fibrosis on chromosome 7 • Lasers first used to clean out clogged arteries (USA)	• L. Kunkel and colleagues discover the gene defective in Duchenne muscular dystrophy • The electronic microscope shows the AIDS virus for the first time (USA) • New vaccine against hepatitis-B is produced by genetic engineering	• Discovery by K.P. Campbell and R. Coronado of the calcium release channel, a protein used to regulate the passage of calcium into and out of muscle cells • Discovery of a gene marker for cancer of the colon by W. Bodmer, E. Solomon, H.J.R. Bussey and A.J. Jeffreys	• R. Jaenisch and colleagues succeed in implanting the gene for a hereditary disease of humans in mice, in order to study the disease • World Health Organization (WHO) estimate HIV infected to be over one million, with 120,000 cases of AIDS reported worldwide	• First successful operation, by M. Harrison and colleagues, on a fetus removed from the womb then returned after lung surgery (USA) • May–Jul: Genetically engineered white blood cells are transferred into cancer patients for the first time, in order to attack tumors (UK)
• Gene for rat growth hormone is transferred to mice, the first time a gene from one mammal has functioned in another mammal • J. Deisenhofer, R. Huber and H. Michel study the structure of the bacterial protein complex (FRG)	• A.W. Murray and J.W. Szostak create the first artificial chromosome • W.J. Gehring and colleagues discover the homeobox, a sequence of genes found in many organisms directing the development of the organism	• Successful cloning of sheep by S.A. Willadsen • Development of DNA fingerprinting by A. Jeffreys (UK) • A. Wilson and R. Higuchi are first to clone genes from an extinct species, the quagga, a kind of zebra (USA)	• J. Deisenhofer, R. Huber and H. Michel succeed in determining the exact arrangement of the more than 10,000 atoms composing the protein complex (FRG) • Swedish scientists succeed in cloning the DNA of 2400-year old mummies	• R.A. Weinberg and colleagues discover the first gene known to inhibit growth, in this case of the cancer retinoblastoma (USA)	• H. Fricke uses a submersible to study coelacanths in the Indian Ocean • The gene for maleness discovered by D.C. Page • Genes for human growth hormone inserted into goldfish and loach, and result in much quicker growth (China)	• Apr: Patent issued to Harvard University for a strain of genetically-engineered mice, developed to discover the causes of cancer (USA) • First forensic use of DNA fingerprinting • Beginning of a major project by US scientists to map human genes	
• B. Cabrera claims to have detected the single magnetic monopole, predicted in theory, but this is not corroborated by further experiments	• First observation of W and Z particles by C. Rubbia and colleagues at CERN, confirming the electroweak theory	• D. Shechtman, I. Blech, D. Gratias and J.W. Cahn discover the first quasicrystal (USA) • C. Rubbia and colleagues at CERN announce their evidence for the top quark and for monojets	• Tevatron particle accelerator at Fermilab begins operation (USA) • Testing of laser-fired nuclear fusion for energy generation (Jap/USA)	• First "high temperature" superconductors, at 30 degrees K, found by K.A. Müller and G. Bednorz • Individual quantum jumps in individual atoms observed for the first time (USA/FRG)	• M.K. Moe, A.A. Hahn and S.R. Elliot observe the double beta decay of selenium-82 • Ching Wu Chu and colleagues create a material superconducting at −196 degrees celsius, the temperature of liquid nitrogen	• First image obtained from a positron transmission microscope • New warm-temperature superconductors developed based on bismuth (Jap) and thallium (USA)	• Aug: First tests conducted with the new LEP particle accelerator (Fr/Swi) • S. Pons and M. Fleischmann claim to have succeeded in achieving "cold fusion", but their experiments are not independently corroborated (UK)
• 29 Aug: Creation of a single atom of element 109 (FRG)	• Aspartame is approved for use as an artificial sweetener in soft drinks	• Creation of three atoms of element 108 (FRG)	• Information on lanxides, crosses between ceramics and metals, is released when the US Defense Department declassifies this subject		• H. Naarman and N. Theophilou develop a form of polyacetylene that is doped with iodine to become a better conductor of electricity than copper	• Development of embryo cloning of dairy cattle (USA)	• G. Winter produces new types of antibodies capable of being used as biodegradable pesticides (UK)
	• Method developed for dating ancient objects based on chemical changes in obsidian • Japanese scientists discover a dust circle around the sun	• Two more rings of Saturn discovered by J.C. Bhattacharyya and colleagues (Ind) • Partial ring around Neptune discovered by European Southern Observatory (Chile)		• Discovery that the Milky Way and galaxies of the local group, as well as parts of the local supercluster of galaxies, are moving toward a point in the direction of the Southern Cross	• B. Tully announces his discovery of the Pisces-Cetus Supercluster Complex • Confirmation of the existence of an ozone hole by scientists flying over Antarctica	• US Senate is first to ratify a treaty intended to reduce the use of CFCs	

Datafile

High-power computers have developed rapidly and are now used extensively in many areas of business and leisure. Computer screens now dominate the financial world. Word processors have revolutionized the office. Desktop and laptop computers bring individual computing power to managers, sales people and technical staff. The average word processor now has more memory capacity than the computers that were used on the Apollo spacecraft of the late 1960s and early 1970s. The period since the oil crisis of 1973 has seen a greater awareness of the problems of energy production and the disposal of waste. Much concern has been expressed about the possible exhaustion of nonrenewable energy sources. Effort has been put into examining sources of renewable energy, such as wave, tide and wind power. Nuclear power has continued to grow in importance but the accident at Three Mile Island in the United States (1979) and the explosion at Chernobyl in the Soviet Union (1986) have set back programs and reduced public confidence.

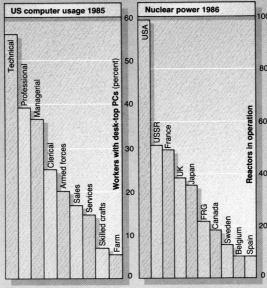

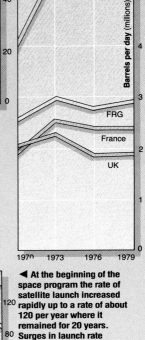

◀▲ **Computer usage has increased greatly during the 1980s throughout Europe, the United States and Japan with the introduction of personal desktop machines. A highly successful marketing effort has also introduced computers at home, although a survey in the United States in 1984 showed that only half of the owners actually used them.**

▲▶ **The rapid growth in oil consumption in the industrialized world was dramatically curtailed in 1973 by the decision of the exporters of the Arab world and their supporters to increase their prices by 70 percent and cut production. This turned interest toward nuclear power; many countries developed reactor-building programs.**

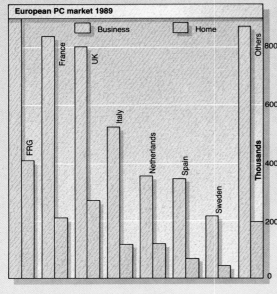

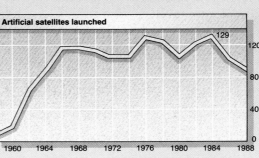

◀ **At the beginning of the space program the rate of satellite launch increased rapidly up to a rate of about 120 per year where it remained for 20 years. Surges in launch rate occurred in 1975 and 1981 with the start of the Skylab and Shuttle programs, but fell after the destruction of the Space Shuttle *Challenger*.**

▶ **Global warming is of great concern. The gas carbon dioxide is produced from fossil-fueled power stations and by various industrial processes. In the atmosphere carbon dioxide retains the Earth's heat. Carbon dioxide is consumed by plants in photosynthesis but the continuing destruction of vast forest areas in South America reduces the planet's capacity to cope with the carbon dioxide produced. The temperature of the Earth could slowly rise.**

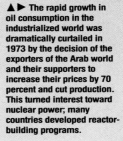

▶ **In 1987 Japan spent the same proportion of its national income on R&D as did the United States (3.3 percent). However, in the USA and in other high-spending nations (Soviet Union, UK, Germany and France) between 40 and 50 percent of this was from government funds whereas in Japan only 20 percent was from public money. In Japan industrial companies are encouraged to compete for government contracts based on their internally funded R&D.**

THE COMPUTER-DRIVEN WORLD

The oil crisis of 1973

Energy conservation

Underwater exploration for important minerals

Science and technology in Japan

The impact of computers

Space technology: the development of the American Space Shuttle

In the late 1970s and 1980s three main issues dominated the international scene: the energy crisis of 1973 and its effects; the environment; and the phenomenal developments in "chip"-controlled electronic devices.

The energy crisis arose in 1973 when several Arab states, working through the Organization of Petroleum Exporting Countries (OPEC), severely curtailed the export of oil to Western Europe, Japan, and the USA and simultaneously marked up the price of crude oil severely in retaliation for some countries' support of Israel against an Arab attack. The economic onslaught was sudden and in the short term the industrialized countries, which had failed to provide adequate safeguards against their excessive dependence on supplies of oil not under their control, were quite literally held over a barrel. Four years later, in 1977, President Carter of the USA still had to warn the American people that they must pursue economy in the use of oil as the "moral equivalent of war".

▼ One of the major problems of the nuclear industry is the disposal of radioactive waste. Here liquid waste is being disposed of in double-walled steel tanks at Hanford, USA, each holding 4.5 million liters.

In the medium term there were three main lines of reaction: first, to encourage the economical use of fuel, by avoiding waste, which was often prodigal, and by devising more efficient "lean burn" engines for automobiles and trucks; second, to exploit new oilfields as quickly as possible (for example, the 1,300km Alaska pipeline was opened in 1977) and to reopen others which had become uneconomic when the price was low; third, to explore the possibility of making greater use of alternative sources of power, in particular nuclear energy and renewable sources of energy such as wind and tide. Practical arguments were reinforced by moral ones: it was widely urged that fossil fuels, as nonrenewable assets, should be conserved for the benefit of future generations.

Not altogether surprisingly, these moral arguments carried less weight as the crisis eased and when it finally blew itself out (when world oil prices crashed at the beginning of 1987).

▶ The oil crisis of 1973 accelerated the exploitation of oil under the North Sea, with consequent developments in the technology and techniques needed for work in deep waters. Here the supports for a drilling platform are under tow off Norway.

▼ The Chernobyl disaster of April 1986, when a Soviet nuclear reactor exploded, released a vast quantity of highly radioactive material and created contamination of an unprecedented scale.

▼ The 1973 oil crisis focused attention on energy conservation and on ways of improving the insulation of buildings. This thermograph shows heat loss from a house. The color coding ranges from white to orange for the warmest areas through to green and blue for the coolest. The roof is well insulated but the greatest heat loss is through the single-glazed windows.

Energy conservation

Energy conservation could be effected at two levels. In the short term, some savings could be made simply by general exhortations – promoted by government publicity campaigns – to use less energy in everyday life. Typical examples were to reduce domestic heat losses by double-glazing and roof and pipe insulation; to use less hot water for baths and other household uses; to restrict the use of automobiles to essential journeys; to avoid unnecessary lighting. While much could be achieved by voluntary cooperation, this could be reinforced by some forms of rationing, by price increases, or – conversely – by grants toward the cost of installing energy-saving equipment. It was, indeed, the traditional stick and carrot policy.

All this proved helpful, but long-term strategies had also to be developed. These took many forms, from improving the performance of automobile and truck engines so that they required less fuel, to developing detergents effective in cool water and improving the basic design of buildings to reduce heating requirements. These measures were the easier to introduce because they were in tune with the ideas of the increasingly influential environmentalist lobby. If there was, for example, less demand for electricity, the emission of destructive waste gases from power stations would be reduced, with corresponding benefit to forests suffering from acid rain and to the protective ozone layer in the upper atmosphere.

But perhaps the most valuable consequence of this appraisal of energy policy was recognition that in some way or another it must figure in the balance sheet for a wide range of industrial operations. In agriculture, for example, the energy cost of spreading fertilizer does not stop with the fuel directly consumed. The tractor itself and associated equipment represent a measure of energy input, as does the manufacture and transportation of the fertilizer. Again, in house-building there are many hidden energy inputs distinct from those involved on site. These include the firing and transportation of bricks; the felling and sawing of timber; the leveling of the site, and so on. In the interests of conservation, such inbuilt energy factors had to be considered when alternative procedures were evaluated.

Mineral resources

Rather surprisingly, while much concern has been expressed about the depletion of fossil fuels, there has been much less about the equally steady depletion of other mineral resources. Essentially, these fall into two categories. First, minerals used as such in bulk. These include sand, gravel, and stone essential for building and civil engineering, and phosphates and potassium salts essential as agricultural fertilizers. Worldwide, these are so abundant that serious shortage need not be considered but there could be strategic considerations. The USA and USSR are self-sufficient in phosphates, but most European countries have to rely on the rich deposits of North Africa and certain Pacific islands.

The situation is very different, however, in respect of minerals which are a source of metals. Here again there are important strategic considerations. The bulk of the world's chromium ore, for example, comes from Turkey, Africa, the Soviet Union, and the Philippines: the world's greatest single resource lies in the Great Dyke of Zimbabwe, one of the world's most unusual geological features. But wherever located, these mineral resources are not renewable, and their ultimate exhaustion is just as certain as that of coal and oil. True, the two situations are not strictly comparable, in that metals can to some extent be recovered by recycling processes, but there is nevertheless a steady erosion. First, much waste is simply discarded and permanently disposed of in tips. Secondly, recycling processes are not 100 percent efficient, and in every cycle something is lost: moreover, recycling always requires energy.

In the nature of things, mines become exhausted and have to be closed, so new sources of minerals are constantly being sought, but not necessarily for immediate development. Traditionally, the search has been made on land, but in the 1970s and 1980s increasing attention was directed to possible submarine resources. Apart from large quantities of sand and gravel dredged up in shallow water, cassiterite (tin ore) was recovered from the seabed off Indonesia and Thailand; sulfur in the Gulf of Mexico; ironsands, containing rare elements such as zirconium, off the Philippines; and diamonds off Namibia.

Additionally, many other large deposits of interesting minerals have been located. Thus deposits of phosphorite suitable for fertilizer manufacture, totalling 100 million tonnes, lie at a depth of 400m (1,300ft) on the Chatham Rise off the east coast of New Zealand. More impressive are the vast deposits of manganese nodules, totalling perhaps a trillion tonnes, lying at depths of up to 4,000m (13,100ft) in the southwestern Pacific: these also contain nickel, copper and cobalt. Such deposits need sophisticated capital-intensive techniques to mine them, and the viability of exploitation depends on the market price of similar ores obtained from land sources.

Science in Japan: Isolation to Integration

In the late 20th century Japanese science has been closely identified for many with the country's impressive technological achievements and economic prosperity. The actual historical record in much of the century has shown, however, a constant struggle by Japanese scientists to justify their work as creators of knowledge in an atmosphere of relative public indifference and substandard funding. Scientists have constantly been on call as consultants on foreign technology. Except in favored research domains (especially those related to energy needs) they have had to make do with meager resources. Attitudes have greatly changed, particularly since the energy crisis of 1973, but their effects linger and continue to influence policy, which is to produce more discoveries and inventions of Nobel prize caliber and eliminate the debilitating, long-term pattern of isolation in science.

The two world wars had a profound effect on Japanese science. The first (1914–18) broke the earlier pattern of extreme dependence on foreign science (which had resulted from the opening of the country in the 1850s). A new imperial university (Hokkaido) was built in 1918 (four such institutions had been established earlier, beginning with Tokyo University in 1877). The prestigious Research Institute for Physics and Chemistry (RIPC) was established in 1917, and a systematic program of grants for all forms of scientific and technical research was inaugurated in 1918. Progress slowed down in the 1920s, a period of relative economic stagnation for Japan;

but militarism and growing international isolation in the 1930s were beneficial to science, particularly through the creation in 1932 of a major funding mechanism for large-scale projects, called the Japan Society for the Promotion of Science. Military defeat in 1945 naturally brought changes (the RIPC's cyclotron was destroyed by the US army). There was some resurgence of the earlier attitude of dependence on foreign science and technology, and applied research was given explicit priority over basic research, even in policy declarations by US military officials. Moreover, many Japanese scientists (including the country's first Nobel laureate, Hideki Yukawa) became relatively estranged from the postwar government and from private business interests which they considered to be tainted by association with wartime policies.

Despite many obstacles, Japanese researchers have managed to win five Nobel prizes, all since World War II. Hideki Yukawa (1949), Shin'ichirō Tomonaga (1965) and Reona Esaki (1973) won in physics; Kenichi Fukui (1981) in chemistry and Susumu Tonegawa (1987) in medicine. Leaving aside the work of Tonegawa on antibody production, which was done outside Japan, the dominant feature of this work is its highly theoretical character, displaying throughout a considerable virtuosity in mathematics. Indeed, mathematics and theoretical physics have been the fields of greatest strength in Japanese science, a fact which may well be related to the historical pattern of modest research funding.

▲ A technology fair at Tsukuba, near Tokyo, 1985. Since World War II business has been a powerful force in the development of science and technology in Japan. In the 1950s business, in pursuit of its economic goals, successfully pressured the government to increase university admission quotas for engineering and science. In the early 1990s Japanese business provided about two-thirds of research funds. It has also promoted new patterns of cooperation with academic scientists, in such ambitious research schemes as the Fifth Generation Computer Project and a program of research in superconductivity.

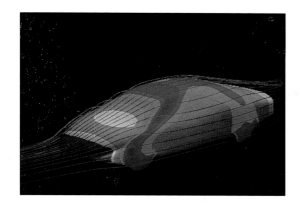

The depths of the Red Sea also contain some unusual mineral deposits. These are hot brines, rich in salt, which have welled up through the sea bed but then remained in deep pools stratified, because of their relatively high density, from the main body of the sea water. They contain manganese, iron, zinc and copper. Again, a decision to exploit them depends on economic as well as technical considerations, but feasibility tests by a specialist German mining company suggested that – particularly in view of their high zinc content – they could become an important resource. For such purposes samples can often be collected by remote-controlled devices. In 1970, for example, the USA introduced a Remote Underwater Manipulator (RUM), a vehicle with caterpillar tracks capable of being controlled from the surface at depths up to 3,000m (9,840ft).

Computers and information technology

One of the most significant, but scarcely noticed, events in the history of computers took place in 1976: the German firm of Keuffel and Esser made their last slide-rule (for 350 years the popular individual calculator) and presented it as a museum piece to the Smithsonian Institute in Washington. The reason was a dramatic change of scale in the computer industry. Until the late 1960s the scene was dominated by mainframe systems which often cost several million dollars and were operated by a considerable staff to whom users (customers) initially had to present their programs for processing, usually in punched-type form. More flexibility was introduced with the introduction of extension (remote job entry) points to which local users could bring their programs, followed by terminals plugged into the ordinary telephone network.

All this was changed with the advent of the mass-produced microcomputer in the 1970s, first for the amateur and later – such as the American Apple II in 1977 – increasingly for professionals. With equipment costing a few hundred dollars operators now had literally at their fingertips computer power for which previously they had had to wait hours or even days. Where extra power was needed the microcomputer could still be keyed into a mainframe system. Miniaturization was further developed in the early 1980s with the appearances of battery-operated laptop portables: for around a thousand dollars a wide range of facilities was available in a package the size of a briefcase. Provided a telephone terminal was available these computers could feed data acquired in the field into mainframe systems.

In the late 1980s and early 1990s emphasis was on the so-called fifth generation computers, nonnumeric supercomputers which can be addressed in a natural language. They embody artificial intelligence and can communicate directly with the human mind on a logical basis. Although interest is widespread, the present leaders in this field are the Japanese, who started an intensive 10-year FGCS (Fifth Generation Computer Systems) research programme in 1982. It is a joint enterprise between the Ministry of International Trade and Industry (MITI) and eight leading electronic manufacturers, designed to meet the needs of the 1990s. The aim is to promote the internationalization of society, and of Japanese society in particular, by cooperation through a global computer network embodying, among other functions, machine translation from one language to another and the processing of other nonnumeric data such as speed, pictures and graphical representations. Such supercomputers will have an important role, for example, in controlling nonstandarized operations in tertiary industry; developing educational, medical and other support systems; and in minimizing energy consumption.

Japan (and also, of course, China) has a particularly difficult problem in communicating with other countries because of the nature of its written language, but the European Community (EC) countries also have their problems even though most use a common alphabet.

At its present size the number of combinations of pairs of EC languages has risen to 72, and hundreds of translators have to be employed at the Community's centers in Luxembourg and Brussels to deal with a million pages of text that have to be processed each year. There is, therefore, keen interest in the possibility of machine translation, and this led, in 1983, to the launch of the Eurotran project, a sophisticated scheme working not on a one-to-one language basis but from a number of source languages into a number

◄ Computerized design: General Motors in the USA uses a supercomputer to simulate aerodynamics.

▼ Photons replace electrons. Glass fibers can carry light signals with less interference and greater capacity than possible with electric cables.

▲▼ Computers controlling machine tools enable a remarkable variety of industrial operation to be performed automatically. Above, "chips" are tested; below, robot welders are uniting components in a car assembly line.

of product languages. Although much needs to be done, it is currently the world's most advanced machine translation system. It grew out of an earlier Systran system developed in the United States in the late 1950s when the Soviet conquest of space made it necessary to monitor all the relevant Soviet radio communications. It was later developed to work with other pairs of language drawn from French, English, Dutch, German and Italian, but no more than 80 percent intelligibility has been achieved: the final texts therefore still need much polishing.

A reusable space vehicle: the US Shuttle

In the early days of space explanation, when the objectives were as much political as scientific and technological, expense was no object. Ultimately, the United States Apollo Program cost around 25 billion US dollars. It landed 21 men on the Moon; the return fare was over a billion dollars each. Not unnaturally, there were serious misgivings as to whether such expenditure was justifiable: ex-President Dwight Eisenhower described the project as "nuts" and the launch of Apollo 2 was picketed by a "Poor People's March".

Many factors contributed to this enormous cost: the research and development program required to solve problems never before encountered; the ground-based hardware for launching and communication; and the recovery process on return to Earth.

In retrospect, the latter was extraordinary cumbersome. When the tiny command module of the Apollo space system reentered the stratosphere a giant parachute opened and the capsule splashed down in the Pacific Ocean, where a small flotilla of naval vessels, including an aircraft carrier, awaited it. The whole process has been likened to sinking a trans-Atlantic liner after its maiden voyage and building a new one for the next. At the start of every Apollo flight 200 million dollars worth of Saturn launching rocket disappeared into the ocean depths. The Soviets operated a similar procedure, save that their capsules parachuted onto land.

August 1977 saw the beginning of a new era in space travel, with the advent of the US Space Transportation System (STS), better known as the Shuttle. It took ten years to develop and cost about 7 billion dollars, but it can effect great

▲ It is hard to believe that less than quarter of a century separate Vostok 1, in which Yuri Gagarin made the first space flight in 1961, and the American Shuttle, put into service in 1984. In the former the single occupant was strapped into a tiny capsule. By contrast the Space Shuttle *Challenger* – here seen launching a satellite over California – had comfortable accommodation for a crew of seven, a 20m (65ft) cargo hold, and a carrying capacity of 40 tonnes.

◀ **Flight deck of the Shuttle *Columbia*. The pilot sits on the right and the commander on the left, facing a highly complex instrument panel. Between them are the navigation aids console and the flight computer.**

▼ **The launching of the Shuttle (seen here leaving its hanger) – and, indeed, of any spacecraft – demands a complex launching pad and an elaborate prelaunch countdown to check that every system is working satisfactorily. On many occasions flights have had to be aborted within seconds of firing.**

economies. It has been calculated that the Explorer 1 satellite cost 225,000 dollars for one orbit of a kilogram load; the corresponding figure for the shuttle is around 225 dollars – a thousand times less.

This dramatic saving must be attributed to many factors, such as the writing-off of overheads, more efficient technology and so on, but the major one is that the space craft can be recycled, supposedly at least 100 times. Like earlier craft, it is launched by rockets, but when their fuel is exhausted the booster rockets are detached and are parachuted down to the Atlantic where they are recovered for reuse. Only the relatively inexpensive fuel tank is lost. Once in space the Shuttle continues in orbit like any other space craft, but it is equipped with stubby wings and on reentry can glide back to its base and land – on a very long airstrip – like a conventional aircraft, at around 300km (200mi) per hour.

Although separated by only a quarter of a century from Vostok 1, in which Yuri Gagarin made the first-ever manned space flight, the contrast with the Shuttle is remarkable. It has comfortable quarters for a crew of seven – up to ten if necessary – and can carry loads of up to 40 tonnes in a cargo hold 20m (65ft) long. Apart from making government-sponsored flights to put military observation satellites into orbit, the original Shuttle craft earned big fees for launching communica-

tion and meteorological satellites into space for private companies.

Sadly, the Shuttle suffered a severe setback in January 1986, when one of the then four craft in use (the *Challenger*) exploded just after takeoff, killing all seven on board. Flights were immediately suspended pending a detailed inquiry and they were not resumed until 1989. Technically, the fault seems to have been very simple: low ambient temperatures had caused a plastic gasket to fail, releasing large quantities of fuel which were ignited by the booster rockets. But the inquiry also revealed serious defects in the organization of NASA and so a radical reorganization was ordered.

The three-year shut-down of Shuttle flights seriously disrupted many programs involving the launch of massive satellites, for no other available launcher approached its capacity. After a hesitant start the unmanned European launcher "Ariane" (developed with strong support from France) proved its reliability but its payload was only around 2,000kg (4,400lb). China and India also developed small launchers, which were available to a limited extent to other countries.

Japan, too, developed an ambitious space program in the 1970s and 1980s. At first they had to rely heavily on US technology but increasingly they developed their own resources through the National Space Development Agency, (NASDA). Their initial need was for small application satellites for which they used the N-1 rocket, based on US technology, with a payload of 130kg (290lb), this was phased out in 1982 after seven unsuccessful launches.

It was followed in 1986 by the H-1 launcher, a three-stage rocket embracing much indigenous technology and capable of carrying a payload of 550kg (1,200lb). Already, however, they had initiated in 1985 a much more ambitious H-11 rocket, expected to become operational in 1992, to be economically competitive and capable of carrying a two-tonne payload. Looking further ahead still, there were plans for a small recoverable shuttle-plane (HOPE) to ride into space atop an H-11 rocket. Work was also being done with a Japanese Experiment Module (JEM) to link up with an international space station.

Datafile

The 1980s were a decade of contrasts in the biomedical sciences. Research budgets in many countries remained large but their rate of increase on occasions fell below that of inflation. On the other hand there was new public skepticism of science and technology, which gave rise to ecological concerns. Ecology itself was the subject of ambitious research programs on a global scale, concerned especially with climate and the atmosphere. Further discoveries in the life sciences, as in genetic engineering, prompted increased controls on experimentation and more discussion of the ethical and legal implications. The advent of AIDS in the 1980s provided a new challenge to medical research.

AIDS cases

▲ The origins of AIDS are obscure. The syndrome was first diagnosed in San Francisco. Improved diagnosis has revealed increasing numbers of cases in certain parts of Africa, where the condition is now spreading the fastest.

US hospital care of AIDS patients 1984–86

◄ There is a high incidence of AIDS among young people in their 20s and 30s – an age group which otherwise has low mortality rates. In the USA the expenditure on treating young AIDS patients has increased rapidly.

Causes of mortality

USA 1980

Sweden 1980

Japan 1980

- Heart disease
- Cancer
- Influenza/pneumonia etc
- Other

◄ These death rates in three countries enjoying high prosperity illustrate similar trends on different continents. While the cancer (tumor) rates are very much the same, differences in the incidence of heart diseases have been ascribed to such dietary factors as the Japanese eating more fish. It is also of interest to note that these countries have very different forms of health-care provision, Sweden's being the most highly socialized and the United States' very dependent on private funding.

Israeli university R&D

- Medicine
- Nat. sciences/maths

Israeli R&D funding 1985–86

Total 911,100,000 shekels

- Industry/agriculture
- Universities
- Government/nonprofit institutions

◄▲ The state of Israel, founded in 1948, has been one of the inheritors of the high value placed on learning by Jewish culture. From 1970 until 1983 expenditure on medical research increased 2.7 times and the substantially greater expenditure on natural sciences more than doubled (above; in New Israeli Shekels). National expenditure on civilian research (left) came in roughly equal proportions from government and universities on the one hand, and from industry and agriculture on the other.

From around 1980 the immensely broad spectrum of the physical sciences – ranging from the ultimate atomic particles to the remote reaches of the universe – began to be echoed in the biological sciences. While molecular biology remained a fertile field for new and useful discovery, there was increasing interest in the environment, originally at the local and regional level but latterly on a global basis. Fears grew that the technological revolution was creating changes in the atmosphere sufficient to alter the world's climate, with far-reaching biological consequences.

At the molecular biological level scientists built on Stanley Cohen and Herbert Boyer's advance of 1973, when they spliced together fragments of DNA present in some bacterial cells – notably *Escherichia coli* – outside the cell nucleus. Their "recombinant" DNA could be introduced into an *E. coli* strain, thereby conferring on it new functional powers. As the bacteria are easily grown in quantity this technique could be used to produce, relatively cheaply, important products such as insulin, hitherto available only from biological sources such as pancreas from abattoirs.

But from the early 1980s it was the environment that excited real public interest in the biological field. The concern was both aesthetic and practical. Polluted rivers and beaches, unsightly waste tips, destruction of wildlife, ill-planned industrial and urban development, were all unpleasing evidence of man's neglect of the environment. But more serious, in the long term, was the so-called "greenhouse effect" on the global climate.

Of the atmospheric gases concerned, one of the most important is carbon dioxide, released by animals in respiration, by plants during the hours of darkness, and in every form of combustion. The latter has been a particularly important factor since the Industrial Revolution, with the increasing use of coal as a fuel. This was followed at the turn of the century by the use of oil, most particularly for automobiles, but also for power stations and other industrial installations. As carbon dioxide builds up in the atmosphere it prevents the escape of reflected radiation from the Earth. The feared overall effect is a steady rise in the average atmospheric temperature, which could lead to some thawing of the polar ice caps, which in turn could raise the average sea level. Estimates cannot be precise, but an increase of 2.5°C – possible by 2030 AD – could raise the level of the oceans by as much as 30cm (12in). This would be disastrous for low-lying countries such as Bangladesh. Additionally, there would be far-reaching effects on seminatural and natural vegetation, and on organized agriculture and forestry.

If true, these are very real perils for the human race but they must be viewed in the context of great fluctuations in the world climate which

THE LIMITS OF PROGRESS

occurred before any of these man-made factors existed at all. Apart from the great Ice Ages of geological time, there was a significant warming of the atmosphere beginning some 10,000 years ago, followed by a little ice age lasting from 1550 to 1850. The present task is to distinguish such variations – which are, of course, none the less serious because they are natural and uncontrollable – from those resulting from human activities, which might be controlled. To try to clarify the situation various international initiatives have been launched. They include WCRP (World Climate Research Program), a 20-year project launched in 1980 to determine how far climate can be predicted and man's influence on it; TOGA (Tropical Oceans and Global Atmosphere) a 10-year program from 1985, to study the interaction between the atmosphere and the oceans; and IGBP (International Geosphere-Biosphere Program) with the ambitious aim of "describing and understanding the interactive physical,

chemical, and biological processes that regulate the total Earth system".

Other changes in the gaseous context of the upper atmosphere – collectively referred to as the "greenhouse gases" – may also have long-term effects on the climate. The most important, and certainly the most intensively studied, is ozone. This is a form of oxygen in which the molecule contains three atoms instead of two. It is highly toxic and very reactive. It is a normal constituent of the lower atmosphere but – except in the vicinity of electrical discharges – in concentrations too low to be harmful. Its climatological significance lies in the fact that there is a layer of ozone in the stratosphere at a height ranging from 10 to 50km (6–31mi). It is a powerful absorbent of ultraviolet light, which in excess can be harmful, and there is concern that the ozone layer has been becoming thinner over the last ten years, especially over the Antarctic and probably also over the Arctic. This can have two effects. First, exposure to the more

▼ Atmospheric pollution comes from many sources, but worldwide some of the worst offenders are power stations burning fossil fuel. Their chimneys and towers discharge soot, carbon dioxide and oxides of sulfur.

Pragmatism and Coherence: Science in Israel

Since 1989 Israel has been recognized as a technological mini-superpower. That year it launched a satellite using its own ICBM. It is also credited with the possession of nuclear weapons.

From the outset, the Zionist movement was aware of the importance of science to the creation of a Jewish state in Palestine. Before 1948 Jewish science in Palestine focused on agriculture, the medical sciences and the basic technological requirements of the Zionist armed forces. Three major institutions were established before 1948: the Hebrew University of Jerusalem, 1925; the Israel Institute of Technology, 1925; and the Weizmann Institute, 1934. Very soon after the state of Israel was established in 1948, the Ministry of Defence launched its Nuclear Weapons Program. This was an early expression of a continuing trait of Israeli science: to give heavy emphasis to military applications.

The Israeli political, military and scientific communities have set their priorities in a coherent and pragmatic fashion which has led to an orderly evolution of the scientific establishment and the pursuit of stable national scientific policies. The extensive participation of Israeli scientists in the world's "invisible colleges" has made it possible for them to mobilize international science in the service of national scientific policies.

A new phase was ushered in when the recommendations of a government committee (the Katchalski Committee) began to be implemented after 1967. These focused on the promotion of civilian science-based industries with a view to deriving economic returns. The military-industrial complex (MIC) that had emerged by then continued to expand in parallel; by 1985 the MIC employed 10 percent of the national labour force or 25 percent of industrial labor. More than one quarter of Israel's industrial exports were either weapons or security services.

Subsidies from the United States government (for the Lavi-fighter and the Arrow projects, and others) went a long way to supporting the expansion of giant "companies" such as the Israel Aircraft Industries and Koor. Another indirect form of support was the purchase by the US government of Israeli services and products in its military foreign aid programs in the Third World. But by 1989–90, there was a crisis situation in Israel's MIC's and associated industries. This was largely due to Israel's small economic base, changes in the fortunes of its clients in the Third World, and the inability of Israeli nuclear and space technologies to compete with the more advanced products developed elsewhere.

Israeli scientists undertake research work in all front-line fields of science and technology and have made notable contributions. There is a heavy emphasis in the civilian research programs on the agricultural, biological and medical sciences. Applied research programs emphasize electronics and aerospace.

There are about 5,000 publishing scientists in Israel. The Hebrew University of Jerusalem accounts for a quarter of the total, thus making Jerusalem the largest center of scientific activity in the country.

▼ Rain forests are important regulators of the world's climate and their wholescale destruction has serious long-term effects, though these cannot at present be quantified. In Brazil much land is cleared by "slash and burn" methods, as seen here, and crops are raised for a few years before the land is exhausted and eroded. The nomadic farmers then move on and repeat the process elsewhere. In other parts of Brazil much land is being eroded by the use of hydraulic methods for mining gold.

intense radiation can substantially increase the risk of skin cancer. It may also cause genetic effects, because it is absorbed by DNA. Second, if more ultraviolet light reaches the surface of the Earth and at the same time increasing amounts of carbon dioxide prevent heat escaping, the greenhouse effect will be enhanced.

There is evidence that the changes in the ozone layer are, to some extent at least, a consequence of human activity. In particular, it is linked to the use of chemicals known as chlorofluorocarbons, (CFCs) which are widely used in aerosol sprays, refrigerators, dry cleaning and blowing foamed plastic. They accummulate high in the stratosphere and there form chlorine and oxides of chlorine which react with ozone and destroy it. In 1987, 24 countries signed a protocol which will limit the expansion of production of CFCs, aiming at a 50 percent cut by 1996. Many fear that this will be inadequate, especially as many countries are likely to ignore the protocol.

The environmental lobby

These developments were influenced by the increasingly strong environmental lobby, which manifested itself in various ways. There were those who, very properly, were concerned that technological development should recognize society's need to have its health and the quality of its environment properly safeguarded. In this there was, of course, nothing new. Britain, for example, brought in the first Alkali Act in 1863 specifically to limit the severe air and water pollution resulting from the manufacture of soda by the Leblanc process. But in the second half of the 20th century a vociferous minority was demanding far more restrictive, and sometimes quite unrealistic, legislation. One manifestation of this was a new political phenomenon: the emergence of "green" parties who appealed to the electorate largely on the single issue of environmental conservation. They found no lack of targets: agrochemicals, leaded petrol, pharmaceuticals, radioactive waste disposal and whaling.

Such conservationists directed their attention particularly to the power industries, which they attacked on two fronts. On the one hand, they attacked conventional power stations burning fossil fuel on the ground that their waste gases produced "acid rain" which could cause severe damage to trees and other plant life at great distances, and also poison rivers. These waste gases – notably carbon dioxide and nitrogen oxides – were later identified as a contributory factor in the so-called "greenhouse effect". On the other, they organized resistance to any form of nuclear power and in Sweden persuaded the government to abandon its nuclear power program altogether. Their arguments tended to be emotive rather than informed, qualitative rather than quantitative, but nonetheless persuasive: the major disaster at Chernobyl in the Soviet Union (when a nuclear reactor exploded) in 1986 greatly strengthened their hand. Attention was drawn, too, to the risk of climatic changes inherent in the reckless felling of the rain forest.

Automobile exhaust gases were similarly identified as a serious and increasing risk, with the

▲ Whales rank high among the world's endangered species and many international agreements have been made to conserve them. But enforcement is difficult, especially when, in view of the economic importance of the whaling industry, loss of income and unemployment are the inevitable results. A loophole allowing whales to be taken for research purposes is widely exploited. This picture, taken in the Greenpeace Antarctic Expedition of 1988–89, shows a Minke whale taken by a Japanese catcher.

additional danger that their lead content was a particular hazard to children, impairing their mental ability. As a consequence many countries introduced legislation designed gradually to reduce the lead content of petrol, with the ultimate aim of abolishing it altogether.

Concern for the environment was expressed also in a changed attitude towards wildlife, especially endangered species. The whaling industry, for example, came under increasing pressure on the ground that the oil was no longer essential and some species faced extinction. The oil companies faced increasingly hostile public criticism when major slicks destroyed birds and other marine life.

The fight against smallpox

Smallpox is almost alone among major infectious diseases in having been totally eradicated, but the battle has been a long one. In ancient China and in India a crude form of immunization was practiced by sniffing up the powdered crusts of dried pox marks and in the 18th century Lady Mary

▼ ▶ **The AIDS virus has been intensively investigated and much has been learnt about its nature. Nevertheless, in spite of the discovery of so much detailed knowledge there is still no cure for the infection once it is established. AIDS appears to be particularly prevalent in central Africa. This electron microscope (false color) picture (right) shows AIDS virus particles inside a stricken T4-lymphocyte (white blood cell), the destruction of which makes the victim, such as this patient in a Mexican hospital, exceptionally liable to fatal secondary infections. It is a retrovirus: that is, its genetic program is in the RNA (red cores of particles). It is thence translated to DNA inside the T4 cells and infiltrates the cell's own DNA. Once activated, the viral DNA replicates and progeny bud from the cell wall, killing the host cell.**

Wortley Montagu introduced a similar method into Britain from Turkey, scratching the skin with a needle dipped in pus. This practice of vaccination resulted, in effect, in a mild attack of smallpox which gave lasting immunity to virulent infections. But the results were hazardous: the reaction was not always mild and could be severe, or even fatal. In 1798 Edward Jenner announced an alternative form of immunization resulting from his observation that milkmaids infected with cowpox were immune to smallpox. Although Jenner's technique of vaccination with cowpox was widely adopted, and he received many honors and awards, there was also much opposition to it on religious grounds. Nevertheless many countries made vaccination compulsory – for example Sweden (1814) and the United Kingdom (1853). With the emergence of bacteriology as a disciplined science in the latter part of the 19th century more dependable and readily administered liquid vaccines became available.

The results were highly satisfactory in terms of reducing the incidence and severity of the disease, but largely in northern climes. The liquid vaccine rapidly lost its potency in the heat of the tropics and in these regions progress was made only when stable freeze-dried vaccines became available after the World War II. Use of this, together with careful monitoring of active foci of the diseases, has led to a truly remarkable result. In 1980 the World Health Organization was able to announce that smallpox had been eradicated throughout the world and was never likely to recur.

This confidence that there has been absolute eradication rests on the fact that as there are no animal reservoirs of infections which might fuel a new epidemic, it can be assumed that the virus simply no longer resists. The only exception are a few cultures maintained under high security in a few laboratories around the world. It is thought that the risk of mutation from a wild form of virus is negligible.

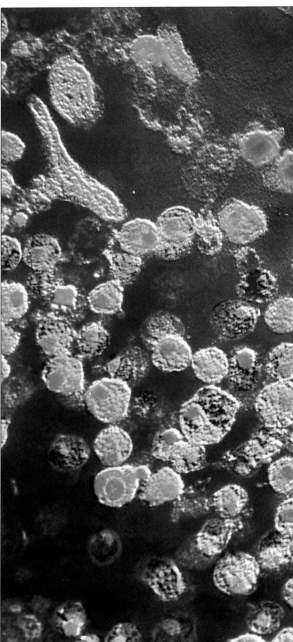

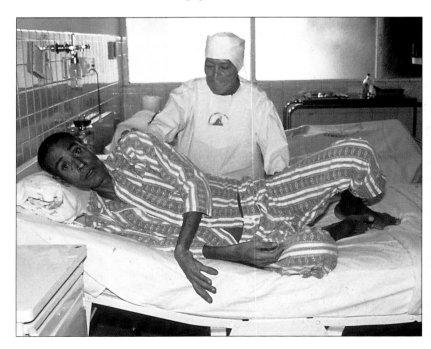

The war against AIDS

In 1981 it seemed that medicine was advancing steadily on most fronts, and that the most serious diseases could be contained if not cured. Complacency was shattered by the first diagnosis of AIDS (Acquired Immune Deficiency Syndrome) caused by HIV (Human Immune Deficiency Virus). It is an insidious disease, prevalent worldwide, with a high mortality and incurable.

HIV attacks cells of the immune system and some other cells, including those of the nervous system. It can remain latent for a period, and may be activated by other infections, but apparently never disappears spontaneously. It is infectious from person to person, primarily through the body fluids but particularly through every form of sexual intercourse. It may also be contracted, quite unwittingly, through transfusion with infected blood, or among drug addicts by sharing

an infected needle. Infected mothers can pass it on to their babies in pregnancy, at birth, or possibly through breast feeding. The general effect of the virus is to destroy the natural mechanisms that normally resist infections such as pneumonia, tuberculosis and herpes: resistance to cancer is also reduced, and as many as one-third of sufferers may become demented through neurological disorders. The time-scale of the infection is very variable, but on average those infected with HIV show symptoms within eight years. The total incidence of HIV infection is uncertain, but in 1988 WHO estimated it to be over one million. By January 1990 over 215,000 cases had been reported from 130 countries.

AIDS has suddenly emerged as a serious medical and social problem and much research has been devoted to containing it. So far, conventional methods have been unsuccessful. Vaccines developed in the late 1980s showed little promise, and the drug AZT is expensive and seems at best only to extend survival time. Faced with this, governments have had to resort to massive publicity campaigns to inform the public how the risk of infection can best be avoided. Short of total sexual abstinence, the best defense is the unfailing use of the contraceptive condom. This is an interesting repetition of history, for the condom was first introduced not for prevention of conception but for protection against venereal disease.

The advent of AIDS had some remarkable repercussions. To be effective, publicity about the sexual aspects had to be very explicit, embracing both heterosexual and homosexual activities. The vocabulary of sex was aired in a way which would have been unthinkable even 20 years earlier. The word condom did not even appear in the massive *Oxford English Dictionary* of 1933: in the 1980s the public had its use carefully explained on television screens. There was also a notable change in sexual morality. Libertarian attitudes about the desirability of many sexual partners had to be revised in the light of medical advice that this enhanced the risk of infection.

Medical developments in the 1970s and 1980s

In 1975 C. Milstein and G. Köhler, at Cambridge in the UK, produced monoclonal antibodies (MCAs). These are totally homogeneous antibodies produced by "cloned" cells: that is, cells which are genetically identical. Because they are chemically pure, having a unique amino acid sequence, MCAs find many uses. For example, in typing blood used for transfusion and treating pregnant women with Rh-negative blood. Milstein and Köhler shared a Nobel prize in 1984.

Cloning in mammals was achieved in 1975 by Derek Brownhall at Oxford in the UK. Removing the egg cell of a rabbit, he replaced its nucleus with the nucleus of a cell taken from the animal's body. Replacing the egg cell, it grew to form a perfectly normal animal with – of necessity – exactly the same genetic makeup as its mother. In normal reproduction, of course, the offspring derives its gametes from two parents, and is thus different from both.

More dramatic, because of its human connotations, was the birth of the first "test-tube baby"

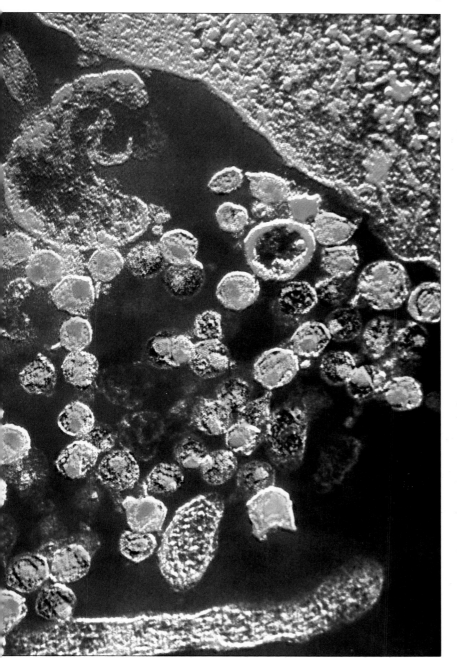

◀ Smallpox has been a scourge of mankind throughout recorded history. In Europe and other temperate regions its incidence was much reduced by vaccination but in tropical regions conventional vaccines quickly lose their potency. After World War II the World Health Organization launched a worldwide eradication campaign using stable freeze-dried vaccine. In 1980 they were able to announce that the disease had been eradicated.

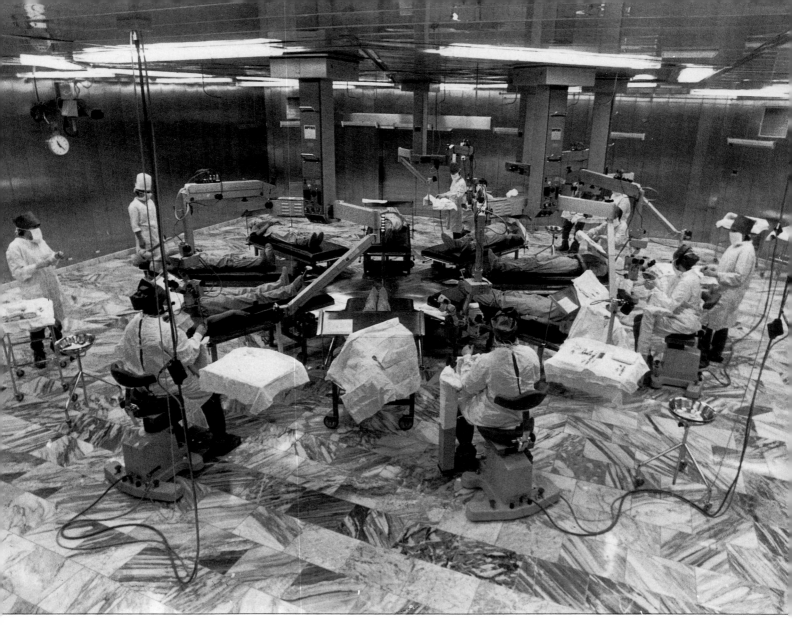

Louise Brown, in England in 1978. Produced by a technique pioneered by P. Steptoe and R. Edwardes, the ovum was fertilized outside the mother's body, kept in a test-tube for two days, and then planted in the uterus. Within a few years this method was being widely used to treat certain forms of infertility. The thousandth test-tube baby was born in 1988.

Two important advances in the treatment of viral infections were made at the end of the 1970s. The drug company Burroughs Wellcome developed a process for manufacturing the antiviral protein interferon, discovered by Alick Isaacs in 1957. This made possible a systematic evaluation of its therapeutic value, including the treatment of certain forms of cancer. At the same time P. Provost and M. Hilleman, at the Merck Institute in America, succeeded in growing, in kidney cells of the Rhesus monkey, the virus that causes infective hepatitis. This paved the way for the introduction of a vaccine in 1981. All such advances in the treatment of disease are welcome, but they pose increasingly difficult economic problems. Many of the new forms of treatment are so demanding of human and technical resources that

the cost of treating a single patient can run to tens of thousands of dollars. In Britain, the cost of setting up an in-vitro fertilization clinic capable of treating 400 patients a year has been estimated at around £700,000 exclusive of premises, running costs and drugs. Even in the rich Western nations, the provision of adequate health services along conventional lines has stretched resources. In the developing countries, representing a majority of the world's population, the most sophisticated medical equipment could not even be considered: there, few have any regular access to any sort of organized health service. The infant mortality figures stress the point. In Western countries mortality in the first year of life is now around 2 percent: in Afghanistan it is 25 percent.

In 1978 WHO and the UN Children's Fund launched an initiative to encourage all governments to devote more resources to promoting health rather than treating disease. Millions of deaths, it was pointed out, could be cheaply avoided by systematic campaigns against insect vectors of major diseases, such as the malaria-spreading mosquito and the tsetse fly. In the modern world, the incidence of serious illness,

▲ In 1980 the USSR announced a radical new approach to eye surgery, instigated by Professor S. Fyodorov. Named "Daisy", it involves a conveyor-belt approach. To perform all the stages of a complex operation the patient is passed to a succession of surgeons, who each contribute their own particular skills. The first such clinic was opened in Cheboksary: by 1990 fifteen more had been opened at other centers. This "Daisy" eye clinic is at Cheboksary.

such as certain forms of cancer and cardiac diseases, can be reduced by reducing indulgence in alcohol, tobacco and certain kinds of food. Very modest expenditure on anthelmintics (drugs which destroy worms) could dramatically increase the expectation of life in Third World countries where these parasites are prevalent.

Ethical problems of medical developments

Until comparatively recently most of our knowledge about the early development of mammals came from experiments with mice, but following the advent of in-vitro (test-tube) fertilization (IVF) in the 1970s a great deal of human material has become available for study. How far, and in what way, this can properly be used for research purposes has aroused much controversy.

The facts are very simple. In Britain alone it is estimated that half a million couples could benefit from IVF but not only are the available facilities far too limited to cope with more than a fraction of this, the success rate is less than 10 percent. Both to improve the success rate and to have a deeper understanding of the causes of infertility more research using human material is essential. In some cases, for example, the fault lies with the sperm, which is unable to penetrate the membrane (zona pellucida) surrounding the egg. Possibly this might be corrected by drilling a hole in the membrane or injecting the sperm beneath it.

Nearly half of all child deaths are due to some kind of genetic disorder. The best known is Down's syndrome but there are at least 3,000 others, including hemophilia (absence of blood clotting), cystic fibrosis (gland malfunction), and thalassemia (a form of oxygen deficiency in blood). For adults who are, or might be, carriers of such diseases there are at present only two alternatives: either to have no family or to have tests done late in pregnancy with the prospect of abortion; for many genetic disorders, however, there are still no tests.

With IVF this situation can be avoided. Using a technique known as gene amplification, specific genes can be picked and then multiplied until there is sufficient to test for defects. This involves taking a cell from the developing embryo within a few days of conception, when it is known as a conceptus.

The ethical dilemma lies on the fact that many people believe passionately that life begins at the moment of conception and that the conceptus is therefore sacrosanct. The validity of this must not be judged only by the pragmatic, nonreligious standpoint of much of the Western world but also by that of the hundreds of millions whose beliefs are guided by their faith.

The scientific argument is that to manipulate the conceptus is permissible. In the first few days after fertilization the cells simply divide – from one to two, two to four, and so on – and the growing group remains microscopically small. On the fifth day a blastocyst forms, a hollow ball with a distinct inner and outer layers of cells, some 30–120 in all. If a cell is moved at this stage to a new part of the conceptus it will assume a new role, depending on where it ends up. It is impossible to specify any one cell that will go on

to form the embryo itself, or indeed whether an embryo will result. To remove a single cell is thus not damaging. If an embryo does result most of the cells will in fact go to form the placenta and the various membranes that will surround the embryo.

After a week the blastocyst becomes attached to the wall of the womb but an identifiable embryo does not appear until about the fifteenth day, in the form of a column of cells known as the primitive streak. Most medical research workers believe that experimental work is permissible at least up to this stage.

A problem with embryo research is devising a legal frame work. In Britain a Voluntary Licensing Authority was set up in 1985 to approve all clinical and experimental research on human embryos, but it has no statutory powers. Italy has no restrictions but Belgium, France, Holland and Sweden allow research only with strict guidelines. Germany allows no research. The USA has no national policy but in many states existing laws effectively ban all research. However, public support for experiment is indicated by the millions of couples worldwide who pin their hopes on IVF to start their families.

Where research is permitted, some clear benefits have been defined. Experiments up to the sixth day would facilitate improvements in IVF techniques with detection of genetic diseases. Ninth-day cells have something in common with cancer cells and may be expected to improve the recognition and treatment of this disease. Experiments at the stage of attachment to the womb (7–14 days) will increase knowledge of the implantation process, significant for natural conception as well as IVF. Experiments at the stage of the primitive streak (15 days) will throw light on the way in which the embryo gains its nourishment and on the initial stages of spina bifida and anencephaly, a fatal defect in the development of the skull.

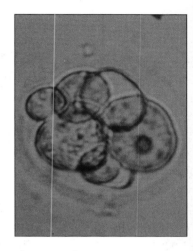

▼ Following fertilization, the human egg cell develops by division as a zygote. By day five it is still no bigger than a pin-point and the cells are still undifferentiated: that is, they have no assigned function in the embryo. This picture shows an eight-cell zygote.

▼ The 1980s saw the advent of a new scanning system for medical diagnosis: positron emission tomography (PET). It is based on introducing positron-emitting radioisotopes (produced in a cyclotron) into the patient's blood stream and then systematically scanning the emitted radiation to detect metabolic anomalies. A PET scanning system costs roughly 3.2 million US dollars.

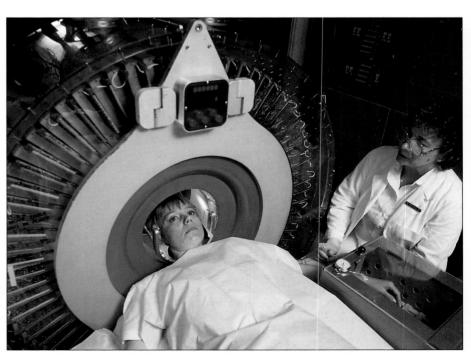

Biotechnology and genetic engineering

Life, in the biological sense, is the expression of the immensely complicated biochemical changes that go on ceaselessly in living tissue. These changes do not occur at random, but are strictly programmed by the genes which can be regarded as a succession of different beads strung on immensely long necklaces within the chromosomes. One gene, or perhaps a small number of genes working in combination, may control the color of eyes in humans or the length of stalk in a cereal. In sexual reproduction, the offspring inherit one gene from each parent, and thus while resembling their parents have an individuality of their own.

In traditional methods of improving crop plants and domestic animals, breeders choose for further breeding individuals which have in an enhanced degree some desirable quality – such as high milk yield in cattle or late flowering in fruit trees. They are, however, limited to the genes available within a particular species, though they may improve their chances by including in their program the different genes of wild varieties exhibiting some desirable trait. It is because they contain these reserves of new genes that the preservation of wild varieties is so important. Nevertheless, breeders are essentially in the position of the gambler who throws four dice: with great luck he may throw four sixes but more likely he will throw a less remarkable combination. By no means can he throw sevens or eights, for this would require a different, octahedral, die.

Genetic engineering removes the inherent limitation of conventional breeding. The essence of this is that a gene will express itself specifically wherever it finds itself. Thus in the animal body one gene codes for the hormone insulin, lack of which causes diabetes in humans. This connection was recognized in the 1920s and diabetes is now effectively controlled by means of insulin extracted from animal pancreases. Plants and lower

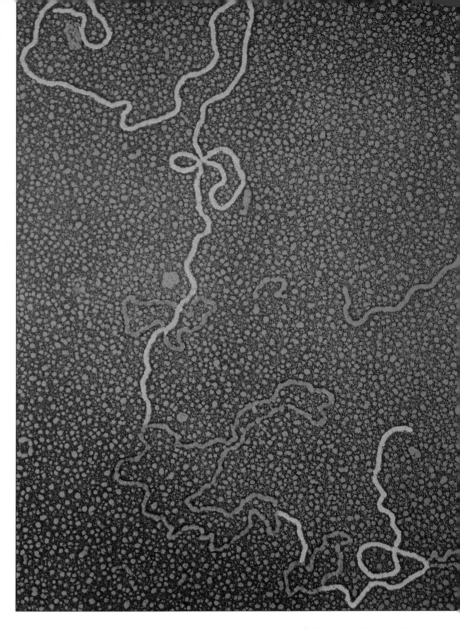

▲ Genetic engineering allows DNA – the carrier of the genetic code – to be modified. This color-coded electron micrograph (magnified X65,000) shows how DNA from two related viruses have been spliced together to form a heteroduplex DNA molecule (substitution loops red; deletion loops blue).

▶ Over the years conventional techniques of plant breeding have led to the introduction of important new varieties – better, for example, in terms of yield, disease resistance, and soil nutrients. The method is necessarily slow, however, and commercial availability is measured in years. Modern cloning techniques make it possible to speed up this breeding process enormously, as here in a commercial cloning operation.

The European Molecular Biology Laboratory

The European Molecular Biology Laboratory (EMBL) was formally inaugurated in May 1979 with John Kendrew, a British Nobel laureate, as its first Director General. The concept of EMBL first emerged in 1962 when Kendrew and James D. Watson visited the international nuclear physics research center (CERN) in Geneva on their way back from the Nobel ceremonies in Stockholm. They met with Leo Szilard, a nuclear physicist turned molecular biologist, who proposed that European governments should be persuaded to support an international laboratory for molecular biology following the CERN model. Within Europe at that time there was both concern that the initiative in molecular biology was passing from Europe to the United States and a conviction that some pooling of European resources would be necessary.

The European Molecular Biology Organization (EMBO) was formed in 1964. At the end of the 1965 the German Volkswagen Foundation awarded EMBO a grant of DM 2,748,000 to support a 3-year program of courses and research fellowships and to further efforts to establish

the laboratory. After some political debate, Heidelberg was selected in 1972 as the main site with outstations in Hamburg and Grenoble.

In the early 1990s EMBL had nearly 500 staff, both permanent and temporary, representing most European countries. Its work was recognized to be of the highest standard and encompassed studies of protein structure, metabolism, evolution, genetic engineering, embryology, carcinogenesis, virology and computers in biology. Through a program of long-term (1–2 year) fellowships, costing around DM 7 million in 1989, EMBO's influence extended far beyond Heidelberg: over 100 fellowships awarded per year allowed young researchers to work at laboratories outside their own country, a system which particularly benefited southern European countries like Spain, Italy and Greece. The fellowship scheme gained the highest possible scientific acclaim when in 1984 George Köhler was awarded a Nobel prize for the discovery of monoclonal antibodies in César Milstein's Cambridge laboratory under an EMBO fellowship.

forms of animals do not produce insulin, because their cells do not contain the essential gene. Now, however, it is possible to insert the insulin-producing animal gene into the cell of a micro-organism, which then acquires the power to produce insulin. The microorganism, typically *Escherichia coli*, can then be grown in fermentation vessels, producing insulin much as yeast produces alcohol in brewing. This is a radical departure, with far-reaching consequences: for the first time a bead can, as it were, be removed under human control from one necklace and strung on another. This opens up an important general possibility: the power to mass-produce any particular protein at will.

There are also other potential medical applications of this genetic engineering. Thus some congenital diseases – such as the blood disease thalassemia – are due to the absence of an essential protein or abnormal production of a harmful one. It may be possible in the future to diagnose these conditions before birth and introduce corrective genes into the bone marrow of affected children, allowing them to develop normally.

For Paul Ehrlich, in 1911, Salvarsan was a "magic bullet" which singled out the syphilis spirochaete as its target. Unfortunately, although a considerable advance, it had very serious side-effects. In the 1940s penicillin – and later other antibiotics – seemed to have assumed this role, but again adverse reactions have been encountered and, more seriously, pathogenic organisms show themselves capable of becoming resistant, rendering the drug ineffective. Then, in 1975, César Milstein and George Köhler invented something much more nearly deserving the title. These are monoclonal antibodies (MCAs) which are specific not merely for a particular organism but for a specific antigen. They promise to be the guided missiles of medicine, able to home in on targets within particular viruses, bacteria, and cancer cells.

Hazards of genetic engineering

Genetic engineering has opened up entirely new possibilities for creating new strains of plants and animals far more diverse, and more quickly, than was feasible with the traditional methods of breeders. It has also, however, created misgivings in the mind of the public-at-large. Clearly, genetic engineers could create forms of life which had no counterpart in nature. Their intentions are no doubt good and the results – such as bacteria which could produce cheap insulin for diabetes sufferers – are demonstrably useful. But proverbially the road to hell is paved with good intentions, and could it be guaranteed that some of the new creations might not be dangerous? Discounting imaginative tales of Frankenstein monsters that might rule the earth, there are real fears that at the microorganism level, viruses and bacteria might be created, in pursuit of other characteristics, that could be deadly and resistant to any existing form of treatment. Genetic engineering might conceivably produce a devastating viral or bacterial strain which would rage out of control.

Another aspect of genetic engineering raises potential ethical questions. It has long been

◀ Genetic "fingerprinting" (ie analysis of DNA) allows family relationships to be established with virtual certainty. In 1975 Modoris Ali applied for his family in Bangladesh to join him in the UK, but his claim of family relationship was disputed. In 1988 it was proved by DNA analysis.

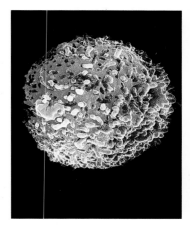

▼ False-color scanning electron micrograph of a hybridoma cell producing a monoclonal antibody to cytoskeleton protein. Such monoclonal antibodies are produced by injecting a mouse and then harvesting its antibody response by removing the B lymphocytes.

known that certain closely related species can mate and produce distinctive offspring, though these hybrids are sterile and cannot themselves reproduce. But modern techniques of fusing the cellular nuclei of different species make it possible to take this hybridization a good deal further and to interbreed disparate species. Clearly, there are dark possibilities open to irresponsible researchers. By the same technique it might be possible, for example, to cross a human with an ape or even a quadruped, with gruesome results. With such possibilities in mind most countries have devised codes of practice, and in some instances legislation, to regulate such experiments and the release of genetically engineered organisms into the environment.

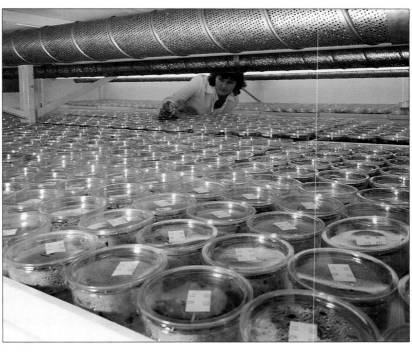

MONITORING THE ENVIRONMENT

The rise of environmental consciousness in the 1980s brought science and technology back into the public eye in new ways. In some respects they were implicated with the threat to the environment – agrochemicals had for example been used without proper consideration of their effect on wildlife – but in this and many other cases the cause of the problem has been as much economic as technical. The means of ending the pollution were at hand but were prohibitively expensive or might produce even more serious side effects.

Of all the environmental problems, that of the so-called greenhouse effect, in which an increasing proportion of carbon dioxide in the atmosphere is preventing heat from escaping from the Earth's atmosphere, and thus causing a general rise in temperatures across the globe, is by far the most considerable. As scientists study the phenomenon in its entirety, it is becoming more, not less, complex. Initially it seemed as if these climatic changes, like others observed in the 1980s, were solely the consequence of human imprudence – burning enormous amounts of fossil fuels, using CFC aerosols and refrigerants which endangered the ozone layer of the Earth's atmosphere, failing to purify automobile exhaust gases and so on. If all these causes were controlled, the changes would be arrested and gradually reversed. But hard information on the scale of the problem, as well as its causes, is only gradually being built up. As scientists accumulate more data, debate continues to rage over the speed at which global warming is in fact taking place, and over the extent to which local climatic variations can be attributed to this cause.

Meanwhile, as scientists study the world's climate in more detail, it has been suggested that the changes observed are part of huge historic swings that have wrought cataclysmic changes in the past. For example, nitrous oxide, one of the gases implicated in the greenhouse effect, is produced in every electrical storm. In 1989 the research vessel *Charles Darwin* reported after a three-year circumnavigation that huge quantities of nitrous oxide are produced by bacteria in the northwest Indian Ocean. If, as this study suggests, natural forces are indeed the major contributory factor there is little that can be done to control them; the introduction of legislation reducing greenhouse gases caused by human agencies such as power stations and automobiles might retard the rate of change, but not reverse it.

But science and technology still have a vital role to play. Supercomputers are being employed to produce sophisticated models of the world's weather systems. Scientists can monitor environmental changes and give advance warning of growing hazards; technology holds out the possibility of mitigating the effects in many ways, from building embankments to protect lowlying land threatened by a rising ocean level to breeding new varieties of crop plants appropriate for changed climatic conditions.

▼ The smog which daily envelopes modern conurbations such as Mexico City requires careful monitoring, before action can be taken to mitigate its effects or remove its cause.

▲ The nuclear power station explosion at Chernobyl in the USSR in 1986 led to a massive cleanup campaign in the surrounding countryside. Residents were evacuated until radiation had fallen.

▲ The ozone layer in the Earth's atmosphere, which is threatened by chemicals used in refrigerants and aerosols, is monitored regularly from space, as here by the US NIMBUS-7 satellite over the North Pole in January 1989. Changes in its depth are studied, and then the environmental threat can be analyzed in detail.

▼▼ Chemical pollution of the natural environment, whether it occurs on land or sea, or as here, in the Antarctic, is studied in detail so that action can be taken against offenders.

▼ Underground detectors allow scientists to monitor minute changes in rock formation that could assist in the prediction of earthquakes. As yet scientists can only give a few minutes warning.

▲ Satellites can assist in pollution control, as here in New Jersey where the effects of water (blue) pollution on agriculture (yellow and green) or urban (red and purple) activities are monitored.

Datafile

In the 1970s and 1980s investigations into some areas of the physical sciences required such large investments that they were beyond the resources of a single nation. International collaborations were set up, particularly within the European Community, to construct the instruments required for research into elementary particle physics, nuclear physics and space exploration. The European Organization for Nuclear Research (CERN) and the Joint European Torus (JET) are among the world leaders in their fields. Experimental work at CERN has confirmed theories which predict that the same physical processes are responsible for the electromagnetic force and one of the nuclear forces. Present work includes the search for evidence of the sixth quark, predicted by current theories of nuclear particles. American space research has received a boost with the successful relaunch of the Shuttle following the *Challenger* disaster, and from the magnificent television pictures of the outer planets of the Sun's solar system which were sent back by the Voyager 2 space probe.

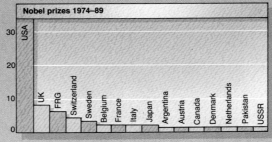
Nobel prizes 1974–89

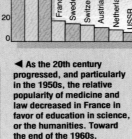
Nobel prizes 1901–89

◀ Since 1974 the domination of the Nobel prizes by the United States has been spectacular (including a clean sweep in 1983). Of the 106 individual scientists awarded prizes, 62 have been from the United States. Scientists from 15 other countries have been honored but many of these actually work in American laboratories.

◀ From the inception of the Nobel prizes in 1901, exactly 400 scientists received awards up to 1989. Of these, nine were women, three of whom were awarded the prize jointly with their husbands. The youngest winner was William Bragg at 25. Marie Curie and John Bardeen won two prizes. Of the 400, 153 were from the USA.

Chronology: JET project

1973–78
Design team established at Culham (south Oxfordshire, UK). Approval given to construct JET tokamak fusion experiment.

1978–83
Construction Phase. Construction of buildings (for approx. 600 staff) and JET tokamak machine; installation of computer systems for control and data acquisition.

1983–85
Operation Phase I. Establishment of first plasmas using hydrogen gas, to temperature of 30 million degrees.

1985–88
Operation Phase II. Heating studies with hydrogen and deuterium plasma routinely reaching 75 million degrees and occasionally temperatures as high as 250 million degrees.

1989–91
Operation Phase III. Studies using tokamak at full power.

1991–92
Operation Phase IV. Full-power operation using tritium plasmas.

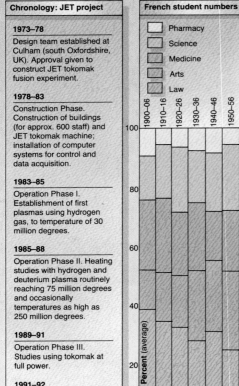
French student numbers

▲ The establishment of large international collaborations, such as the JET project, requires planning ahead for 20 years of work and is dependent on continuing funding from many governments. In that time administrations can change. The administrative difficulties can be as formidable as the technological problems.

▶ The centrally planned economies of the GDR and USSR have shown the largest fraction of population involved in research and development. However, if the number of Nobel prizes awarded in relation to a nation's population is taken as a measure of creativity, it is the UK and Switzerland which head the list calculated for 1974–89.

◀ As the 20th century progressed, and particularly in the 1950s, the relative popularity of medicine and law decreased in France in favor of education in science, or the humanities. Toward the end of the 1960s, however, the study of law and the arts was increasing while the proportion of students choosing to study science declined.

▼ Government funding of research and development in France showed a steady decline from the mid 1960s to the end of the 1970s. After the election of François Mitterand as president in 1981, French R&D was increasingly well supported. Strenuous efforts were made to promote science to schoolchildren and the public.

▼ The CERN nuclear research facility occupies an area of 5.6 sq km (2.2 sq mi), straddling the Swiss–French border. Its annual budget of 810 million Swiss francs (US$520M) supports the employment of 3,500 people and more than 50 experiments either in progress or preparation. Over 200 universities and other institutes are involved.

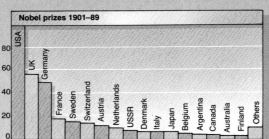
French investment in research & development

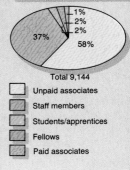
CERN staff 1988

Total 9,144
Unpaid associates
Staff members
Students/apprentices
Fellows
Paid associates

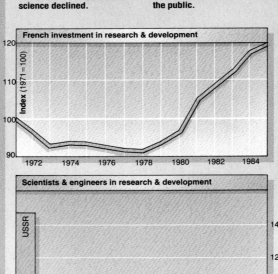
Scientists & engineers in research & development

SCIENCE: NECESSITY OR LUXURY?

The new generation of
particle accelerators

Attitudes to the cost of
scientific research

Nuclear fusion research
projects

Earth studies from space

French science

Astronomical research

On 14 July 1989, Bastille Day, all France celebrated the bicentenary of the start of the French Revolution. At 1630 hours on the same day physicists of CERN, the international research center for particle physics in Geneva, were celebrating the commissioning of LEP (Large Electron Positron Collider), the largest scientific machine ever built. Lodged in a circular tunnel some 27km (16.8mi) in diameter – most of it under French territory – it is a particle accelerator in which particle beams are fired against each other to obtain very high collision energies. It creates the sort of conditions that existed a fraction of a second after the big bang in which the Universe was supposedly originated and creates particles and effects no longer seen. In particular, physicists were seeking to create Z particles, whose existence had been predicted in the 1960s in the theory that unites electromagnetism and the weak nuclear force. These particles are carriers of weak force. The first Z particles had been seen by mid-August and the evaluation of the first results had been made by the end of October.

LEP was the culmination of nearly 10 years of planning and construction, at a cost of around

▼ At Fermilab in Illinois, USA (below), a service road marks the 6km (4mi) circumference of the underground ring of magnets in the laboratory's particle accelerator. In 1983 Fermilab upgraded its machine by installing superconducting magnets, and in 1990 it was still providing the world's most energetic man-made proton beams.

£400 million. As LEP was commissioned, the USA was planning to build in Texas an even bigger machine, the Superconducting Supercollider (SSC) with a circumference of 84km (52mi): the estimated cost was over one billion dollars. If this project proceeds, however, it may be the end of the line, as physicists are now directing their attention to new techniques using linear rather than circular machines.

CERN, founded in 1953, was a cooperative venture by 14 European countries. Physicists from other countries including the Soviet Union, Japan and the United States later took part in its research program. It was an indicative of a new pan-European movement reflected also in the economic and political sphere. Europe was not lacking in scientific talent, as demonstrated by continuing success in gaining Nobel prizes, but individual countries could not in many fields afford to compete with the United States.

It was not only a matter of finance, but also of skilled scientific manpower. Lacking opportunity at home, European scientists naturally gravitated toward the United States, with its higher salaries and better facilities. This trend was, of course,

particularly noticeable in the physical sciences, the domain of the "big science" projects. Collaborative scientific ventures in Europe gained new impetus in 1973, with the admission of the UK, Ireland, and Denmark to the European Economic Communities. New initiatives included the European Space Agency (established in 1975) and the multidisciplinary EC research center (ISPRA) in Italy.

But fashions and needs change in science as in other human activities and strategies have to be changed accordingly. In the UK, for example, the great atomic energy research laboratory at Harwell – a source of national pride in the postwar euphoria and an important bargaining counter in the exchange of information with the United States – had to be reorganized and to some extent earn its keep by doing contract work for industry. Contrariwise, the JET (Joint European Torus) experiment project – aiming at producing power by the fusion of light nuclei, the energy source of the Sun – became operative in 1983 at nearby Culham. Even for this public enthusiasm waned as the green movement – opposed to nuclear power in any form – gathered strength, the more so as its program is measured in decades rather than years.

The first great scientific event of the 1990s was the launching of the Hubble Space Telescope in April 1990, after two decades of planning, but its reputed ability "to see ten times deeper into the Universe than ever before" did not impress all those – including many scientists struggling with modest budgets – who weighed this against a cost of $1.3 billion. At the same time the Supercollider program began to be reassessed.

While exploration of the innermost parts of the atom and the remotest parts of the Universe continued to capture the popular imagination, there was much activity in other fields of physical science. Indeed, progress in both these major fields would have been impossible without advances made in many others. Even fields of classical physics proved capable of providing new surprises.

In magnetism, known through the lodestone since ancient times, the advent of liquid magnets opened up new prospects. These consist of minute particles of magnetic materials, such as some oxides of iron, dispersed in a liquid: as in ordinary colloids the particles do not separate from the liquid. Each acts as a small permanent magnet and can confer remarkable properties on the magnetic liquid, commonly known as a ferrofluid.

Particle physics in the 1970s and 1980s

Following the announcement of the concept of quarks by Murray Gell-Mann and George Zweig in 1963 a worldwide search was begun to demonstrate their existence, and the lack of success encouraged growing skepticism about the whole idea. However, 1973 saw a significant advance in the theory of the strong force necessary to bind quarks together. This was a theory akin to QED (quantum electrodynamics) called quantum chromodynamics (QCD): it postulates that quarks are held together by force carriers called gluons. More importantly, it postulates that individual quarks

cannot escape from particles, a plausible explanation of why the search for them had proved uniformly fruitless. Subsequently (1979–80) various experiments confirmed predictions made on the basis of QCD and established it as part of the "standard model" for interpreting particles and forces.

Meanwhile, more and more powerful accelerators, and computerized analysis of millions of collisions, led to the recognition of still more particles. In 1972–73 the first experiments were conducted with a new type of accelerator at the Stanford Linear Accelerator Center, USA. In this, two beams of electrons and positrons respectively, collide while traveling in opposite directions. One of its first major achievements, in 1974, was the discovery of a new kind of quark and of a heavy electron. Yet another new kind of quark was discovered in 1977 at Fermilab in Illinois in the USA.

By this time a clearer and simpler picture of nuclear structure was beginning to emerge. On the one hand there are particles built from quarks, capable of interacting strongly with other particles. On the other, there are leptons, which do not react strongly with other particles. Leptons and quarks both appear to be fundamental in that they have no structure.

As part of the move toward ever more powerful accelerators the proton synchrotron at CERN was converted in 1981 to the world's then most powerful proton–antiproton collider. This led two years later to the discovery of the W and Z particles with masses as predicted by Abdus Salam, Steven Weinberg and Sheldon Glashow. In 1989 CERN commissioned its giant LEP (Large Electron Positron) collider, 27km (16.8mi) in diameter. As was hoped, this quickly produced thousands of Z particles and firmly established that there can be no more than the six quarks and six leptons currently accepted.

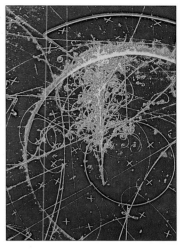

◀▲ In the 1970s bubble chambers remained important particle detectors, revealing ionized tracks as trails of bubbles in a superheated liquid (seen in false color above). But electronic detectors are the only possibility at machines where beams collide head on to produce sideways sprays of particles. Left, a technician stands where the beams in LEP, at CERN in Europe, would eventually pass through the detector known as Opal.

▼ The LEP machine accelerates beams of electrons and positrons traveling in opposite directions and makes them collide head-on. Its large ring keeps the particles on a gently curving path and so minimizes the energy they lose by radiating.

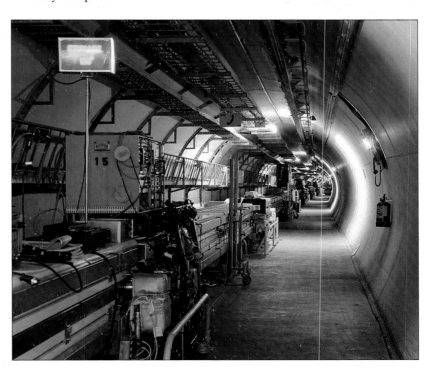

The search for fusion

The enormous potential of fusion as a source of energy is demonstrated daily by the Sun whose seemingly inexhaustible energy comes from a continuous process of nuclear fusion. On Earth the devastating power of the hydrogen bomb is produced by fusion. Since the early 1950s scientists have been attempting to create fusion reactions on Earth to make a new and safe way of generating electricity.

The bulk of the world's electricity comes from burning coal and oil but these fuels will not last forever. Their by-products cause atmospheric pollution and the resulting greenhouse effect. Similarly, nuclear fission power stations are not universally popular for a variety of reasons even though they could provide all electricity needed. Can nuclear fusion therefore be an acceptable long-term solution to the world's energy problem?

The great advantage of fusion is the plentiful supply of fuel. Reactions take place between the nuclei of light atoms such as deuterium and tritium – the isotopes of hydrogen. Deuterium is extremely plentiful – there being 34g of it in every cubic meter of water – and as only small amounts are required there is enough to last for millions of years. Tritium is not plentiful and is thus made within the reactor using the light metal lithium, widely distributed in the Earth's crust.

To get deuterium and tritium nuclei to fuse together requires heating the fuel to enormous temperatures so that the nuclei are traveling fast enough to overcome the electrostatic force between them. The optimum temperature needed to create more energy than is required for heating is in the region of 100–200 million °C – ten times hotter than in the core of the Sun. At such extreme temperatures the deuterium and tritium

▲▼ Scientists have long known of unusual phenomena at temperatures close to absolute zero, such as the superfluid behavior of liquid helium (above). Superconductivity also seemed to be restricted to very low temperatures until 1986, when ceramics with the type of structure shown below were found to be superconducting at higher, more accessible temperatures.

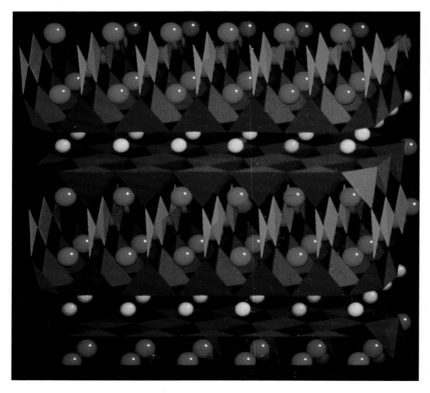

gas is fully ionized and is called a plasma. The only way to contain high-temperature plasmas is to use magnetism. Many different heating methods and different types of magnetic cages have been tried in an attempt to contain enough high-temperature plasma for long enough to yield a net gain of energy. The favored shape of the container for plasma is a ring-shaped vessel called a torus.

The most successful fusion experiment to date is the 14-nation Joint European Torus (JET) based in the UK at Culham, south of Oxford. JET is not a reactor although it has near-reactor dimensions. Its aim is to establish the scientific feasibility of fusion. Experiments started in 1983 and by the end of 1989 plasmas up to 250 million degrees at the required density had been routinely produced. The magnetic field confined the plasma energy sufficiently well but the peak temperatures and densities were obtained only for a second or so.

▲ For nuclear fusion to occur readily, hydrogen nuclei must be in a hot, dense plasma (a gas so energetic that the atomic electrons move independently of the nuclei). Such plasma cannot be allowed to interact with anything, so it must be contained within a vacuum vessel and restricted by magnetic fields from touching the walls. The most promising design is the "tokamak", with its doughnut-shaped (toroidal) containment region, as used here in the European JET project.

and also to make tritium from lithium for fueling the reactor.

The fusion reactor will be very safe as there will only be enough fuel in the reactor at any given time to last for a few tens of seconds; it cannot therefore get out of control. Unfortunately, the neutrons also make the reactor radioactive but with careful selection of construction materials, storage of the radioactive structure may be limited to about 100 years or so. The quest for fusion power is a long and difficult task and it will be well in the 21st century before commercial fusion reactors will be generating electricity.

New research in superconductivity

In 1911 Heike Kamerlingh Onnes, working at Leiden Univeristy, discovered the phenomenon of superconductivity. At temperatures very near absolute zero metals suddenly show near zero electrical resistance: once induced, a current will flow indefinitely. It took almost a half a century to produce a satisfactory theory for the phenomenon. In 1957 J. Bardeen, L.N. Cooper, and J.R. Schrieffer devised the so-called BCS theory (for which they received the Nobel Prize for Physics in 1972). This revived interest in superconductivity but it still appeared to be a phenomenon exhibited only at very low temperatures – at the liquefying temperature of helium. It was, therefore, difficult to make practical use of it. Then, in 1986, J.G. Bednorz and K.A. Müller, at the IBM Research Laboratory in Zurich, created a sensation by announcing the discovery of new kinds of superconductors which were effective at temperatures above 30°K. Soon afterward, at a crowded meeting of the American Physical Society in New York in 1987 other researchers described superconductors effective at 90°K. This was sensational, for temperatures of this kind – above the boiling-point of liquid nitrogen – are relatively easily attainable. The rapid achievement of this considerable increase in the critical temperature encouraged the belief that superconductors could be found that are effective at ordinary ambient temperatures.

Such a development would open up tremendous prospects, especially in the electrical power industry. For example, the considerable losses in overhead power transmission lines could be virtually eliminated if they were superconducting. Equally, the power requirements of electric motors and electromagnets would be minimal if their windings could be made of superconducting wire. Superconducting levitation promises to have practical application: for example, in rail transport. If a magnet is placed on a surface which, by lowering its temperature, is made superconducting, it will rise up and float above it.

However, these dreams are yet to be realized, for no room-temperature superconductor has yet emerged. However, scientists in Japan have made a notable advance. The new superconductors are ceramic in nature and as such not easily converted into wire form. In 1988 the Nippon Steel Company devised a "melt-processing" technique which yields a satisfactory superconducting wire. Japan is, indeed, the leader in this very promising new field of technology.

This was because impurities coming from the torus walls spoiled the plasma's performance. New experiments were planned to provide a solution to the problem. When this had been done the next step was to build an experimental reactor to address all the engineering aspects required to generate electricity in a continuous manner.

Design teams have already been set up to prepare for the next step. One such team involves experts not only from Europe but also from the USSR, USA and Japan, who were designing a reactor called ITER – International Thermonuclear Experimental Reactor. In such a reactor deuterium and tritium nuclei fuse together producing a helium nucleus (alpha-particle) and a neutron. The alpha-particle's role is to continue heating plasma to make the reaction self-sustaining. The neutron heats a surrounding blanket of lithium to provide the heat source for generating electricity

▼ Photography from space has greatly enhanced our ability to follow changing phenomena on Earth. Imaging at infrared in addition to visible wavelengths, as well as highly-developed techniques in imaging processing provide a wealth of information. This false-color infrared image shows Mount St Helens in Washington in the USA brewing up to its big eruption of 1980.

Earth studies – by satellite

Satellites – either orbiting at a height of about 1,000km (620mi) or geostationary at 36,000km (22,400mi) – provide not only a unique means of continuous weather surveillance but also of monitoring many other terrestrial features such as military installations, the pattern of land utilization and the distribution of minerals. New techniques of image intensification make possible pictures with an almost unbelievable precision of detail despite the distance. Additionally, "optical subtraction" makes it possible quickly to detect quite small changes – for example in forest clearings – which might escape routine inspection. In this new technique photographs are compared electronically with others of the same scene taken earlier: any changes can then be automatically highlighted.

One serious limitation to satellite surveying, especially in temperate and humid tropical areas, is that cloud-cover may obscure a particular area, sometimes for days on end. To overcome this, new kinds of sensors have been developed to work on microwave or radar wavelengths, to which clouds are transparent.

Overall views of the pattern of land utilization and the way in which it is changing are essential for formulating agricultural policy. However, conventional surveying techniques – including photography from high-flying aircraft – are too laborious and expensive, except for areas of particular importance. Satellites have transformed the situation. For example, the US Landsat satellite can (clouds permitting) survey any given area regularly and in detail every 16 days.

For maximum utility, satellite pictures must be able not only to make gross distinctions between arid or barren areas and those with continuous plant cover but to distinguish between different kinds of crops. As the latter often look very similar in color some fine tuning is necessary. This is done with the aid of multispectral sensors which can distinguish, for example, between different shades of green by analyzing the wavelength of the light they emit.

An important feature of vegetation surveys from space is in land enforcement. Many countries, such as Brazil, now have laws restricting forest clearance but in remote parts of the country it is impossible to discover and punish infringement until the damage has been done. Satellite pictures can reveal an illicit operation within days, sometimes hours, of its start.

What is true of land utilization is also true of minerals. Many of the latter lie undetected in areas too remote and inaccessible to be surveyed by traditional field methods. Surveying from space, however, can detect outcrops and surface indicators of minerals located in the ground.

Astronomy in the late 20th century

Escalating costs of research have also encouraged international collaboration in the field of astronomy. A major cooperative European venture is the European Southern Observatory (ESO), founded as long ago as 1962: its observatory is located in La Silla in Chile. In the 1980s this organization developed plans for a Very Large Telescope (VLT) project. This will have an objective 16m (52ft) in diameter. To cast a conventional mirror of this kind – as used at Mount Palomar, for example – would be tremendously expensive and technically very difficult, perhaps impossible. Instead, the mirror will be built up of a combination of smaller units of manageable proportions, though this involves formidable design problems in keeping the smaller mirrors properly aligned. Two similar telescopes, though rather smaller 10m (52ft) and 15m (49ft) respectively – are being developed in the USA.

Such instruments will extend the observable limits of the Universe by a factor of between two and five, and will provide a valuable complement to astronomical observatories being made from satellites. It will be possible, for example, to map star surfaces, seek other planetary systems, and resolve galaxies.

The VLT project will also complement another southern hemisphere telescope – the UK Infrared Telescope (UKIRT) situated on Mauna Kea, Hawaii, at an altitude of 1,200m to provide good observational conditions. The biggest of its kind in the world – with an aperture of 3.8m (12.5ft) – it

From Neglect to National Priority: Science in France

▲ In the 1980s and early 1990s France showed a public commitment to the importance of science and technology for its economic and cultural future perhaps stronger than that of any country. In 1986 it founded the Museum of Science and Industry at La Villette, Paris. In part it is a traditional science museum, in part an exciting complex of displays – many interactive – designed especially to capture the attention and interest of children, as here. Funding for science in France is directed through the Centre National de la Recherche Scientifique, which in 1990 had 1,300 laboratories and research groups, 17,000 researchers and an annual budget of 10,000 million francs.

A high degree of centralization has been a characteristic of French science since the 17th century. More than in most countries, therefore, the course of French science and technology has been colored; and continues to be colored, by national policy.

The most distinctive institutions are the technical Grandes Écoles, specialized vocational schools, of which the École Polytechnique (founded in 1794 and originally a school for military engineers) has always been the most prestigious. It is characteristic of the growth of advanced technical education in France since the 18th century that, as new specialities have emerged, schools have been created to provide the appropriate specialized instruction. By the late 20th century there were over 150 such schools in science and engineering.

By comparison, the universities have tended to occupy a subordinate position. Until after World War II the Sorbonne in Paris dominated the university sector in science. At the Sorbonne the emphasis was traditionally on pure science, and in this area the university invariably had a distinguished professoriate, notably in mathematics. But even the Sorbonne suffered from modest levels of funding.

In these circumstances, research suffered. Until the 1960s specialities requiring a significant investment in laboratories were poorly supported, and even expenditure on the three national research institutions – the Musée National d'Histoire Naturelle (for the life and earth sciences), the Paris Observatory (for astronomy), and the Collège de France (for subjects across the whole range of the sciences,

humanities and social sciences) – was modest by international standards.

In the long process of economic reconstruction after World War II the claims of science and technology for a greater share of the national budget were gradually heeded. Important improvements were made in the provision for research, through the founding of new bodies, such as the National Atomic Energy Commission (Commissariat à l'énergie atomique), and the consolidation of the National Center for Scientific Research (Centre National de la Recherche Scientifique – CNRS), founded in 1939.

The fostering of these bodies gathered pace with the presidency of Charles de Gaulle (1958–69) and provided France with a new focus for its research effort. From the early 1960s science and technology were made national priorities, a position conspicuously reinforced in the aftermath of the election of President François Mitterrand in 1981 and the succession of socialist governments between 1981 and 1986. So from the 1960s scientific research was well funded. Most of the enhanced funding, however, was directed to designated priority areas, ranging from molecular biology to nuclear power technology. Increasingly, too, the emphasis was on cooperation with industry.

Twentieth-century France has never lacked its great men and women of science: Marie and Pierre Curie, Paul Langevin, Jean Perrin, and Frédéric Joliot-Curie were figures of international standing even in the days of poor funding and neglect. In the 1990s, however, France's scientific community had a strength in depth that placed it firmly along the leaders of European science.

operates at wavelengths from submillimeters to the near infrared, although it can also be used with visible light.

It can be used to study some of the cool giant stars of our own Galaxy, the nuclei of other far distant galaxies, and the clouds of interstellar dust that permeate the Universe. As with VLT, this telescope has been achieved by using novel forms of design – notably cutting the mirror thickness by half without sacrificing stability. In this case the mirror weighs only 6.5 t against 15 t required by conventional design.

Space probes and laboratories

From the earliest astronomical studies in the pre-Christian era until the 1960s observations of space were limited by the presence of the intervening atmosphere. Cloud, of course, completely prevented observations with optical equipment but even in clear weather normal irregularities in the atmosphere caused distortion, as evidenced by the familiar twinkling of stars. More seriously, however, the atmosphere absorbs most of the wavelengths of the electromagnetic spectrum so that observations at ground level with these are impossible. Large optical telescopes are commonly erected on mountains to reduce the effect of atmospheric aberrations, but for other wavelengths observation must be made from points beyond the atmosphere. While sounding balloons and rockets made possible very limited progress in this direction, it was the advent of the space laboratory and research probes which completely transformed the situation.

Such craft are of three main kinds. First, there are unmanned orbiting or geostationary satellites packed with scientific instruments which relay observations over long periods to ground bases. Second, there are unmanned probes similarly equipped, which – elaborately programmed – explore the solar system. Finally, there are the big manned space stations – such as the Russian Salyut 1 (1971) and the American Skylab (1973), in which scientists can in person conduct experiments for weeks or months on end. Both Salyut and Skylab were outclassed in 1986 with the launching of the enormous Russian M-1. The cost of such laboratories is measured in billions of dollars.

Typical of the first class of space laboratory was COS-1, launched in 1975 by a group of European countries. Its task was to make a complete gamma-ray chart of the sky, a task which it successfully completed before being shut down in 1982. Incidental to its main task, COS-1 discovered new stars, characterized by their intensive emission of gamma radiation, most of which are invisible at other wavelengths. This matter is uncertain: they may be young pulsars, the remains of supernovae, or evidence of black holes. A similar satellite was NASA's Solar Maximum Mission satellite (SMM). Its task was to investigate solar activity, especially solar flares. When launched in 1980 its designed life-time was only one year, but thanks to the development of techniques for making repairs in space it continued in operation until 1989.

Of the many scientific probes, the one that most captured the public imagination was undoubtedly the American Voyager 2. Launched in August 1977, it passed in turn close by Jupiter, Saturn and Uranus, sending back to Earth a

▲ To see detail, radio telescopes need large dishes to compensate for the long wavelengths at which they operate. One alternative is to use the effect of the Earth's rotation on several small dishes to build up the same coverage as a large dish. Above are some of the 27 dishes in the biggest system of this kind, the Very Large Array in New Mexico, USA.

◄ One of the major astronomical events of the 1980s was the return of Halley's Comet in 1986. Probes sent to encounter the comet sent back detailed photographs and analyses of the gas that envelops the nucleus. The biggest surprise concerned the nucleus itself, which proved to be unexpectedly dark.

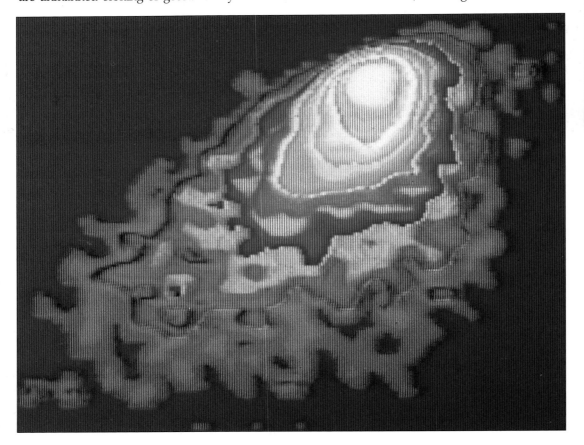

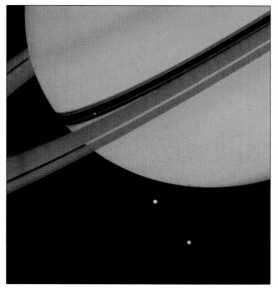

► The USA's two Voyager probes captured the public imagination world-wide in the late 1980s with their amazing pictures of Jupiter, Saturn, Uranus and Neptune. Their many discoveries included volcanoes on one of Jupiter's moons, an unexpectedly rich structure to Saturn's icy rings (right), a nitrogen atmosphere to Neptune's moon Triton, and for all the giant planets many new moons, too small to see from Earth.

remarkable series of pictures of the planets and their moons (many of which it discovered).

Finally – 12 years and more than six billion kilometers later – it reached Neptune, the most distant of the large planets. By then it was so distant that radio signals took more than four hours in transmission even at the speed of light (300 million km per second). Its task completed, Voyager 2 still travels on into outer space, where its transmitters may remain active for another 25 years. It is hoped that it may be able to transmit signals back when it reaches the heliopause, the limit of the Sun's influence in space.

While manned laboratories can make many measurements of radiation in space, they also provide a unique opportunity to make experiments not possible on Earth. Because of the condition of weightlessness the effect of gravity can be eliminated. Experiments conducted under these conditions have, for example, thrown new light on crystal growth.

THE ORIGIN AND FUTURE OF THE UNIVERSE

In 1929 the American astronomer Edwin Hubble showed that each galaxy is receding from every other one and that the Universe is expanding. Recent measurements indicate that the expanding Universe originated in a hot "big bang" ten to twenty billion years ago. Studies of the origins of the Universe involve both astronomical observation and exploration of subatomic particles. It is hoped that the many links between big bang cosmology and particle physics may eventually lead to a complete theory of space, time, matter and the cosmos.

According to the "standard model", particles and antiparticles formed out of high-temperature radiation during the first millionth of a second after the big bang, when time itself began. As the Universe expanded and cooled, particles and antiparticles annihilated each other, but a tiny excess of matter over antimatter ensured that a residue of protons, neutrons and electrons survived – these form the substance of the Universe as it is now known.

About a hundred seconds later neutrons and protons combined to form helium in the ratio relative to hydrogen that we see today. After a hundred thousand years, space became "transparent" and radiation was released, which is detectable even now – 13–15 billion years later – as a weak microwave background. Galaxies formed some time later, and have been moving apart ever since.

Whether the Universe will expand forever, or eventually cease to expand and then collapse into a "big crunch" depends on whether or not its mean density exceeds the minimum, or "critical", value. Although there is not enough luminous matter to halt the expansion, there may be enough as yet undetected "dark matter" to tip the scales.

All matter in the Universe is controlled by four forces – the strong and weak nuclear forces, the electromagnetic force and gravity. Many physicists believe that at progressively higher energies, these forces unite into three, then two, and finally, perhaps, just one fundamental force. Although the first stage of unification – in which the electromagnetic and weak nuclear forces are the same – has been verified experimentally, to test Grand Unified theories (GUTs) which unite the electroweak and strong forces would require energies a million million times greater than the most powerful particle accelerators can deliver. However, because these colossal energies would have existed throughout the Universe during the first 10^{-35} seconds after the Big Bang, physicists may be able to test their theories by observing the Universe.

GUTs can explain why there was a small excess of matter over antimatter, but the explanation implies that protons eventually decay, so that after some 10^{33} years matter as we know it will cease to exist. GUTs also imply the existence of a plethora of elusive and so far undetected elementary particles. They may provide sufficient mass to cause the Universe to collapse in the predicted "big crunch".

▼ During the first 10^{-43} seconds all four forces were probably unified and behaved as a single force. This period is known as the Planck era. Conditions during this era cannot be described by known physics. At the end of this era, gravity and the GUT force became distinct.

▼ After about 10^{-35} seconds the GUT force began to split into the strong nuclear force and the electroweak force. While this change was happening, the Universe inflated, doubling in size every 10^{-34} seconds. By the end of this epoch, all distances had increased by at least 10^{50}.

▼ While the temperature was very high, radiation transformed into particle-antiparticle pairs and colliding particles and antiparticles transformed back to radiation. About 10^{-6} seconds after the initial event, fundamental particles (quarks) formed protons and neutrons.

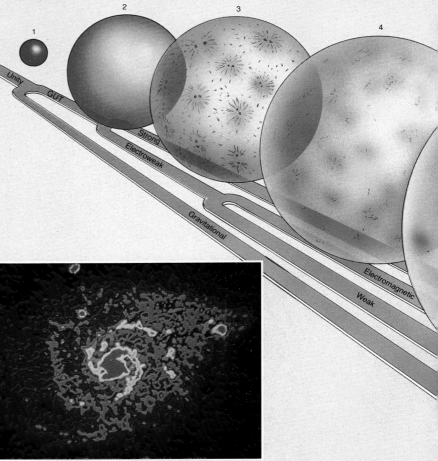

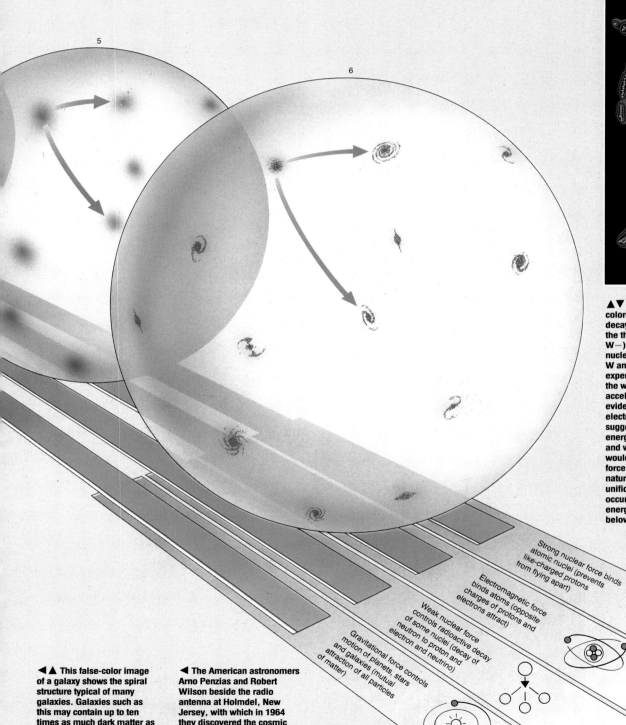

▼ After about 100 seconds protons and neutrons fused to produce helium nuclei. The resulting ratio of helium to hydrogen was 25:75 (by mass). A hundred thousand years later, atoms formed, space became transparent, and the cosmic background radiation was released.

▼ Eventually, galaxies began to form, perhaps a billion years after the initial event. Possibly, very large clumps of matter formed first and then fragmented into individual galaxies. Alternatively, small galaxies may have formed first, then coagulated into larger galaxies and clusters.

▼ As the Universe expands, galaxies move further apart. Each galaxy "sees" every other galaxy receding with a speed proportional to its distance but no galaxy can claim to be the center. Every galaxy is receding from every other one, like dots on the surface of a balloon.

▲▼ This photographically-colored image shows the decay of a Z particle — one of the three particles (Z, W+ and W−) which carry the weak nuclear force. The detection of W and Z particles in 1983 in experiments conducted in the world's largest particle accelerators provided vital evidence in favor of the electroweak theory which suggested that at high enough energies the electromagnetic and weak nuclear forces would unite into a single force. The four forces of nature, and the proposed unifications which should occur at progressively higher energies, are illustrated below.

Strong nuclear force binds atomic nuclei (prevents like-charged protons from flying apart)

Electromagnetic force binds atoms (opposite charges of protons and electrons attract)

Weak nuclear force controls radioactive decay of some nuclei (decay of neutron to proton and electron and neutrino)

Gravitational force controls motion of planets, stars and galaxies (mutual attraction of all particles of matter)

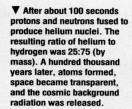

◄▲ This false-color image of a galaxy shows the spiral structure typical of many galaxies. Galaxies such as this may contain up to ten times as much dark matter as luminous matter. In total 90–99 percent of all the mass in the Universe may be in the form of dark matter.

◄ The American astronomers Arno Penzias and Robert Wilson beside the radio antenna at Holmdel, New Jersey, with which in 1964 they discovered the cosmic microwave background radiation. This discovery provided crucial evidence in favor of the big bang theory.

BIOGRAPHIES

Adams, Walter Sydney 1876–1956

US astronomer. He graduated in arts in 1898 and then read mathematics. He gained his first astronomical experience under G.E. Hale at the Yerkes Observatory, and returned there from Europe in 1901 before moving on in 1904 to the Mount Wilson Observatory, where he remained, latterly as director (1923–46). He was a skilful observer with interests ranging from sunspots to interstellar gases and planetary atmospheres. He is chiefly remembered, however, for the first observation of a "white dwarf", and for devising a method for measuring the distance of stars.

Adrian, Edgar Douglas 1889–1977

British neurophysiologist. He entered Cambridge University in 1908: graduating in natural science, he was a fellow from 1913, professor of physiology (1937–51) and master (1951–65). His first research was on responses in muscle-nerve preparations, but the special problems created by World War I led him to examine the clinical problems of nerve injury and war neuroses. In 1925 he began a study of the encoding of sensory information to the brain by impulses in nerve fibers. Using the cathode-ray oscilloscope, he began in 1930 to investigate the "brain waves" discovered by Hans Berger. Finally his research extended to the brain itself. In 1932 he shared a Nobel prize with C.S. Sherrington. He was later president of the Royal Society (1950–55).

Alfvén, Hannes Olof Gösta 1908–

Swedish theoretical physicist. After graduating in physics at the University of Uppsala (1934), he held professorships in the Royal Institute of Technology, Stockholm (1940–73). In 1967 he was appointed visiting professor in the University of California, San Diego. Much of his research was concerned with plasma, neutral gases which contain a high concentration of positive and negative ions in equilibrium. His interest was twofold: in the plasma that exists in the far reaches of the Universe and in the contained artificially produced plasma which is potentially a source of fusion energy, largely free of the radiation hazards of fission processes. In 1942 he predicted that magnetohydrodynamic waves must exist in plasma; the existence of such so-called Alfvén waves was subsequently demonstrated experimentally. In 1970 Alfvén shared the Nobel Prize for Physics with L.E.F. Néel.

Alpher, Ralph Asher 1921–

US physicist. With George Gamow and Hans Bethe he formulated a "hot big bang" theory of the origin of the Universe, based on interpretation of the relative distribution of its known elements, and involving thermonuclear process. This convincingly explained the universal abundance of helium, resulting from fusion of four hydrogen nuclei, with intense emission of energy. A

corollary of this theory – known as the alpha, beta, gamma theory – was that there should be a background of radio "noise" throughout space corresponding to a temperature of 5°K. Such radiation was detected in 1965.

Aston, Francis William 1877–1945

English physicist. A graduate in chemistry, he entered the Cavendish Laboratory, Cambridge, in 1910 to work under J.J. Thomson, investigating the streams of positively charged particles emitted when electricity is discharged through a tube filled with gas at low pressure. Using electric and magnetic fields it was possible to separate the particles according to their mass:charge ratio. From this it appeared that neon consisted of two isotopes. In 1913 he succeeded in enriching the gas in respect of the rarer isotope, neon 22. On his return from war service he devised, on the same principle, his mass spectrograph (1919) by which the mass of isotopes could be deduced from a photographic image. With this he investigated the isotopic composition of some 50 elements. He was awarded a Nobel prize in 1922.

Baekeland, Leo Hendrik 1863–1944

Belgian-US chemist, whose name is commemorated in "Bakelite", the first major industrial plastic. He settled in the USA in 1889, initially as chemist to a photographic manufacturer. While there, he developed a fast-printing (Velox) photographic printing paper which he made and sold through his Nepara Chemical Company: this he sold to George Eastman for 1 million dollars in 1898. He then turned to electrochemistry and, in seeking a substitute for shellac, a widely used insulator, studied the reaction product of phenol and formaldehyde, normally a useless conglomerate. He discovered that if the reaction is conducted at raised pressure an easily molded resin is obtained, eminently suitable for electrical equipment. This he manufactured worldwide through his Bakelite Corporation, subsequently merged into Union Carbide and Carbon Corporation (1939).

Baird, John Logie 1888–1946

British pioneer of television. Ill health prevented his completing an electrical engineering course. After several unsuccessful business ventures, and a complete breakdown, he retired to Hastings. There he became interested in the possibilities of television and on 26 January 1926 gave the world's first public demonstration of it, using a crude electromechanical system. In 1927, using a telephone line, he transmitted from London to Glasgow, and in the following year to New York. From 1929 to 1935 he gave regular television broadcasts with the British Broadcasting Corporation (BBC). Baird's system was then superseded, and abandoned.

Baltimore, David 1938–

US molecular biologist. After graduating at Swarthmore College he did postdoctoral research at Massachusetts Institute of Technology, the Einstein College of Medicine, and the Salk Institute (1963–68). In 1968 he was appointed associate professor of biology at MIT, becoming full professor in 1972 (from 1982 also director of the Whitehead Institute). Working on the polio virus, he discovered how it propagates itself and how the DNA and protein which are its main constituents are formed. In 1970, however, he announced a discovery of more fundamental importance. According to accepted belief DNA was converted to protein by way of an intermediate RNA (ribonucleic acid). The first stage of the conversion was known as transcription, the second as translation. It was further believed that this was a one-way, irreversible process. Baltimore showed, however, that, at least in some tumor viruses, the first-stage, or transcription process was reversible: RNA can be converted to DNA by the enzyme reverse transcriptase. In 1975 Baltimore shared a Nobel prize with H.M. Temin, who had discovered reverse transcriptase independently.

Banting, Frederick Grant 1891–1941 and Best, Charles Herbert 1899–1978

Canadian medical research workers who first isolated insulin and successfully used it for the treatment of diabetes. After war service, Banting practiced as a surgeon in London, Ontario. In 1921, with his research student Best, he succeeded in preparing a pancreatic extract from dogs which was active against diabetes. With the aid of a chemical colleague, J.B. Collip, and the Professor of Physiology, J.J.R. Macleod, they succeeded in preparing much larger quantities of active extract from animal pancreas (sweetbread). Clinical trials were highly successful and a pharmaceutical firm began manufacture in 1922. Banting and Macleod shared a Nobel prize in 1923: believing that the contributions of Best and Collip had been gravely underestimated they insisted on sharing it with their younger colleagues.

Bardeen, John 1908–

US physicist, co-inventor of the point-contact transistor. A graduate in electrical engineering, he served in the Naval Ordnance Laboratory, Washington (1941–45). He then joined Bell Telephone Laboratories, where he developed – with W.H. Brattain and W. Shockley – the transistor (1947) which rapidly superseded the thermionic tube (valve) which had dominated electronics for nearly half a century. For this work all three shared a Nobel prize in 1956. Bardeen gained a second Nobel prize in 1972 – with his colleagues L.N. Cooper and J.N. Schrieffer – for the first satisfactory theory of superconductivity, now generally known as the BCS theory.

▲ Johannes Bednorz

▲ Susan Bell

▲ Hans Bethe

Barnard, Christiaan Neethling 1922–

South African surgeon. After studying medicine at the Universities of Cape Town and Minnesota he set up in private practice. After various medical posts he became a specialist cardio-thoracic surgeon in the University of Cape Town (1958) and subsequently (1961) head of cardio-thoracic surgery. He became interested in the technical problem of open-heart surgery and the possibility of replacing diseased hearts with healthy hearts from accident victims. He successfully performed a heart transplant operation in 1967, although the patient died 18 days later from postoperative complications. Later, he met with increasing success and pioneered also the replacement of diseased heart valves with artificial ones.

Beadle, George Wells 1903–

US biologist, pioneer of biochemical genetics. After graduating in 1925 he did postdoctoral work on maize genetics at Cornell before working (1935) on the genetics of eye color in the fruit fly. As professor of biology at Stanford University (1937–46) he collaborated with the microbiologist E.L. Tatum in research on the biochemical genetics of the bread fungus *Neurospora crassa*, using X-rays to produce mutations. Tests on these led to formulation of the "one gene/one enzyme" concept, which postulates that the role of a gene is to control synthesis of a specific enzyme. In 1958 Beadle and Tatum shared a Nobel prize with J. Lederberg, pioneer of bacterial genetics.

Becquerel, Antoine Henri 1852–1908

French physicist remembered chiefly for his discovery of radioactivity. He entered the École Polytechnique in 1872, received his doctorate in 1888 for a thesis on the absorption of light and in 1895 was appointed professor. From 1882 to 1892 Becquerel investigated the phenomena of phosphorescence and fluorescence and took an immediate interest in the X-rays discovered by W.K. Röntgen in 1895. He supposed that strongly fluorescent substances, like certain uranium salts, might emit X-rays and found that they did in fact produce penetrating rays that blackened a photographic plate, but these were not X-rays. They were in fact not strictly waves but streams of electrically charged particles. In 1898 Marie Curie named this phenomenon radioactivity. In 1903 Becquerel shared the Nobel Prize for Physics with Marie and Pierre Curie.

Bednorz, Johannes Georg 1950–

Swiss physicist, co-inventor of an important new category of superconductors. He graduated in physics from the University of Münster in 1976 and then joined the IBM Research Laboratory at Rüschlikon, working under K.A. Müller. The main subject of their research was superconductivity. In the years after World War II superconductors found many practical applications, for example in operating large electromagnets, but all had the disadvantage of working only at very low temperatures. In 1986 Bednorz and Müller announced the discovery of a new kind of superconductor effective at temperatures above 30°K: later this was raised to 90°K. As this can be attained without too much difficulty with the aid of liquid nitrogen, this enormously extended the potential of semiconductors. Bednorz and Müller shared a Nobel prize in 1987.

Bell, Susan Jocelyn 1943–

British astronomer. A Glasgow University graduate, she did postgraduate research at Cambridge, later held a research fellowship at the University of Southampton (1968–73) and was research assistant in the Mullard Space Science Laboratory, University College, London (1974–82). At Cambridge she worked with Antony Hewish, a radio astronomer who had completed in 1967 a radio telescope specially designed to observe the scintillation of stars. In the summer of 1967 they observed an unusual signal at a wavelength of 3.7m (12.1ft) – unusual in that it corresponded to a sharp burst of radio energy at regular intervals of about one second. This was the first known pulsar. Pulsars are believed to be rapidly rotating neutron stars whose radio emission is perceived much as one perceives a light signal from the rotating lantern of a lighthouse.

Bergström, Sune 1916–

Swedish biochemist. He graduated in medicine at the Karolinska Institute in Stockholm in 1943, having also studied at London University (1938) and Columbia University, New York (1940–41). He was subsequently professor of biochemistry at the University of Lund (1947–58), and then returned to the Karolinska as professor of biochemistry. His research has included work on the blood anticoagulant heparin, the bile acids and cholesterol but he is particularly distinguished for his research on prostaglandins. Chemically, these are fatty acids which produce a range of physiological effects, notably inducing contraction of smooth muscles. Their existence had been discovered in the 1930s but Bergström was the first to isolate them in pure crystalline form (1957). In 1962, together with R. Ryhage, B. Samuelsson and J. Sjovall, he showed that prostaglandin molecules contain five-membered carbon rings.

Bethe, Hans Albrecht 1906–

German-US physicist, noted for contributions to the theory of nuclear reactions and the production of energy in stars. After graduating in physics at the Universities of Frankfurt and Munich, he did research work in Frankfurt, Stuttgart, Cambridge and Rome. In 1932 he was appointed assistant professor at Tübingen, but in 1933, when the Nazis came to power, he left for England, working in Manchester and Bristol. In 1935 he moved to Cornell University in the USA, where he remained as professor of theoretical physics. During the war he was seconded to the Radiation Laboratory of MIT, to work on microwave radar, and to Los Alamos to work on the atomic bomb project. His main research was concerned with nuclear reactions, developing Bohr's theory of the compound nucleus. This interest led him to consider the source of stellar energy and he concluded that this is basically the result of the condensation of four hydrogen nuclei to form one helium nucleus, with emission of fusion energy. With R.A. Alpher and G. Gamow he formulated in 1948 a theory of the distribution of the chemical elements consistent with the "big bang" theory of the origin of the Universe and the abundance of helium in it. In 1967 Bethe was awarded the Nobel Prize for Physics.

Birkeland, Kristian 1867–1917

Norwegian inventor (with Samuel Eyde) of the first commercially successful process for the fixation of atmospheric nitrogen. After studying physics in Paris, Geneva and Bonn he was appointed professor of physics at Christiania (Oslo) University in 1898. Research on the aurora borealis led him to experiments with electric discharges through gases. In 1784 Henry Cavendish had showed that the nitrogen and the oxygen of the air combine in an electric spark and in 1901 Birkeland and Eyde, an engineer, joined forces to develop this as a viable industrial process. They began manufacture in 1904 and in the following year founded the now well-known Norsk Hydro-Elektrisk Kvaelstofaktieselskab. They continued manufacture until the early 1920s when the process was displaced by one invented by Fritz Haber.

Bjerknes, Jacob Aall Bonnevic, 1897–1975

Norwegian meteorologist. He entered Christiania (Oslo) University in 1914 and in 1920 was appointed director of the Weather Forecasting Center in Bergen and subsequently (1930) professor of meteorology in the university there. He established an international reputation and in 1940 – happening to be in America at the time of the German invasion of Norway – became professor of meteorology in the University of California, Los Angeles. Under his direction UCLA became one of the world's great centers for teaching and research in meteorology. From the records of a network of observing centers established in Norway during World War I he developed in 1919 the concept of vast air masses – cold polar ones and warm tropical ones – which keep their identity for long periods and are separated by fronts. A lifetime of experience was summarized in his *Dynamic Meteorology and Weather Forecasting* (with C.L. Godske) (1957).

▲ Wernher von Braun

▲ Melvin Calvin

▲ Alexis Carrel

Black, Sir James Whyte 1924–

British pharmacologist. After graduating in medicine at St Andrews University in 1946 he held academic positions there and in the Universities of Malaya and Glasgow. He subsequently held industrial positions with ICI (1958–64), Smith, Kline and French (1964–73) and Wellcome Research Laboratories (1978–84) and was then appointed professor of analytical pharmacology in London University (1984). His development of beta-blocking drugs (notably propranalol, in 1964) was based on the fact that some hormones act on the heart by attachment to certain sites known as beta-receptors. By blocking their sites, the effect of the hormones can be diminished and the load on the heart reduced. His second major discovery (1972), of drugs for the treatment of gastric ulcer, had a similar basis. The wall of the stomach contains receptor sites to which histamine can become attached, stimulating acid production. Drugs such as cimetidine block the receptors, and so encourage the healing of the ulcers. Black was awarded a Nobel prize in 1988.

Bloch, Felix 1905–83

Swiss-US physicist. After graduating in physics in Zurich (1927) and doing postgraduate research in Leipzig under W. Heisenberg he spent three years teaching in Germany before emigrating to the USA in 1933 when the Nazis came to power. He spent the rest of his working life as a professor of physics at Stanford. In 1939 he measured the magnetic movements of the neutron, discovered by James Chadwick in 1932. In 1946 he developed the technique of nuclear magnetic resonance (NMR) spectroscopy, a widely used method of analysis. This was developed independently by the American physicist E.M. Purcell, with whom Bloch shared a Nobel prize in 1952.

Bohr, Niels 1885–1962

Danish physicist who applied the quantum theory of Max Planck to atomic structure. He studied physics in Copenhagen and in 1912 went to England to work with Ernest Rutherford at Manchester University. The latter had proposed a model of the atom in which negative electrons circled a positive nucleus. But this failed to explain why the electrons did not simply spiral down into the nucleus, losing energy. Bohr proposed that the orbital momentum of the electrons is quantized, and that radiation is emitted only when an electron jumps from one orbit to another. In 1916 he returned to Copenhagen and became director of a new Institute for Theoretical Physics which became a leading international center. He was awarded a Nobel prize in 1922. In 1943 he left German-occupied Denmark and went to Britain, and then to the USA, to advise on the Manhattan atomic bomb project. In the 1950s he was concerned with establishing CERN.

Boyer, Herbert Wayne 1936–

US biochemist. After graduating in the University of Pittsburgh he held various research posts until appointed professor of biochemistry in the University of California, Berkeley. By 1970 it was well established that the synthesis of protein is controlled by the helical, ladder-like DNA molecule. In theory, therefore, new powers of synthesis could be conferred on a cell if a DNA fraction derived from some other species could be, as it were, spliced into the normal DNA. In 1973 Boyer turned theory into practice by grafting into the DNA of *E. coli* some extrachromosomal DNA present in the form of plasmids. Within a few years this sort of genetic engineering with recombinant DNA was being generally practiced. For example, the genes of simple organisms such as *E. coli* can be manipulated to include foreign DNA capable of synthesizing valuable biological products such as insulin.

Bragg, William Henry 1862–1942 and William Lawrence 1890–1971

British physicists, father and son, who pioneered the application of X-ray diffraction to the determination of crystal structure. After graduating at Cambridge, W.H. Bragg was appointed professor of mathematics and physics at Adelaide University (1886). While he was there he began to investigate the newly discovered X-rays and alpha particles. In 1912, after Bragg had taken a chair in physics at Leeds, Max von Laue announced his discovery that X-rays can be diffracted by crystals. Bragg immediately saw that this could be used to work out the exact positions of atoms and ions in a crystal lattice. He and his son, W.L. Bragg (then a research student at Cambridge), used X-ray diffraction to reveal the structure of many single crystals. In 1915 they shared the Nobel Prize for Physics. In 1923 W.H. Bragg was appointed director of the Royal Institution, London, a prestigious appointment to which his son succeeded (1953–66).

Brattain, Walter Houser 1902–87

US physicist. He graduated in physics at the University of Minnesota in 1929 and then joined the Bell Telephone Laboratories. There he did research on the surfaces of semiconductors, initially copper oxide but later silicon and germanium, which had potentially more interesting properties. With J. Bardeen and W. Shockley he developed the point-contact transistor (1947), which had some similarities to the cat's-whisker detector of the early days of radio. It incorporated a thin germanium crystal and had the rectifying properties of a thermionic tube (valve). After 1967, after he left Bell to go to Whitman College, he turned his attention to the properties of the surfaces of biological membranes. In 1956 he shared a Nobel prize with Bardeen and Shockley.

Braun, Wernher von 1912–77

German-US pioneer of rocket propulsion. After studying engineering in Zurich and Berlin, he began rocket research for the German army in 1932, initially with solid fuel propellant but within two years using liquid fuel. In 1938 he was appointed technical director of the German rocket research establishment at Peenemünde on the Baltic. This already had a staff of 3,000, which rose to 20,000 by the end of World War II. He there developed the highly destructive V2 rockets. Afterward he and his team surrendered to the American forces. He then played a leading role in the American space program which ultimately succeeded in landing men on the Moon. He resigned from NASA in 1972.

Brenner, Sydney 1927–

South African-British molecular biologist, pioneer evaluator of the genetic code. After graduating in the University of the Witwatersrand he spent some time in Oxford before joining the British Medical Research Council's Molecular Biology Laboratory in Cambridge in 1957: he was appointed its director in 1980. By 1953 Francis Crick and J.D. Watson had demonstrated how genes reproduce themselves and carry genetic information: specific amino acids that become linked to make up proteins are coded by triplets of bases, known as codons, in the DNA chains. Brenner showed that these triple codon sequences do not overlap. He also determined that there were no breaks (punctuation points) in the sequence of bases.

Broglie, Prince Louis-Victor de 1892–1987

French physicist, pioneer of the wave theory of matter. He originally studied history and turned to physics only after radio service during World War I, enrolling in the Sorbonne: he was professor of theoretical physics in the Henri Poincaré Institute (1928–62). In his doctoral thesis of 1924 – later published at length in the *Annales de Physique* – he suggested that in addition to waves sometimes behaving as particles, particles such as electrons could also behave as waves. This was demonstrated experimentally by G.P. Thomson in 1927. This particle-wave duality was the basis of the quantum mechanics of E. Schrödinger. De Broglie was awarded a Nobel prize in 1929.

Burnet, Frank MacFarlane 1899–1985

Australian virologist. After graduating in medicine at Melbourne University he joined the Walter and Eliza Hall Institute of the Melbourne Hospital where he remained, apart from two brief spells in London. From 1944 to 1965 he was director. Until 1957 his research was concerned largely with viral and rickettsial infections, especially influenza; thereafter with immunology and graft-versus-host reactions. P.B. Medawar's work complemented his, and they shared a Nobel prize in 1960.

◄ Ernst Chain

◄ Subrahmanyan Chandrasekhar

◄ John Cockcroft (left)

Butenandt, Adolf Frederick Johann 1903–

German biochemist. After studying at Marburg and Göttingen, he was professor of organic chemistry in Danzig before being appointed director (1936–72) of the Kaiser Wilhelm Institute for Biochemistry, Berlin-Dahlem (postwar, Max Planck Institute for Biochemistry, Munich). His research was mainly on the chemistry of sex hormones. He isolated oestrone in 1929 and androsterone in 1931, followed by progesterone in 1934. Later he worked on insect hormones, including ecdysone, which controls the molting process. He also did research on the physiologically extremely active sex attractants of insects, the pheromones. He was awarded a Nobel prize in 1939 but the German government forbade him to accept it.

Calmette, Léon Charles Albert 1863–1933 and Guérin, Camille 1872–1961

French bacteriologists who developed the BCG (Bacille Calmette-Guérin) vaccine for protection against tuberculosis. Calmette qualified in medicine in Paris before serving in the French Navy (1883–90) to study the incidence of malaria and sleeping sickness in Gabon. In 1889 he investigated bubonic plague in Oporto before moving to Saigon to establish a Pasteur Institute there (1891), developing vaccines against plague and snakebite. In 1895 he returned to France to found another Pasteur Institute in Lille. In 1917 he was appointed administrative head of the Pasteur Institute in Paris, where he spent the rest of his working life. His most important work was done over the years 1906–24 when, with Guérin, he developed an effective vaccine against tuberculosis. BCG was dramatically effective, reducing the incidence of the disease by 80 percent and giving protection for up to ten years.

Calvin, Melvin 1911–

US biochemist. A chemistry graduate, he went to the UK as a research fellow of Manchester University (1935–37). He spent the rest of his working life (except for wartime work on the atomic bomb) in the University of California, Berkeley, latterly as professor of chemistry. He was also director of the Laboratory of Chemical Biodynamics (1960–80) and associate director of the Lawrence Berkeley Laboratory (1967–80). Most of his research concerned photosynthesis, the process by which plants utilize carbon dioxide from the air and convert it, through chlorophyll as an intermediary, into starch and oxygen. He developed the technique of using radioisotopes to follow chemical reactions, pioneered by G. de Hevesy, and he identified a cycle of reactions now known as the Calvin Cycle. In this the atmospheric carbon dioxide is fixed by an enzyme and then reduced to form sugar. He was awarded the Nobel Prize for Chemistry in 1961.

Carothers, Wallace Hume 1896–1937

US industrial chemist. In 1928 he became head of the organic chemistry department of the giant chemicals company Du Pont. Charged with investigating substances of high molecular weight, he produced neoprene (1932), one of the first satisfactory synthetic rubbers. He then proceeded to study the fiber-forming properties of polymers based on esters and amides. Finding the first unpromising, he concentrated on polyamides. This led to nylon, a highly successful synthetic fiber first marketed in 1938, the year after Carothers' suicide.

Carrel, Alexis 1873–1944

French pioneer of blood-vessel surgery and organ transplantation. A qualified surgeon, he joined the Rockefeller Institute, New York, in 1906, and examined problems of organ transplantation. Here a major task was to reconnect the blood vessels of the transplanted organs. With new micro techniques for suturing the severed vessels, he successfully removed organs and replaced them in the same animals. Transplantation from one animal to another was unsatisfactory, however, as the new organ was usually rejected. His suturing techniques found further application in vascular surgery. Carrel is also remembered for his pioneer research on tissue culture. He was awarded a Nobel prize in 1912.

Chadwick, James 1891–1974

British physicist. A Manchester University physics graduate, he continued there to do research on radioactivity under Ernest Rutherford. In 1913 a scholarship took him to the Technische Hochschule, Berlin, to work with H.W. Geiger. There he was interned for the duration of the war. Afterward he rejoined Rutherford, now director of the Cavendish Laboratory, Cambridge; Chadwick was appointed assistant director in 1923. In 1932 he proved the existence of the neutron, a hitherto unknown particle with approximately the mass of a proton but without charge. He was professor of physics at Liverpool University (1935–48) and was closely involved in the British contribution to the development of the atomic bomb.

Chain, Ernst Boris 1906–79

German-born biochemist. Chain, a Jew, emigrated to England in 1933, after graduating in Berlin. After working briefly in Cambridge, he joined Florey in Oxford in 1935. He worked for a time on the antibacterial enzyme lysozyme, and then in 1939 embarked on a joint project with Florey to make a general study of the antagonisms between microorganisms. Fortuitously, this included penicillin, abandoned by Fleming, its discoverer. By May 1940 its unique properties had been demonstrated. After the war Chain became director of the International Research Center for Chemical Microbiology in Rome (1948–61). In 1954

he became associated with the Beecham Group and collaborated with them in developing a range of semisynthetic penicillins. In 1961 he returned to England as professor of biochemistry at Imperial College, London (1961–73). In 1945 he had shared a Nobel prize with Fleming and Florey.

Chandrasekhar, Subrahmanyan 1910–

Indian-US astrophysicist. He studied physics at the Presidency College, Madras, and then did research at Cambridge, UK (1931–37). In 1937 he was appointed professor of physics in the University of Chicago, where he remained. His particular interest was in white dwarfs – stars in the last stage of development. Such stars have collapsed inward under their own weight to form an exceedingly dense shell – roughly equal to the size of the Earth – sustained by the outward pressure of plasma within it. Only certain stars can undergo this change. The limiting factor is that all must conform to the so-called Chandrasekhar Limit, according to which no white dwarf can have a mass greater than 1.4 times the mass of the Sun. Chandrasekhar received a Nobel prize in 1983.

Cockcroft, John Douglas 1897–1967

British physicist. After a year at Manchester University, and war service, he read mathematics at Cambridge. In 1924 he joined Rutherford's team of atomic physicists at the Cavendish Laboratory. It was there that, with Walton, he succeeded in splitting atoms of lithium and boron by bombardment with protons (1932). For this they were jointly awarded a Nobel prize in 1951. Later he was associated with the development of radar and other aspects of air defense, concurrently serving as Jacksonian Professor of Natural Philosophy at Cambridge (1939–46). After the war he was successively director of the Canadian Atomic Energy Commission (1944–46); director of the UK Atomic Energy Research Establishment, Harwell (1946–58); and master of Churchill College, Cambridge (1959–67).

Coolidge, William Davis 1873–1975

US industrial scientist, remembered for the development of ductile tungsten and the hot-cathode X-ray tube. He studied electrical engineering at MIT, physics at Leipzig and chemistry once again at MIT. In 1905 he joined the General Electric Research Laboratory at Schenectady, New York, eventually becoming its director (1932–61). He quickly had a major success. An urgent need of the day was to draw the very refractory metal tungsten into wires fine enough to use as filaments in electric lamps. This he succeeded in doing in 1908. In 1914 the invention reduced the cost of electric light in the USA alone by around two billion dollars. His second major invention, in 1913, was an X-ray tube in which current and voltage can be varied independently.

Crick, Francis Harry Compton 1916–

British molecular biologist. A physics graduate, after postgraduate research in Cambridge he spent the war years as a scientist with the Admiralty (1940–47), before returning to Cambridge as part of a Medical Research Council (MRC) Unit. He was a member of the MRC Laboratory of Molecular Biology, Cambridge (1949–77), and then joined the Salk Institute in San Diego. The MRC unit was investigating the structure of large biological molecules by means of X-ray crystallography. In 1951 the group was joined by J.D. Watson, an American biologist with a particular interest in genetics. The two complemented each other, both personally and professionally. It was known by then that cellular DNA is the basic material that carries the genetic code. Its chemical structure was broadly understood and there was a suspicion of a helical configuration, but the details of the molecular geometry had still to be worked out. With their own X-ray pictures of DNA, and some others supplied by M.H.F. Wilkins, a New Zealand biophysicist working in London, they eventually constructed a model containing a double helix – connected at intervals by rungs like those of a ladder – which was fully consistent with the observed X-ray patterns. For this original and highly significant work Crick, Watson and Wilkins were jointly awarded a Nobel prize in 1962.

Curie, Marie 1867–1934 and Pierre 1859–1906

French pioneers in the phenomenon of radioactivity. Marie Sklodowska, daughter of a Warsaw physics teacher, worked as a governess before entering the Sorbonne in 1891 to study physics and mathematics. There she met, and soon married (1897), Pierre Curie, director of laboratory work of the newly founded École Municipale de Physique et Chimie. Both were intensely interested in the newly discovered phemonena of X-rays and radioactivity. They decided to collaborate in a search for substances with properties similar to those of uranium. The discovery in 1898 of polonium and radium brought fame to both. In 1904 Pierre was appointed professor of physics in the Sorbonne but, tragically, was killed in a street accident in 1906. Marie succeeded to his chair, and was awarded a Nobel Prize for Chemistry in 1911.

Cushing, Harvey Williams 1869–1939

US physician, founder of modern neurosurgery. After qualifying in 1895 he practiced general surgery before going to work with the Swiss surgeon E.T. Kocher. This aroused his interest in neurosurgery and he spent some time with C.S. Sherrington in Oxford before returning to the USA. There he spent 30 years developing new techniques of brain surgery, in which there had hitherto been a depressing lack of success. They involved meticulous attention to preoperative diagnosis and readiness to carry out operations lasting many hours. He made a study of the pituitary gland, inaccessibly situated at the base of the brain, and showed a particular form of wasting disease (Cushing's Syndrome) to be associated with a tumor of the pituitary.

Dam, Carl Peter Henrik 1895–1976

Danish biochemist. He graduated in the Copenhagen Polytechnic Institute in 1920 and, after training in veterinary medicine, became a lecturer in the Physiological Laboratory of Copenhagen University. He worked in the Institute of Biochemistry (1929–41) before becoming professor of biochemistry at the Polytechnic. At the time of the invasion of Denmark he was in the USA: he returned to Denmark in 1956 to join the Danish Fat Research Institute. In 1929 he had observed that chicks on low-fat diets develop hemorrhages and their blood fails to clot normally. At first he thought that this was a form of scurvy, but later he identified it with a then unknown vitamin which he called Vitamin K. This he showed to be present in many plants and also in certain animal organs, notably the liver. Although he prepared highly active concentrates of Vitamin K, he failed to isolate it in pure form. This was achieved in 1939 by the American biochemist E. Doisy, with whom he shared a Nobel prize in 1943.

Diesel, Rudolf 1858–1913

German engineer, remembered for the internal combustion engine that bears his name. He grew up in Paris and England, was later sent to an uncle in Augsburg and attended the Munich Technische Hochschule, where he studied thermodynamics. In 1880, while working in Paris, he experimented, unsuccessfully, with an expansion engine based on ammonia. About 1890 he conceived the idea of an engine in which ignition would be effected by the heat generated by highly compressing a fuel/air mixture. Having patented this (1892) he published a detailed account of its theory and design. On the strength of this, two German companies supported its development, Maschinenfabrik of Augsburg and Krupp of Essen. It was exhibited in Munich in 1898 and aroused worldwide interest. Diesel was soon a millionaire but did not live to see his invention fully exploited.

Dirac, Paul Adrien Maurice 1902–84

Swiss-British theoretical physicist, who made major contributions to quantum theory and predicted the existence of the positron and other antiparticles. After studying electrical engineering and mathematics in Bristol he took his doctorate in Cambridge in 1926. In 1932 he was appointed Lucasian Professor of Mathematics, a post he held until retiring in 1969. A talk by Heisenberg in 1925 aroused his interest in what was to be quantum mechanics and in the winter of 1927–28 he formulated the celebrated "Dirac Equation", a relativistic theory of the electron. This led on to his prediction of a positvely charged "anti-electron", experimentally observed by C.D. Anderson in 1932. In 1933 he shared the Nobel Prize for Physics with E. Schrödinger.

Domagk, Gerhard 1895–1964

German industrial chemist, discoverer of the sulfonamide drugs. A graduate, in medicine, from the University of Kiel, he worked first as a pathologist. In 1927 he became director of research in experimental pathology and bacteriology in the new company I.G. Farbenindustrie, seeking chemical agents capable of destroying infective bacteria within the body. In 1932 he discovered that Prontosil Red, a dye developed by I.G., could control streptococcal infections in mice. This was published in 1935, and it was soon discovered elsewhere that the antibacterial activity resided not in the whole molecule but in a moiety known as sulfanilamide. As this had been known of since 1908 (although its antibacterial activity was unsuspected) I.G. were unable to patent the discovery. The Nazis would not let him accept the 1939 Nobel Prize; in 1947 he received the medal.

Dulbecco, Renato 1914–

Italian-US microbiologist. After studying medicine at the University of Turin and working in its anatomy institute, he went to the USA in 1947, first to the University of Indiana, and then the California Institute of Technology (1954–63). He was resident fellow at the Salk Institute, California (1963–72) and was appointed distinguished research professor in 1977. He discovered the photoreactivity of phages inactivated by ultraviolet light, and he devised the patch test for the recognition of animal virus mutations. His research advanced knowledge of oncogenic viruses, polioma and simian viruses, and their action in cellular transformation. He was awarded a Nobel prize in 1975.

Eckert, John Presper 1919–

US electronic engineer. A science graduate from the University of Pennsylvania (1941), he was appointed research associate in the University's School of Electrical Engineering (1941–46). In his first year he was concerned with the design of radar equipment but in 1942 began a long collaboration with J.W. Mauchly, initially on the giant ENIAC machine (Electronic Numercial Integrator and Calculator) completed in 1946. In 1947 he and Mauchly founded the Eckert-Mauchly Computer Corporation, absorbed into Remington Rand in 1950. He was concerned with the development of UNIVAC (1952), the first commercial electronic computer, and with a wide range of electronic digital computers.

Edison, Thomas Alva 1847–1931

US inventor of great originality and versatility. He had little formal education, being regarded as retarded, and had a series of casual jobs as a youth. While working as a telegraph operator during the American Civil War, he read Michael Faraday's *Experimental Researches in Electricity* and gained a technical insight into the principles of electrical communication. In 1869 he invented the ticker-tape machine used to distribute stock exchange prices nationwide. This he sold for 30,000 US dollars, which he used to fund a research laboratory where he would devote himself to invention. In all he lodged over 1,000 patents. His major inventions included the carbon-granule microphone, the phonograph, and (independently of Joseph Swan in Britain) the incandescent filament lamp. He also discovered the Edison effect (the one-directional passage of electricity between filaments in a vacuum lamp), which was to be the basis of the thermionic tube (valve).

Ehrlich, Paul 1854–1915

German pioneer of hematology (the study of blood and blood forming organs) and founder of chemotherapy. He studied medicine at several German universities, finally graduating at Leipzig (1878). While a student, with the help of amline dyes, he discovered all the different types of white blood cells (1877–81). In 1881 he began to use methylene blue as a stain and the specificity of this for certain bacteria suggested to him that such substances might be used to destroy infective organisms without harming their host. In Liverpool, UK an arsenical drug had been used to treat trypanosomiasis without success. By systematically ringing the chemical changes in this he eventually discovered a compound effective against the organism that causes syphilis. This was marketed as Salvarsan.

Einstein, Albert 1879–1955

German-Swiss-US mathematical physicist who conceived the Theory of Relativity. In 1896 he entered the Polytechnic Academy, Zurich, after some delay because of his weakness in mathematics. On completing his studies he acquired Swiss citizenship and got a junior post in the Berne Patent Office. After the publication of the first of his remarkable papers on relativity, he held senior academic appointments in Zurich and Prague and in 1913 became director of the Kaiser Wilhelm Institute of Physics, Berlin. He traveled widely, and when Hitler came to power in 1933 he was in California, and never returned to Germany, taking up a position at the Institute of Advanced Study, Princeton. In 1905 he published three seminal papers on the Special Theory of Relativity: his General Theory was published in 1915. The latter gained him a Nobel prize in 1921. In 1952 he was offered, and declined, the presidency of Israel.

Enders, John Franklin 1897–1985

American microbiologist. He entered Harvard as a student of languages but found that he preferred microbiology: on graduation he entered Harvard Medical School and remained there, latterly as professor in the Children's Hospital. Particularly interested in virology, he succeeded at last in maintaining viral cultures by using live whole animals, such as chicken embryos. Using penicillin to prevent bacterial infections he found – with F.C. Robbins and T.H. Weller – that simple tissue cultures would suffice; it was then possible to culture the viruses responsible for mumps, polio and measles (1948–51). For measles they developed in 1951 a vaccine which was widely used in the 1960s. They shared a Nobel prize in 1954.

Esaki, Leo 1925–

Japanese physicist. He did postgraduate research on semiconductors and then worked for the Sony Corporation (1956–60). He then joined IBM at their Thomas J. Watson Research Center, of which he was appointed director in 1962. From the outset of his career he investigated the so-called tunnel effect by which, in accordance with wave mechanics, a current carrier can penetrate a potential barrier which, in accordance with classical physics, it could not surmount. In 1960 he invented the tunnel (Esaki) diode, an important variation of Shockley's junction diode. Such diodes can be used as low noise amplifiers or as oscillators, up to microwave frequency. In 1973 Esaki was awarded a Nobel prize jointly with B.D. Josephson and I. Giaever.

Euler, Ulf Svante von 1905–83

Swedish physiologist. He graduated in medicine from the Karolinska Institute, Stockholm. In 1930 the award of a Rockefeller Fellowship enabled him to do research in the UK, Belgium and Germany. In London he worked with H.H. Dale who had established that in the nervous system impulses are transmitted from one nerve fiber to another by chemical intermediaries, notably acetylcholine. This raised questions: how are such highly active substances synthesized, stored and released? How are they disposed of, literally in a flash, once the nervous impulse is triggered? In 1946 von Euler identified noradrenaline as a major neurotransmitter and thereafter closely investigated its mode of action. With N.Å. Hillarp he discovered that within the cells noradrenaline is synthesized and stored in minute granules. In an independent line of research he identified in 1935 the first of the prostaglandins, substances with a very powerful physiological effect, especially on blood pressure and muscle contraction. From 1939 to 1971 he was professor of physiology in the Karolinska Institute. In 1970 he shared a Nobel prize with Bernard Katz and J. Axelrod. In 1966 he was elected president of the Nobel Foundation.

Fermi, Enrico 1901–54

Italian-US physicist. After graduating at Pisa, he did postgraduate research in Germany and Holland, returning to Italy to be professor of theoretical physics in Rome (1927). There his brilliant papers established for him an international reputation. In 1938 he was awarded a Nobel prize and left Fascist Italy for Columbia University, New York. There he built an atomic pile in 1941, but it was too small to become self-sustaining. He then moved with his research group to Chicago and there, on 2 December 1942, set in train a self-sustained atomic reaction that could be started and stopped at will: the Atomic Age had begun. From 1943 he took part in the Manhattan Project for the production of atomic bombs. After the war he resumed his research at Chicago, working mainly on radioactivity.

Fischer, Emil 1852–1919

German organic chemist. He was successively professor in Erlangen (1882), Würzburg (1885) and Berlin (1892–1919). The last was the major chair of chemistry in Germany and with the appointment went the promise of a new laboratory, completed in 1899. There he built up a flourishing school of chemistry with a particular interest in natural products. His earliest work was on nitrogenous substances known as purines -- which include guanine and adenine, identified years later as constituents of DNA – and sugars. Later, he turned his attention to proteins, breaking them down into amino acids, and achieving some success in recombining these to form polypeptides. He was awarded a Nobel prize in 1902.

Fleming, Alexander 1881–1955

British bacteriologist. A small legacy allowed him to qualify in medicine at St Mary's Hospital where, save World War I, he spent the rest of his working life as a bacteriologist. He was appointed professor in 1928 and on retiring in 1948 continued as principal of the Wright-Fleming Institute of Microbiology. During these years he made two major discoveries. The first (1922) was lysozyme, an enzyme present in nasal and other bodily secretions which has the power of digesting (lysing) certain types of bacteria. This demonstrated the existence of substances which could destroy bacteria and be harmless to human tissues, and he soon observed the similar lytic action of a mold which had accidentally contaminated a staphylococcal culture (1928). He rightly attributed this to the production of an antibacterial substance, penicillin. However, he failed to recognize penicillin as a uniquely powerful chemotherapeutic agent, and by 1934 had lost interest in it. It was left to H.W. Florey and E.B. Chain to establish penicillin in clinical medicine in the early 1940s. All three shared a Nobel prize in 1945.

Howard Walter Florey

Sigmund Freud (right)

Kenichi Fukui

Florey, Howard Walter 1898–1968

Australian-born physician. After qualifying in Adelaide in 1921 he went to Oxford as a Rhodes Scholar and then held a succession of medical appointments in Britain, finally as professor of pathology in Oxford (1934–62) subsequently becoming provost of The Queen's College, Oxford. Early in his career Florey took an interest in the antibacterial enzyme lysozyme, discovered by Fleming in 1922. This led him in 1939 to undertake, with E.B. Chain, a general study of antimicrobial substances. Fortuitously, one of those they selected was penicillin. By May 1940 the unique chemotherapeutic properties of this substance had been demonstrated, but manufacture in wartime Britain was impossible. With his colleague N.G. Heatley he went to the USA and was instrumental in getting government agencies and industrial firms to launch a crash program to produce penicillin for all military casualties. In 1943 he went to North Africa to demonstrate the use of penicillin to army doctors there. He shared a Nobel prize with Fleming and Chain in 1945 and was president of the Royal Society (1960–65). He became a Life Peer in 1965.

Ford, Henry 1863–1947

US automobile pioneer. Destined to be a farmer, he followed a mechanical bent, becoming successively a watch-repairer, a repairer of agricultural machinery and chief engineer to the Edison Illuminating Company (1887–99). He devoted much of his spare time to constructing his first automobile and in 1899 joined the Detroit Automobile Company. He left in 1903 to found the Ford Motor Company, with the aim of producing a popular-price vehicle. This he achieved in 1909 with the famous Model T, of which he sold 15 million over the next 19 years. Major factors in its success were his insistence on rigorous standardization and the use of conveyer belts for assembly. His famous dictum that customers could have any color they wanted, provided it was black, was due to the fact that only black paint would dry quickly enough to keep pace with assembly.

Freud, Sigmund 1856–1939

Austrian pioneer of psychoanalysis. After graduating in medicine in Vienna in 1881 he specialized there for a time in neurology before going to work in Paris with the eminent French neurologist J.M. Charcot. Returning to Vienna, he set up in practice as a consultant on nervous disorders but became disillusioned with existing forms of treatment, notably hypnotism and electrotherapy. He developed a technique of "free association" to penetrate the patient's subconscious, thus founding what is now known as psychoanalysis. For a time (1906) he was associated with Alfred Adler and Carl Jung in the International Association for Psychoanalysis. Freud

identified repressions of the subconscious with infantile sexuality. He believed in the significance of dreams and formulated the concept of the "id" – the subconscious drive – and the "ego", the executive force. His theories had a considerable effect on contemporary art and literature.

Fukui, Kenichi 1918–

Japanese chemist. He graduated as an engineer at Kyoto University in 1948: he was appointed professor of physical chemistry there in 1951. The 1930s had seen considerable advances in the theory of valency – the force which binds atoms together to form molecules. From the 1950s Fukui devoted himself to studying the mechanics of the way in which molecules interact, developing (1954) the hypothesis that the site of reaction is the highest electron orbital of one and the lowest electron orbital of the other. This results in a new combined orbital which he called a frontier orbital. Similar ideas were developed by the American chemist Roald Huffmann, with whom Fukui shared a Nobel prize in 1981.

Funk, Casimir 1884–1967

Polish biochemist. After studying in Berne and London he became biochemist in the Pasteur Institute, Paris (1904–06). After appointments with the Cancer Hospital, London, and in New York, he returned to Warsaw as director of the State Institute for Hygiene (1923–27). Returning finally to New York as a research consultant, he founded the Funk Foundation for Medical Research in 1953. He published many papers in the field of nutritional science and discovered that yeast is effective for curing beri-beri. As it appeared that all food factors associated with deficiency diseases belonged to the chemical group known as amines, he suggested the name vitamine. When this assumption was proved wrong the modern form, vitamin, was introduced in 1920.

Gabor, Dennis (Denes) 1900–79

Hungarian-British electrical engineer. After studying in Budapest and Berlin, he worked briefly with the electronics companies Siemens and Halshe in Germany, but with the Nazis' rise to power (1933) he emigrated to Britain, and joined the Thomson-Houston Company (1934–48), then moving to Imperial College, London, latterly as professor of applied electron physics (1958–67). In 1947, literally in a flash of inspiration while waiting his turn for a tennis court, he conceived the idea of holography, originally as a way to improve the performance of electron microscopes. In producing a hologram a beam of monochromatic light is reflected from an object on to a photographic film and a second (reference) beam goes direct to the film. The two beams are combined to form a three-dimensional image. Gabor was awarded a Nobel prize in 1971.

Gajdusek, Daniel Carleton 1923–

US virologist. After studying medicine at the University of Rochester and Harvard Medical School he held a number of overseas research appointments. In 1958 he was appointed director of a program – sponsored by the National Institute of Health, Bethesda – to study child growth and disease patterns in primitive communities. He visited Papua New Guinea and came in contact with the Fore people, among whom a fatal disease called kuru is endemic. He identified the cause as a kind of virus previously unknown. This is the "slow" virus, which takes up to a year to develop. In the case of the Fore people its transmission is possibly linked to their cannibalism, in which the brains of their dead are virtually eaten. Gajdusek was awarded a Nobel prize in 1976.

Geiger, Hans Wilhelm 1882–1945

German pioneer of atomic physics and inventor of the Geiger counter. After graduating in physics in Germany he did postdoctoral research on the discharge of electricity through gases, before going to Britain to work with Ernest Rutherford in Manchester. They devised a method for counting alpha-particles and showed that these have two units of charge. In 1909, with E. Marsden, he demonstrated that, exceptionally, the particles showed a very large deflection when directed at gold leaf. This led to Rutherford's initial concept of the nuclear atom. In 1912 Geiger returned to Germany and in 1925 became professor of physics at Kiel. There, with W. Müller, he perfected his famous counter. This consists of a tube filled with a mixture of argon and halogen gas down which passes a high-voltage wire. If a charged particle passes through the tube it causes a discharge from the wire, which is quenched by an electronic circuit, activating a counter.

Gell-Mann, Murray 1929–

US theoretical physicist, noted for the identification of new atomic particles. He studied physics at Yale University and MIT. In 1956 he was appointed professor of physics (later, 1967, of theoretical physics) at the California Institute of Technology. His research has been almost entirely concerned with the classification and description of elementary particles and with their interactions. As a young man he introduced the concept of "strangeness", a quantum number that must be conserved in strong and electromagnetic reactions between particles. With Yuval Ne'eman he deduced from this that elementary particles can be categorized as multiples of 1, 8, 10 or 27 units. This led to the prediction of the omega-minus particle, experimentally identified in 1964. With G. Zweig he developed the concept of quarks, from which other elementary particles (hadrons) can be derived. He was awarded the Nobel Prize for Physics in 1969.

Glaser, Donald Arthur 1926–

US physicist. He graduated in 1946 and did postgraduate research on cosmic rays at the California Institute of Technology. He was professor of physics at the Universities of Michigan (1949–59) and California (Berkeley; 1959–64). His interest turned increasingly to molecular biology and he was appointed to a new professorship in physics and molecular biology. In the 1950s the detection of new high-energy particles was becoming increasingly difficult, as the cloud chamber often failed to detect their tracks. Glaser therefore developed a particle detector based on a different principle. If particles are shot through a superheated liquid, their track is revealed by a row of tiny bubbles. Bubble chambers are now essential adjuncts to the big new particle accelerators. Glaser was awarded a Nobel prize in 1960.

Goddard, Robert Hutchings 1882–1945

US physicist. He was educated at Worcester Polytechnic Institute and Clark University, where he remained, as professor of physics. His interest in rocketry derived from a desire to investigate the physics of the upper atmosphere. In 1929 he launched a liquid-fueled rocket carrying camera, barometer and thermometer. For the next ten years he worked mostly at a research station in New Mexico. By 1935 his rockets had achieved heights of 2.3km (1.5mi). He lodged over 200 patents, from which he gained little reward, but in 1960 the US government paid his widow a handsome indemnity for the use they had made of them in developing the space research program.

Goeppert-Mayer, Maria 1906–72

German-US theoretical physicist who made major contributions to the understanding of atomic structure. In 1924 she enrolled in the University of Göttingen to study mathematics but, attracted by the exciting new developments in quantum mechanics in which her professor, Max Born, was closely involved, she soon turned to physics. From 1929 she worked at Johns Hopkins University on physical aspects of chemistry including the cause of color in organic compounds. In 1939 she and her chemist husband went to Columbia University, where she worked in the SAM Laboratory on the separation of uranium isotopes. After the war she was appointed professor of physics in Chicago. There in 1948 she began research on the so-called "magic numbers". In the 1920s it had been found that nuclear stability was associated with the number of nucleons in the nucleus, the favored numbers being 2, 80, 20, 50, 82 and 126. This discovery had been made empirically – Goeppert-Mayer's great achievement was to give a rigorous theoretical interpretation. For this she was awarded a Nobel prize in 1963, jointly with J.H.D. Jensen, who had reached similar conclusions.

Goldmark, Peter Carl 1906–77

Hungarian-US physicist. After studying physics in the University of Vienna (1925–31), he joined the television division of Pye Radio in Cambridge, Massachusetts (1931–33) and then went on to New York, first as a consulting engineer and then (from 1936) on the staff of the research organization of CBS, where he became director of engineering research and development in 1944 and vice-president in 1956. With a keen interest in television, he introduced in 1940 the first practicable (field sequential) color system. However, this was a photomechanical system and proved not to lie on the main line of development. He therefore turned to all-electronic systems and in 1946 received the Zworykin Award for the development and utilization of electronic television. In 1948 he invented the long-playing (LP) microgroove record.

Golgi, Camillo 1844–1926

Italian histologist, pioneer of research on the microstructure of the nervous system. After graduating in medicine at Pavia he was for seven years physician in the hospital there. In 1875 he was appointed professor of histology, and later professor of general pathology in Pavia. He was much interested in the dye-stain techniques being developed by bacteriologists and in 1873 discovered that nervous tissue could be differentiated by staining with silver. He classified nerve cells and showed that their fibers did not join directly but were separated by short gaps (synapses). In the 1880s he discovered significant differences between the intermittent and the pernicious types of malaria parasite. In 1898 he described an important feature (Golgi body) present in the cytoplasm of many cells. In 1906 he shared a Nobel prize with S. Ramón y Cajal: he was the first Italian to achieve this honor.

Granit, Ragnar Arthur 1900–

Swedish neurophysiologist. He graduated in medicine at Helsinki University, Finland and was appointed professor of physiology there in 1937. In 1940 he moved to Stockholm as professor of neurophysiology (declining an invitation from Harvard) and subsequently became director of the department of neurophysiology in the Medical Nobel Institute (1945–47). Early in his career he became interested in the visual process, approaching its study initially through psychology, but soon turning to physiology. Using electro-physiological methods he demonstrated that the retina contains three different types of cones differentiated by their spectral sensitivity. From 1947 he turned his attention to the different nerves serving the muscular system and their relationship to the spinal cord. In 1967 he shared a Nobel prize with two other pioneers in the physiology of vision, H.K. Hartline and G. Wald.

Haber, Fritz 1868–1934

German Jewish physical chemist. In the early 1900s there was grave concern about the rapid depletion of natural sources of nitrogenous fertilizers. Processes of fixing atmospheric nitrogen were developed in Norway and Germany, based on blowing air through an electric arc, but they were not cost-effective. Haber, director of the Kaiser Wilhelm Institut für Physikalische Chemie und Electrochemie in Berlin (1911–33) developed a far more efficient process based on the combination of nitrogen and hydrogen, to form ammonia, at high pressure. He was awarded the Nobel Prize for Chemistry in 1918 but the Nazis ended his career; he spent the rest of his life in England.

Hassel, Odd 1897–1981

Norwegian chemist. He did postgraduate research in Copenhagen and Berlin before returning to Oslo, where he later became professor of physical chemistry (1934–64). His main focus was the investigation of crystal structure using X-ray and electron diffraction techniques. He also investigated electrical dipole movements (the segregation of charge within molecules) and the stereochemical configuration of organic molecules.

Hawking, Stephen William 1942–

British theoretical physicist. After reading physics at Oxford he did postgraduate research at Cambridge, where he has remained. In 1979 he was appointed to the Lucasian professorship of mathematics. Despite a disease which has all but robbed him of speech and movement, he has directed his genius to elucidating the structure of black holes showing (1971) that these can issue not only from the collapse of stars but also – as mini-black holes – from the time of the original "big bang". He predicted that when found such mini-black holes will emit "Hawking radiation" at a fixed rate. He is now working on a quantum mechanical theory of gravity which will subsume the four basic forces.

Heisenberg, Werner Karl 1901–76

German physicist, founder of quantum mechanics. He studied physics in Munich and then lectured under Max Born and Niels Bohr (1924–26). Back in Germany, he was professor of theoretical physics at Leipzig 1927–41; director of the Max Planck Institute in Berlin, 1941–45; and professor of physics in Berlin. He considered Bohr's atomic theory to be inadequately supported by experiment and in 1927 formulated the system of matrix mechanics from which the wavelength and intensity of spectral lines – both easily measurable – could be deduced. Also in 1927 he put forward his revolutionary uncertainty principle, according to which it is impossible exactly to determine both the position and momentum of a body at any given moment. He received a Nobel prize in 1932.

▲ Dorothy Hodgkin

▲ Bernardo Houssay

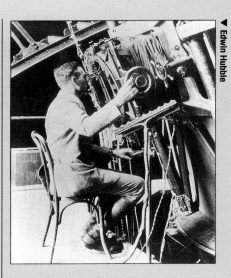

▲ Edwin Hubble

Hess, Harry Hammond 1906–69

US geologist. After graduating at Princeton he joined Loangwa Concession, Northern Rhodesia, as geologist in 1928. In 1934 he returned to Princeton, later becoming professor of geology. From the 1930s he carried out gravity measurements and echo sounding experiments with the US navy which gave him a keen interest in submarine geology and gravitational anomalies. After the war there was growing interest in this kind of research and in 1957 W.W. Ewing demonstrated the global distribution of deep central rifts in the mid-ocean ridges, associated with anomalously young deposits in the ocean depths. Hess suggested that material rose constantly from the earth's mantle to form the mid-ocean ridges and then spread out sideways until it reached the continental margins, sank below the lighter continental crust and eventually found its way back to the mantle.

Hevesy, Georg (György) von 1885–1966

Hungarian chemist, pioneer of the technique of isotope labeling. After studying at Budapest and Freiburg he worked briefly (1911–13) with Rutherford in Manchester. There he was set the task of separating radioactive Radium-D from ordinary lead: as they proved chemically inseparable he concluded that they were in fact isotopes of the same element. As extremely small quantities of radioactive isotopes can be accurately traced with the aid of Geiger counters, or photographic film, he conceived the idea that they might be used to follow the normal (nonactive) atoms of elements through chemical reactions. This idea he worked out with F.A. Paneth in Vienna in 1913, determining the solubility of lead sulphide and lead chromate in water, too low to be measured by existing methods. Subsequently he made extensive use of such marker elements: for example, to trace the absorption of phosphate in human tissue, using radiophosphorus (1934). He was awarded a Nobel prize in 1943.

Hewish, Antony 1924–

British radio astronomer. He graduated in Cambridge in 1948 and did postgraduate research there before going to the Royal Aircraft Establishment, Farnborough (1943–46). Returning to Cambridge, he was a lecturer in physics until appointed professor of radio astronomy in 1971. He designed a special form of radio telescope, completed in 1967, to examine a variation in radio signals from space analagous to the twinkling of stars, noticed after the war. He and his research assistant S.J. Bell picked up signals in the 3.7m waveband which had an unusual characteristic; they fluctuated steadily. The source proved to be a tiny neutron star within our own galaxy. It was the first pulsar. For this discovery Hewish was awarded a Nobel prize in 1974.

Hodgkin, Dorothy Mary Crowfoot 1910–

British chemist, distinguished for elucidating the structure of large biological molecules. After studying chemistry at Oxford (1928–32), she spent two years at Cambridge working wth J.D. Bernal on the structure of sterols. In 1934 she returned to Oxford to work in the sub-department of crystallography, eventually as Wolfson research professor (1960–77). In 1941 she began an investigation, with the aid of X-ray crystallography, of the structure of penicillin, then the subject of intensive research in Oxford. Progress was slow because computational methods were then very laborious but a complete structural analysis was completed in 1945, confirming a result achieved by purely chemical methods. In 1956 she and her group elucidated the structure of vitamin B12. Now assisted by computer technology, in 1972 she announced the structure of insulin. In 1964 she received the Nobel Prize for Chemistry.

Hopkins, Frederick Gowland 1861–1947

British biochemist, discoverer of essential food factors later called vitamins. After qualifying in both chemistry and medicine in London, he became lecturer in chemical physiology at Cambridge (1908–14) and subsequently professor of biochemistry (1914–43). He had a wide range of biochemical interests and isolated the animo acids tryptophan and glutathione. He also showed, for example, that when a little fresh milk was fed to rats fed on a synthetic diet they resumed normal growth. He did not himself succeed in isolating a specific vitamin but the originality of his work earned him a Nobel prize in 1929 (shared with C. Eijkman, another pioneer in this field). Perhaps his greatest contribution was his insistence that the methods of the physical sciences should be applied to the solution of biological problems.

Hounsfield, Godfrey Newbold 1919–

British physicist. After studying in London at the City and Guilds College and Faraday House he served in the Royal Air Force (1939–46). He returned to Faraday House (1947–51) and then joined EMI (Electrical and Musical Industries), becoming initially head of medical systems and later, senior staff scientist (1977). His early work was on radar and computers: in 1958–59 he led the design team for the first large fully transistorized computer to be built in the UK (the EMIDEC 1100). Later he developed the technique of X-ray computerized tomography or CAT (1969–72). If soft tissues are scanned with X-rays the differences in absorption are too small to be very revealing: in the CAT technique these slight differences are accentuated to give an effective degree of contrast by means of a computer. In 1979 Hounsfield shared a Nobel prize with A.M. Cormack, a South African physicist who independently developed a similar scanning system.

Houssay, Bernardo Alberto 1887–1971

Argentinean physiologist remembered for his work on the endocrine glands. When he graduated in medicine from the University of Buenos Aires in 1911 he was already interested in the physiological role of the pituitary gland. This engaged his interest throughout a lifetime of research devoted largely to the endocrine glands. Houssay showed that among the hormonal products of the pituitary is one which increases, and can even induce, the symptoms of diabetes. For his research demonstrating that hormones do not act wholly individually, but are part of a complex interacting system, Houssay was awarded a Nobel prize in 1947. The Argentinean press – controlled by the dictator Juan Perón, with whom Houssay had longstanding differences – was critical, alleging that the award was designed to embarass Perón. Houssay's reputation was restored when Perón was exiled in 1955, and he resumed his career at Buenos Aires University.

Hubble, Edwin Powell 1889–1953

US astronomer. He began life as a lawyer, graduating at the Universities of Chicago and Oxford, but an amateur interest in astronomy led to his making this his life's work, mostly at the Mount Wilson Observatory, California, with its 2.54m (100in) telescope. In 1923 he showed that the nebulous outer part of the Andromeda galaxy in fact consisted of a myriad of individual stars. Turning to other galaxies, he was able by 1929 to measure the speed of recession from the Earth of 18 of them. Analyzing the results, he showed that this speed of recession is proportional to their distance (Hubble's Law). This gave the first positive evidence for the concept of an expanding Universe, advanced some years previously. Calculation of the distance at which the recession reached the speed of light, and calculating backward to the moment of zero speed, led him to put the boundary of the Universe at a distance of 18 billion light years. He estimated its age as two billion years, but later research suggests that this was a tenfold underestimate.

Hubel, David Hunter, 1926–

Canadian-US neurophysiologist, noted for research on the physiology of vision. He graduated at McGill University, Montreal, first in mathematics and physics (1947) and then in medicine (1951). After various appointments in neurology he joined the Harvard Medical School in 1959, where since 1982 he has been John Franklin Enders professor. There he was closely associated with Torsten Nils Wiesel. Using electro-physiological techniques, they investigated the response to light of cells in the cerebral cortex. This led to the identification of regions responsive to specific photo-stimuli, and to a detailed mapping of the whole visual cortex. Hubel and Wiesel shared a Nobel prize in 1981.

▲ Pyotr Kapitsa

▲ Har Gobind Khorana

▲ Hans Adolf Krebs

Hull, Albert Wallace 1880–1966

US electron physicist. He graduated in Greek at Yale and taught French and German for a year, then returned to Yale to graduate in physics and after five years teaching joined the General Electric Research Laboratory, Schenectady, in 1914, working initially on electron tubes (valves). In 1921 he published a classic paper on the motion of electrons in a magnetic field between coaxial cylinders. He called this configuration a magnetron. Also in the 1920s he invented the thyratron, a heavy-duty triode for converting alternating to direct current for long-distance power transmission. Independently of Walter Schottky he devised the tetrode tube (valve). In the 1930s his interest turned to glasses and metallurgy. This led to alloys with expansion coefficients similar to those of glass, making strain-free vacuum seals possible.

Jansky, Karl Guthe 1905–50

US physicist, pioneer of radio astronomy. He graduated in physics at Wisconsin in 1928 and then joined Bell Telephone. One of his assignments was to investigate the sources of the "static" which interferes with short-wave radio reception. Using a directional aerial he identified one constant source as lying in the direction of Sagittarius and suggested in 1932 that this corresponded with some source of radio emission in that constellation. Surprisingly – for the use of radio waves could be useful in astronomy because, unlike light, they penetrate interstellar dust – this observation aroused little interest. For all practical purposes radio astronomy is a postwar field of research. However, Jansky's name is commemorated in the international unit used to measure the power received at a radio telescope from a cosmic source.

Josephson, Brian David 1940–

British theoretical physicist, discoverer of "tunneling" between superconductors (Josephson Effect). After graduating in physics at Cambridge he remained there in the Cavendish Laboratory as, successively, director of research in physics (1967–72), reader (1972–74) and professor (1974). The two discoveries by which he is known was made in 1962 while he was still a research student. If two superconductors are separated at very low temperatures by a narrow insulating gap, current will pass from one to the other without an electric potential being applied. Conversely, if an electric potential is established across such a gap an alternating current will flow in proportion to the potential. The effects find application in various high-speed switching devices as in memory cells and logic circuits. Such applications led Josephson to interest himself in the theory of intelligence. In 1973 he was awarded a Nobel prize jointly with Leo Esaki and Ivar Giaever.

Kamerlingh Onnes, Heike 1853–1926

Dutch physicist. After graduating in physics at Gröningen, he went to Heidelberg in 1871 to work with R.W.E Bunsen and G.R. Kirchhoff. In 1882 he became professor of physics at Leiden and remained there. Influenced by J.D. van der Waals, he embarked on a lifelong study of the properties of gases and liquids over a wide range of pressures and temperatures. Eventually he concentrated on very low temperatures near absolute zero, and founded the famous Cryogenic Laboratory in 1894. In 1908 he succeeded in liquefying helium and he discovered the phenomenon of superconductivity (1911) whereby the resistance of electrical conductors vanishes at near-zero temperature. This is a phenomenon of major significance in solid state physics. He received a Nobel prize in 1913.

Kapitsa, Pyotr Leonidovich 1894–1984

Russian physicist. He graduated in electrical engineering in 1918 at Petrograd (Leningrad) Polytechnic and in 1921, sent by the Soviet government to buy scientific equipment, went to the Cavendish Laboratory, Cambridge, where he remained for 13 years and worked with Rutherford. He investigated the electrical properties of metals in high magnetic fields and at very low temperatures, showing much experimental ingenuity in achieving both conditions. He discovered the phenomenon of superfluidity (complete loss of viscosity) in liquid helium. In 1934 he visited the USSR and was forbidden to return to Cambridge, though the Soviet government negotiated a deal to enable his equipment to be sent to Moscow. During the war he worked on the mass production of oxygen for steel-making. This brought him into conflict with the notorious Beria, head of the KGB. He was dismissed, but later rehabilitated, and did important research in high-temperature physics (plasma) and lightning. Very belatedly, at the age of 84, he was awarded a Nobel prize for his early research on low temperatures.

Kendrew, John Cowdery 1917–

British molecular biologist, noted for his use of X-ray crystallography to determine the structure of proteins. After graduating at Cambridge, he served in the ministry of aircraft production. After the war, he joined Max Perutz in a small unit set up in the Cavendish Laboratory to study the chemistry of large molecules of biological importance. This was to develop into the world famous MRC Laboratory for Molecular Biology, of which he became deputy chairman. From 1975 to 1982 he was the first director general of the European Molecular Biology Laboratory in Heidelberg, and subsequently president of St John's College, Oxford. In 1947 he and Perutz began a project to elucidate the structure of proteins by means of X-ray analysis. They chose hemoglobin, a

substance of intermediate complexity, whose vital role is to carry oxygen in the blood, and the related protein myoglobin, derived from the muscles of sperm whales. By 1953 full structures for both substances were announced. They were jointly awarded a Nobel prize in 1962

Khorana, Har Gobind 1922–

Indian-US molecular biologist, the first to synthesize a gene. After studying science at the Universities of the Punjab, Liverpool, Zurich (ETH) and Cambridge, UK, he did research at Simon Fraser University, Canada (1952–60); then at the Institute for Enzyme Research, University of Wisconsin; and finally at MIT. Much of Khorana's research was devoted to the synthesis of nucleic acids and, ultimately, to the synthesis of a gene. In 1968 he shared the Nobel Prize for Physiology or Medicine with R.W. Holley and M.W. Nirenberg, who also made major contributions to the deciphering of the genetic code.

Kolff, Willem Johan 1911–

Dutch physician. With a small group of colleagues he succeeded, despite the German occupation of the Netherlands, in devising in 1944 a machine that would take over the function of the kidneys. In this, blood was circulated through a cellophane tube immersed in a water-bath. Toxic constituents of the blood diffused out into the water, while protein and blood cells were returned to the patient's body. The machine could provide only temporary relief while damaged kidneys recovered. It was, nevertheless, the prototype of the dialysis machines of the 1960s by which patients with renal defects could remain active indefinitely with the help of regular dialysis.

Krebs, Hans Adolf 1900–81

German biochemist, notable for his elucidation of metabolic pathways. After studying at several German universities, as was then customary, he graduated in medicine at Hamburg in 1925 and then worked under Otto Warburg in Berlin (1925–30) and at Freiburg. With the rise of the Nazis (1933) he emigrated to England, working first under F. Gowland Hopkins at Cambridge, then at Sheffield and finally (1954) as professor of biochemistry in Oxford. Throughout his career the chemistry of metabolic processes had a particular fascination for him, especially those concerned with the utilization of energy. He is remembered for three main developments. First, the synthesis of urea in the liver. Second, the tricarboxylic acid cycle for the oxidation of pyruvic acid to carbon dioxide and water: this is now generally known as the Krebs Cycle. Third, in association with Hans Kornberg, the glyoxalate cycle, important in fat metabolism. In 1953 he shared a Nobel prize with Fritz Lippmann, another pioneer of the biochemistry of metabolism.

Land, Edwin Herbert 1909–

US inventor. Although he enrolled at Harvard he never graduated (though he was awarded an honary degree in 1957). Scientific research often requires polarized light, in which the wave vibrations are in a single plane. Up to the 1930s this was commonly obtained by means of a large (dichroic) crystal of Iceland spar. Crystals of certain organic chemicals are also dichroic but could not be grown large enough. As a student Land realized that the necessary effect could be achieved if easily obtainable small crystals were permanently aligned in a transparent sheet of plastic and he abandoned his studies to develop this, using crystals of quinine iodosulfate, and marketed it under the name Polaroid. It was used not only in scientific instruments but in photography and sunglasses. A second major achievement (1947) was the Polaroid Land instant camera. In this positive and negative paper were combined with developer in a single flat pack. The developer was released by passing the pack through a roller, when the picture appeared within a minute or two.

Landau, Lev Davidovitch 1908–68

Soviet theoretical physicist. He graduated from the physics department of the Leningrad Physico-Technical Institute in 1927. A Rockefeller Fellowship (1929–31) allowed him to do research in Germany, Switzerland, England and Copenhagen (where he was particularly influenced by Niels Bohr). He returned to the Soviet Union where he joined the Physico-Technical Institute, Kharkov (1932–37) and, from 1937, was head of the theoretical department of the Soviet Academy of Sciences Institute of Physics in Moscow. He never fully recovered after an automobile accident in 1962. His interest in theoretical physics was catholic. After P.L. Kapitsa's discovery of the superfluidity of liquid helium he made a prolonged study of the so-called quantum liquids at very low temperatures. This led to the formulation of a detailed theory to explain their behavior. With E.M. Lifshitz he wrote, from 1938, a famous series of textbooks of physics. He was awarded a Nobel prize in 1962.

Landsteiner, Karl 1868–1943

Austrian-US immunologist. He graduated in medicine in Vienna in 1891 and then spent five years studying chemistry at various European centers. He returned to Vienna for some years, before working (1919–22) in Holland. He then accepted an invitation to work in the USA at the Rockefeller Institute of Medical Research. His interests were wide – ranging from morbid anatomy to typhus – but his main contribution was in immunology, especially that associated with the blood. By 1909 he had identified the four main blood groups recognized today (A, B, AB and O). This made blood transfusion – previously

extremely hazardous – feasible as a routine procedure. He continued his research on blood and in 1940, with A. Wienes and P. Levine, announced the discovery of the rhesus (rh) factor in the red blood cells of certain individuals. In certain circumstances this is of critical importance. Landsteiner was awarded a Nobel prize in 1930.

Langmuir, Irving 1881–1957

US industrial chemist, remembered for research on thermionic emission and the properties of surfaces. After graduating in the USA (metallurgical engineering, 1903) and Germany (physical chemistry, 1906) he returned to America and in 1909 joined the research laboratories of the General Electric Company at Schenectady. He remained there for the rest of his working life, later (1932–50) as associate director. He was a versatile research worker, but two particular achievements were outstanding. One was the invention of the coiled-coil filament which much improved the efficiency of electric lamps, especially when coupled with the further improvement of replacing the vacuum with an inert gas. His investigation of the spread of oils and other immiscible liquids to form monolayers on water gave a measure of the size and shapes of molecules. During the two world wars he was associated with the development of submarine detectors, the improvement of smoke screens and other devices. He was also concerned with stimulating rainfall by "seeding" cumulus clouds with chemicals. He was awarded a Nobel prize in 1932.

Laue, Max Theodor Felix von 1879–1960

German physicist. At an early age he abandoned classics for physics and became assistant to Max Planck at the Institute of Theoretical Physics, Berlin (1905–19) where he worked on Albert Einstein's Special Theory of Relativity. Moving on to Munich, he also did research on wave-optics and reached the conclusion that short-wave electromagnetic radiation should be diffracted by crystals. In 1912, using copper sulfate, his assistants Friedrich and Knipping verified this. His discovery was subsequently used by W.H. and W.C. Bragg to investigate the structure of crystals. He was awarded the Nobel Prize for Physics in 1914, as were the Braggs in 1915. After subsequent appointments in Zurich and Frankfurt he became professor of theoretical physics in Berlin, resigning in 1943 in protest at Nazi racial policy. After the war he worked to revive German science.

Lawrence, Ernest Orlando 1901–58

US physicist. After gaining his doctorate at Yale in 1925 he was on the staff of the University of California, latterly (1936–58) as director of the Radiation Laboratory. A.S. Eddington's suggestion that nuclear reactions might occur within stars at very high energies led Lawrence to explore the

possibility of effecting nuclear reactions with high-energy particles generated in the laboratory. With M.S. Livingston he conceived the idea of achieving a very long path within a manageable space by repeatedly accelerating particles along a circular path. This machine they called a cyclotron, and with it a number of radioactive isotopes were prepared. Later a number of synthetic elements were prepared in his Laboratory. In 1961, after his death, element 103 was detected and named Lawrencium (Lw) in his honor. He was awarded a Nobel prize in 1939.

Lee, Tsung Dao 1926–

Chinese-US theoretical physicist. After studying physics at National Zhejiang University and the National Southwest University, Kunming, he went to the USA and took his doctorate at the University of Chicago. After further research at Princeton and Columbia Universities, he was appointed professor of physics at Columbia in 1956. He is particularly identified with the formulation of what are called the parity laws. Conventional physical theory asserted that no fundamental difference exists between left and right – that is, the laws of physics are the same whether expressed in a left- or a right-handed system of coordinates. Lee's highly original contribution was to show that in some circumstances (weak interactions) this assumption is invalid. With his close collaborator C.N. Yang he received a Nobel prize in 1957: they were the first Chinese to be so honored.

Lemaître, Georges Edouard 1894–1966

Belgian cosmologist. He studied at the University of Louvain and was ordained as a Catholic priest. After a period of study in the UK and America he returned to Louvain as professor of astronomy. On the basis of Einstein's relativity theory he deduced – independently of A.A. Friedmann – that the Universe must now be expanding, having begun as a small, very highly compressed unit – the "primal atom". This theory (commonly referred to as the "big bang" theory), was validated experimentally in 1929 by E.P. Hubble.

Levi-Montalcini, Rita 1909–

Italian-Jewish neurophysiologist. A graduate in medicine, after spending World War II in hiding, she went in 1947 to the USA to do research at Washington University on the embryonic nervous system and discovered (1947) that it produces many more cells than are finally required, the number eventually adjusting itself to the volume of tissue to be enervated. She showed that the controlling substance is a specific nerve growth factor (NGF). With the biochemist S. Cohen she discovered that mouse saliva is an excellent source of NGF. Returning to Rome she became director of the Laboratory for Cell Biology, retiring in 1979. In 1986 she and Cohen were awarded a Nobel prize.

Libby, Willard Frank 1908–80

US chemist. After graduating in 1931 at the University of California, Berkeley, he taught there until 1941, when he joined the Manhattan Project. After the war he became a professor of chemistry in Chicago University and then returned to California in 1959 as director of the Institute of Geophysics and Planetary Physics. He is renowned for dating archeological artifacts by measuring their radiocarbon content (1947). The technique is based on the fact that natural carbon contains a small admixture of a radioactive isotope ^{14}C which decays with a half-life of 5,770 years. Living material contains the two isotopes in constant proportion, in equilibrium with atmospheric carbon dioxide. When it dies, the interchange ceases and the ^{14}C begins to decay. By measuring the residual radiocarbon content the age of the material can be calculated. Libby won the 1960 Nobel Prize for Chemistry.

Lorenz, Konrad Zacharias 1903–89

Austrian physician, founder of ethology, the science of animal behavior. After graduating in medicine in Vienna he held a succession of posts there, in anatomy and then in psychology. In 1940 he became professor of psychology in the University of Königsberg. After wartime military service, and capture, he returned to Germany and held various posts in ethology, finally (1961) as director of the Max Planck Institute, Seewiesen. In the 1930s his interest moved from human psychology to that of animals. From a study of birds he concluded that their behavior is not entirely flexible but derives from visual images imprinted in the young. He preferred to draw conclusions from studies in the wild rather than in the laboratory. His attempts to relate the behavior of animals to that of humans had a mixed reception. He was awarded a Nobel prize in 1973.

Lovell, Alfred Charles Bernard 1913–

British physicist. After graduating in physics at Bristol University (1936) he was appointed lecturer in physics at Manchester University. During World War II he was concerned with radar research, afterward returning to Manchester, where observation that radar echoes could be obtained from meteor showers stimulated his interest in the neglected possibilities of radio astronomy. From 1951 until 1981 he was the first professor of radio astronomy there. He was responsible for building at Jodrell Bank a 75m (250ft) steerable dish to seek centers of radio emission in the Universe.

Maiman, Theodore Harold 1927–

US physicist. After military service with the US Navy, he studied engineering physics at Colorado University and joined the Hughes Research Laboratories in Miami in 1955. He left in 1962 to found the Korad Corporation and other industrial firms to develop and manufacture lasers. The principle of the laser (Light Amplification by Stimulated Emission of Radiation) had been illustrated in 1958. In 1960 Maiman perfected such a device, capable of producing pulses of very intense monochromatic light.

Marconi, Guglielmo 1874–1937

Italian commercial radio pioneer. He was educated privately and then at the Leghorn Technical Institute where, in 1894, his imagination was captured by an article discussing the possibility of using for wireless communication the waves discovered by H.R. Hertz in 1888. Within a year he had sent and received signals at distances up to 2 mi (3.2 km). He took out his first patent in London in 1896. His invention interested the British government, particularly the Admiralty, and soon experimental equipment was installed in naval vessels. In 1901 he succeeded in sending signals across the Atlantic. Before World War I he invented a magnetic detector (1902) and a directional aerial (1905): during it he developed short-wave equipment for use over long distances. Thereafter he devoted himself to his business interests. In 1909 he shared a Nobel prize with K.F. Braun.

Matthews, Drummond Hoyle 1931–

British geologist, codiscoverer of anomalies in the magnetism of the ocean bed. After graduating and doing postgraduate research at Cambridge he was appointed geologist to the Falkland Islands Dependencies Survey (1955–57). He returned to Cambridge to do further research in geophysics and was appointed reader in marine geology (1971–82). He then joined the British Institutions Reflection Profiling Syndicate, Cambridge, which uses seismic reflection techniques to study the earth's crust at depths up to 75km (45.6mi). In 1962 H.H. Hess had put forward his seafloor spreading hypothesis. Matthews produced experimental evidence in support of this by showing (with F.J. Vine) that the earth's crust on either side of the ridges is magnetized in different directions in bands running parallel to the ridges. This is in accordance with the fact that the direction the Earth's magnetic field has reversed many times over geological ages and the alternations of remanent magnetism correspond to the magma having solidified at different times.

Matuyama, Motonori 1884–1956

Japanese geologist. After studying at the Imperial University, Kyoto, and at Chicago, he became professor of geology in Kyoto. Studying the remanent (frozen-in) magnetism of basalts he discovered that over approximately the last 5 million years the Earth's magnetic field has reversed in polarity at least 20 times. This total reversal is quite different from the continuous wandering of the north magnetic pole. The cause of this is still uncertain. The main period of change was between one and two million years ago: this is known as the Matuyama reversal epoch.

Mauchly, John William 1907–80

US electronic engineer. He graduated in engineering from Johns Hopkins University in 1927, and after appointments there and at Ursinus College, Pennsylvania, he joined the University of Pennsylvania (1941–43). There, in collaboration with J.P. Eckert, he began to develop, for the US Army Ordnance Department, the first electronic computer. This was ENIAC (Electronic Numerical Integrator and Calculator), completed in 1946. Subsequently he collaborated with Eckert in various business enterprises, setting up in 1948 the Eckert-Mauchly Computer Corporation, incorporated in Remington Rand in 1950. In 1951 they introduced, for the UNIVAC machine, the use of magnetic tape for programming.

Merrill, John Putnam 1917–84

US physician. After qualifying in medicine at Harvard he did postgraduate research in Cambridge, UK, and Paris. Returning to Boston in 1947, he held appointments in the Brigham Hospital and later in the Harvard medical faculty. In 1962 he became associate professor in clinical medicine. His research interest lay in renal physiology and the role of renal pathology in hypertension. He was the first to utilize an artificial kidney in the USA, and this stimulated his interest in kidney transplant surgery. Here the problem of rejection was a major difficulty but in the 1950s he successfully carried out transplants between twins – identical and nonidentical – where genetic makeup is very similar. He also successfully transplanted from accident victims.

Milstein, César 1927–

Argentinean-British molecular biologist. After graduating in chemistry in Buenos Aires he spent three years as a postgraduate student in Cambridge, UK (1958–61). Two years later he returned there permanently to become a member of staff of the Medical Research Council, latterly at the Laboratory of Molecular Biology. His research was particularly concerned with the structure, evolution and genetics of immunoglobulins and phosphoenzymes. In particular, he is identified with the discovery of monoclonal antibodies (MCAs): that is, antibodies produced by a line of cells deriving from a single ancestral cell. Such antibodies are all identical and have a unique chemical structure. They are, therefore, specific for a single antigen. In 1975, with G. Köhler, he developed a technique (hybridoma technique) for producing MCAs in large quantities. They have great therapeutic and diagnostic value. In 1984 he shared a Nobel prize with Köhler and N. Jerne, an immunologist with similar research interests.

Jacques Monod

Thomas Morgan

Hermann Muller

Monod, Jacques Lucien 1910–76

French biochemist. A biology graduate, he did postgraduate research and then moved to the zoology department of the University of Paris (1934–45). After the war he joined the Pasteur Institute, becoming its director in 1971. He shared a Nobel prize in 1965 with F. Jacob and A. Lwoff, with whom he conceived the idea of the operon, a group of genes, rather than a single gene, which controls enzyme synthesis. These are regulated by operators, located near the ends of chromosomes, and the operators are switched on or off by protein units known as repressors. He later speculated about the origin of life, believing it to have been a chance event, subsequently controlled by Darwinian selection.

Morgan, Thomas Hunt 1866–1945

US geneticist and embryologist. A zoology graduate, he spent most of his professional career at Columbia University (1904–28) and the California Institute of Technology (1928–45). A man of wide interests, he favored the direct experimental method over the then prevalent descriptive approach to biology. His research falls into four main phases: embryology, 1895–1902; evolution and heredity, 1903–10; heredity in the fruitfly *Drosophila*, 1910–25; and embryology, again, in relation to heredity and evolution, 1925–45. Originally skeptical about the validity of J.G. Mendel's laws of inheritance, his own experiments with *Drosophila* convinced him of their basic truth and led him eventually to advance the chromosome theory of heredity. He identified several sex-linked characteristics and in 1911 he and his colleagues published the first chromosome map. By 1922 they had mapped more than 2,000 genes on *Drosophila's* four chromosomes. He was awarded a Nobel prize in 1933.

Moseley, Henry Gwyn Jeffreys 1887–1915

British physicist. It had been shown in Mendeleyev's Periodic Table that if the 60 disparate chemical elements then known are arranged in order of their atomic weights, chemically related elements recur at regular intervals. This left certain anomalies, however; that is, that many elements have atoms of different weights, one of which may fit the table, but not the others. In 1914, working at Oxford (where he had graduated in 1910) Moseley showed that the critical index was not the weight of the atom but the charge on its nucleus.

Mössbauer, Rudolph Ludwig 1929–

German physicist, discoverer of the Mössbauer effect. After working briefly in the Rodenstock Optics Factory, he enrolled in the Munich Technische Hochschule in 1949 and spent the years 1955–57 at the Max Planck Institute for Medical Research, Heidelberg, working for a doctorate. He

then returned for three years to Munich before taking up an appointment as professor of physics in the California Institute of Technology (1961). His discovery of the effect named after him – for which he received a Nobel prize in 1961 – was made at Heidelberg. If a free atomic nucleus emits a gamma ray it recoils, like a gun, and this movement affects the wavelength of the emitted radiation. If the nucleus is firmly enmeshed in a crystal lattice, however, it cannot recoil and so the emitted radiation has a slightly different wavelength. This effect can be used to study the electronic environment of nuclei. In 1960 such a change in wavelength was used to verify Einstein's General Theory of Relativity.

Muller, Hermann Joseph 1890–1967

US geneticist. He graduated in biology at Columbia University, and then studied genetics at Columbia; at the Rice Institute, Texas; in Berlin; in Leningrad and Moscow; in Edinburgh; and finally in the University of Indiana. His great interest was in the production of mutations (sports), both natural and artificial, mainly in the fruit fly *Drosophila*. X-rays, for example, were shown to increase the mutation rate 150 times. This led him to draw attention to the potential danger of such radiation to the human race. He was awarded a Nobel prize in 1946.

Natta, Giulio, 1903–79

Italian industrial chemist. Originally a student of mathematics, he enrolled in the Milan Polytechnic Institute to study chemical engineering. He then held several academic appointments in Italy, before being appointed director of the Milan Institute of Industrial Chemistry (1938–73). One of his earliest tasks there was to organize a program to study the manufacture of synthetic rubber – made urgent by the imminence of war. As a complement to this he investigated the use of olefines as agents in chemical synthesis generally, especially when mediated by catalysts. Through this he became familiar with the work done by K. Ziegler in the low temperature–low pressure production of polythene. He advised the Italian chemical manufacturer Montecatini – to whom he was a consultant – to acquire Ziegler's patent rights and on this basis he extended his own research. This led in 1954 to his perfecting a similar process for polymerizing polypropylene. Polypropylene soon became a major new plastic, in both solid and fiber forms. Natta shared a Nobel prize with Ziegler in 1963.

Néel, Louis Eugène Félix 1904–

French physicist. After graduating at the École Normale Supérieure, Paris, he was appointed professor in the University of Strasbourg (1937–45). He then moved to the University of Grenoble as professor in the faculty of science and was

appointed director of the Centre d'Études Nucléaires in Grenoble when it was founded in 1956, retiring in 1971. Throughout his long research career he took a particular interest in the magnetic properties of solids. In 1936 he predicted, on theoretical grounds, that there should be a new form of magnetism, antiferromagnetism, in which the properties were determined by nonsymmetrical ordering of the spins of unpaired electrons. Above a certain temperature (the Néel critical temperature) such materials should become paramagnetic: that is, they should acquire magnetic properties when located in a magnetic field, losing them when this is removed. Such materials were made in 1938, confirming Néel's prediction. He received a Nobel prize in 1970.

Neumann, John (János) von 1903–57

Hungarian-US mathematician, pioneer of computer theory. A child prodigy in mathematics, he retained his brilliance as an adult. After studying in Berlin, Zurich and Budapest he went to America in 1930 and was appointed professor of mathematical physics at Princeton (1930–33), then moving to the newly founded Institute for Advanced Study at Princeton. During World War II he became deeply involved in the development of computers, and formulated the basic von Neumann Concept. This postulates that the two essential features of a computer are a memory in which to store information and a system to transfer memorized information according to a predetermined program. During the war he was consultant on the Manhattan Project. He is remembered also for his contributions to the theory of games. He was perhaps the last of the great mathematicians who were equally at home in both the pure and applied fields.

Northrop, John Howard 1891–1987

US biochemist. A chemistry graduate, Northrop worked from 1925 to 1962 at the Rockefeller Institute, New York. When J.B. Sumner succeeded in crystallizing an enzyme in 1926, showing it to be a protein, Northrop realized that this opened up new vistas in protein chemistry. Proteins could now be manipulated and studied far more easily. By 1930 he had succeeded in crystallizing several other enzymes. In 1946 Northrop shared a Nobel prize with Sumner and W.M. Stanley, the first to crystallize a plant virus.

Ohain, Hans Pabst von 1911–

German aeronautical engineer. Ohain studied aerodynamics at Göttingen, and realizing that the piston engine/propeller system was nearing its limit for high-speed flight, turned his attention to turbines and rockets. In 1936 he joined the Heinkel Company, taking with him a design for a turbo-jet engine. The world's first jet flight – with a Heinkel He 178 – was achieved on 27 August 1939.

▲ Giulio Natta

▲ Ivan Pavlov (center)

However, the first production jet aircraft, the Messerschmitt 262, did not appear until 1944. The delay was due partly to a major change in design. Whereas Ohain, like Whittle in Britain, had used centrifugal compressors, the Jumo gas turbine which powered the Me 262 used a multistage axial flow compressor, which proved to be on the main line of development.

Palade, George Emil 1912–
Romanian-US cytologist. After qualifying in medicine in Bucharest University Palade was appointed professor of anatomy (1940–45). After the war he emigrated to the USA (1946) and did research first in the Rockefeller Institute and later (from 1972) at Yale. Using the increasingly powerful resources of the electron microscope he investigated the fine structure of living cells, notably the minute organelles known as mitochondria in which energy is generated. Later (1956) he identified even smaller organelles (now known as ribosomes) – rich in DNA – which proved to be the site of protein synthesis. He was a Nobel prizewinner in 1974.

Parsons, Charles Algernon 1854–1931
Irish engineer. Son of an astronomer and mathematician, Parsons read mathematics at Trinity College, Dublin, and Cambridge University, and then became junior partner in an engineering firm. The inadequacy of reciprocating steam engines for driving high-speed dynamos led him to develop a steam turbine. He succeeded where earlier inventors had failed by recognizing that the steam pressure must be reduced in stages. His first engine (1884) developed 10hp at 18,000 revolutions per minute. He succeeded in founding a business of his own and turned his attention to marine propulsion. In 1897 his turbine-powered *Turbinia* created a sensation at the Spithead Review of the British Royal Navy's fleet. Within a decade turbines were widely used at sea.

Pauli, Wolfgang 1900–58
Austrian-born pioneer of quantum mechanics. After graduating at Munich (1921) he studied with Niels Bohr in Copenhagen and Max Born in Göttingen. From 1928 to 1958 he was professor in the Federal Institute of Technology, Zurich, except for the war years which were spent at Princeton, USA (when he acquired American citizenship). On the basis of quantum mechanics he formulated (1924) the exclusion principle according to which no two electrons in an atom can be in the same quantum state. He also postulated that in addition to the three quantum numbers assigned to electrons in an atom, a fourth – a positive or negative – was necessary to take account of electron spin. For this highly original concept, verified experimentally in 1926, Pauli was awarded a Nobel prize in 1945.

Pavlov, Ivan Petrovich 1849–1936
Russian physiologist. After graduating in medicine at St Petersburg in 1883 he studied in Germany before becoming director of the St Petersburg Institute for Experimental Medicine (1891–1936). There he began a long series of experiments on the nature of the digestive process, identifying three distinct phases: nervous, pyloric and intestinal. His major work on this subject appeared in 1898, and he was awarded a Nobel prize in 1904. Ironically he is best known for the lesser, though still considerable, discovery of the conditioned reflex. He found that when dogs learnt to associate food with the ringing of a bell they began to salivate at the sound even in the absence of food.

Pedersen, Charles John 1904–89
Norwegian-US chemist. Born in Korea of Norwegian-Japanese parents, in 1921 he went to the USA and studied chemical engineering at the University of Dayton, Ohio, and MIT. In 1927 he joined the company Du Pont, where he worked until retirement in 1969. There he did research in many fields, but his outstanding discovery (1962) was that of a range of new complexing agents, both natural and synthetic, which have the surprising property of making alkali metal salts soluble in nonaqueous solvents such as chloroform. They produce their effect by wrapping themselves round the metal ion, effectively "hiding" it from the solvent. For this reason the complexes have been called cryptates. He was awarded the Nobel Prize for Chemistry in 1987.

Peierls, Rudolf Ernst 1907–
German-British theoretical physicist. After graduating in physics at Berlin University he embarked on a research career just as the advent of quantum mechanics was reshaping the whole basis of theoretical physics. After studying with W.K. Heisenberg in Leipzig and W. Pauli in Zurich he moved to the UK and was appointed professor of mathematical physics in Birmingham (1937–63), Oxford (1963–74) and subsequently Washington, Seattle (1974–77). His early research was largely in solid state physics, but he turned increasingly to atomic physics in the 1930s. In 1940 he was a member of the influential Maud Committee, whose report indicated the feasibility of an atomic bomb. The British government took this up and Peierls became part of an investigative research group, concerned particularly with the separation of uranium isotopes. This became absorbed into the Manhattan Project (1943) which produced the first atomic bomb in 1945.

Penzias, Arno Allan 1933–
US astrophysicist, discoverer of universal background radiation in the microwave band. After studying physics he joined Bell Telephone Laboratories in 1961, eventually being appointed vice-president, research in 1981. Concurrently he held some academic appointments, including that of professor of astrophysics in Princeton University. In the 1960s, with R.W. Wilson, he was investigating radio noise in the 7cm band emanating from the Milky Way. This was more intense than expected; it came equally from all directions; and no terrestrial source could be identified. It corresponded with radiation from a black body at 3.5°K. This was in accordance with, and gave experimental support to, the "big bang" theory of the origin of the Universe. In 1978 Penzias and his colleague Wilson were jointly awarded the Nobel Prize for Physics.

Perutz, Max Ferdinand 1914–
Austrian-British molecular biologist. After graduating in chemistry in the University of Vienna he went to Cambridge, UK, in 1936 to work on X-ray crystallography with J.D. Bernal. In 1937 Perutz turned his attention to the relatively simple, but still complex, blood pigment hemoglobin, whose role is to carry oxygen in the blood. Its crystals proved to give excellent pictures, but this research had to be put aside until after the war. In 1947 he was appointed, with J.C. Kendrew, to head a small Medical Research Council Unit to study the properties of large biological molecules. Under his direction this was to become the world-famous MRC Laboratory of Molecular Biology in Cambridge. In 1953 he published a complete structure for hemoglobin, the first protein to be so described. Kendrew had success with the related protein myoglobin. Perutz and Kendrew shared a Nobel prize in 1962. Their technique has been used to make structure determination for many different proteins, including viruses and enzymes.

Piccard, Auguste 1884–1962
French physicist, famous for his exploration of the stratosphere and the ocean depths. After graduating in physics at the Federal College of Technology, Zurich, he taught there (1907–22) until appointed professor of physics at Brussels Polytechnic Institute. Becoming interested in the physics of the upper atmosphere and in the cosmic rays discovered by V.F. Hess in 1911–12 he decided to make direct observations from a manned balloon at great altitudes, a technique profitably used by Hess. In 1931 he made a record ascent of 15,780m (51,775ft) using an air-conditioned gondola. A year later he achieved 16,200m (53,153ft). In 1937 he made a complete reversal in policy by deciding to explore the ocean depths, using an independent self-propelled bathyscaphe. With his son Jacques he descended 3,099m (10,168ft) in 1953, trebling the record set by William Beebe in 1934. Together they built a new bathyscaphe, the *Trieste*, which they sold to the US navy. This reached a depth of 10,910m (35,800ft) in the Mariana Trench in the Pacific in 1960.

Pincus, Gregory Goodwin 1903–67

US biologist. After graduating in agriculture at Cornell University he studied physiology and genetics at Harvard and in Europe. Becoming interested in the social significance of birth control, in 1951 he turned his attention to reproductive physiology, setting up his own private consultancy. In this context he investigated the contraceptive effects of steroid hormones, such as progesterone, which inhibit ovulation. In 1954 Pincus organized field trials of the new synthetic analogues of progesterone among women in Puerto Rico and Haiti. These were very successful and led to the marketing of the first contraceptive pill in the USA in 1960.

Planck, Max Carl Ernst Ludwig 1858–1947

German physicist. Abandoning an early intention to be a professional musician he studied physics at Munich and Berlin. He held professorships in theoretical physics at Kiel (1885–87) and Berlin (1887–1928). He published a series of papers (1880–92) on thermodynamics, summarizing his conclusions in his *Vorlesungen über Thermodynamik* (Lectures on Thermodynamics) in 1897. His research on radiant heat led him in 1905 to his revolutionary quantum theory, according to which radiation is emitted not continuously but in quanta (packets), the size of which are determined by the frequency of the radiation. He was awarded a Nobel prize in 1919. In 1930 he became president of the prestigious Kaiser Wilhelm Institute in Berlin, but resigned in 1937 in protest at the Nazi persecution of Jewish scientists. He was reappointed after World War II when the institute was reorganized as a Max Planck Institute.

Porter, Sir George 1920–

British chemist. He graduated in chemistry in 1941 and served with the Royal Navy Volunteer Reserve as a wartime radar specialist. In 1945 he went to Cambridge to work with R.G.W. Norrish on the chemical effects of light. He was professor (1955–66) of physical chemistry at Sheffield University and then director (and Fullerian professor of chemistry) in the Royal Institution, London. He was also president of the Royal Society (1986–90). His photochemical research has depended very much on the development of a new technique of flash photolysis in which a brief, intense flash disintegrates the molecules of a gas, leading to the formation of short-life radicals and excited molecules. A second flash, very shortly afterward, is used to analyze the reaction products spectrographically. The principle is similar to that of a photographic flash gun, except that by 1975 the duration of the flash was no more than a picosecond (10^{-12} second). By that time, the technique had been extended to liquids. In 1967 Porter and Norrish were joint recipients of the Nobel Prize for Chemistry.

Prelog, Vladimir 1906–

Yugoslav-Swiss organic chemist. After graduating in chemistry at the University of Prague, he taught there (1929–34), and then became professor in the University of Zagreb (1935–41). From there he went to the Swiss Federal Institute of Technology, Zurich, latterly as professor of organic chemistry (1950–76). His interest was almost wholly in the molecular constitution and stereochemistry of natural products, ranging from alkaloids to antibiotics, enzymes to steroids. Internationally he received many honors. In 1975 he shared the Nobel Prize for Chemistry with the Australian-British organic chemist John Cornforth.

Prigogine, Ilya 1917–

Russian-Belgian physical chemist. He moved to Belgium from Russia in 1929 and graduated in physics in the University of Brussels in 1941. He was appointed professor of physics there in 1951 as well as holding chairs at Chicago (1961–66) and Austin (Texas) (1967–). The equilibria posited in classical thermodynamics are rarely encountered in real life. Living organisms achieve a regular confirmation from disorganized materials: inanimate systems tend to run down and become increasingly disorganized. Prigogine devoted himself to the investigation of nonequilibrium systems and devised mathematical models to account for them. These have significance not only in chemical thermodynamics but in the wider sphere of ecosystems generally. He was awarded a Nobel prize in 1977.

Purcell, Edward Mills 1912–

US physicist. After graduating in electrical engineering from Purdue University he studied physics in Karlsrühe and at Harvard, where he lectured from 1938. After working during the war on radar at MIT, he returned to Harvard as professor of physics (1945–80). In the late 1940s, independently of F. Bloch, he developed the nuclear magnetic resonance (NMR) technique for measuring the magnetic movements of atoms. This led on to the the powerful analytical technique known as NMR spectroscopy. In 1951 he was the first to observe 21cm wavelength radiation emitted by interstellar hydrogen gas. In 1952 he shared a Nobel prize with Bloch.

Rainwater, Leo James 1917–

US physicist. After studying physics at the California Institute of Technology he did postgraduate research at Columbia University, where he remained and was appointed professor of physics in 1952. Although his research ranged over a wide field, it was largely concerned in one way or another with the properties of the atomic nucleus. In the 1950s two theories obtained in this field. According to one its structure was analogous to a drop of water; according to the other it

consisted of a series of concentric shells, rather like an onion. Rainwater formulated a theory which ingeniously subsumed both these concepts and with Aage Bohr and B.R. Mottelson found experimental evidence for it. In 1975 he shared a Nobel prize with Mottelson and Bohr.

Ramón y Cajal, Santiago 1852–1934

Spanish neurohistologist. After graduating in medicine in Madrid in 1877, he held professorships in Valencia (1884–87) and Barcelona (1887–92) before returning to Madrid as professor of histology and pathological anatomy (1892–1922). Using Golgi's staining method of distinguishing nervous tissue Ramón y Cajal demonstrated that though impulses can be transmitted from one to another the nerve fibers always have a minute gap between them. While at Madrid he devoted nearly all his time to research on the mechanism of nervous transmission and the way in which severed nerves regenerate. In 1906 he shared with Golgi the Nobel Prize for Physiology or Medicine, the first Spaniard to achieve this honor.

Robbins, Frederick Chapman 1916–

US virologist. He graduated in science at the University of Missouri in 1936 and in medicine at Harvard. He held various hospital and research appointments associated with infectious diseases and pediatrics. During the war he served with the virus and rickettsial disease section of the US army. From 1948 to 1952 he held appointments in Boston, latterly at Harvard Medical School. In this context he became associated with J.F. Enders and with him and T.H. Weller succeeded in growing cultures of the mumps virus on a homogenate of chicken embryo cells and ox serum to which penicillin was added to prevent bacterial infection (1948). A similar technique led to his culturing of the polio virus in 1949. Robbins also did important research on hepatitis, Q fever and typhus, diseases caused by organisms intermediate between viruses and bacteria. In 1954 he shared a Nobel prize with Enders and Weller.

Robinson, Sir Robert 1886–1975

British organic chemist. After graduating in chemistry in Manchester he did research there for some years before being successively professor of organic chemistry in Sydney, Liverpool, St Andrews, Manchester, University College London, and Oxford (1930–55). In mid-career he was briefly (1920–21) director of research for the British Dyestuffs Corporation (later part of ICI, for whom he was for many years, a consultant). After his retirement from Oxford he became a consultant to Shell and a director of Shell Chemical Co Ltd. He was the last of the great organic chemists in the classical tradition, achieving his results with very simple apparatus. His interests were catholic. His electronic theory of chemical reaction aroused

much interest around 1930 but is now of largely historical significance. His output was prodigious: in a busy life he published over 700 research papers, and his name was on 32 patents. His honors included the presidency of the Royal Society (1945–50); a Nobel prize (1947); and the Order of Merit (1949).

Röntgen, Wilhelm Konrad 1845–1923

German physicist. He entered the Polytechnic Institute in Zurich in 1855 and after a series of academic appointments was elected professor of physics at Würzburg and director of the newly founded Physical Institute there in 1888. His interests were wide and included the mysterious "molecular rays" emitted when eletric discharges are passed through gases at low pressures. In 1895, while experimenting with a Cookes' tube covered in an opaque shield of black cardboard, he noticed that a nearby sheet of paper painted with barium platinocyanide had begun to fluoresce. The cause proved to be a penetrating radiation which he designated X-rays. He immediately announced his discovery in a report to the Würzburg Physical-Medical Society and the great potential diagnostic and therapeutic value of X-rays was quickly recognized. Röntgen was awarded the Nobel Prize for Physics for 1901.

Ruska, Ernst August Friedrich 1906–88

German physicist, pioneer of the transmission electron microscope. After studying engineering in the Technical University, Munich, and working as a research student in Berlin University, he became development engineer for Television Berlin (1934–36), an appointment which stimulated his existing interests in electron optics. From 1937 to 1955 he was with Siemens and Halske AG, and was subsequently appointed director of the Institute for Electron Microscopy. His research career started just as quantum mechanics were establishing a duality between waves and particles. In 1928, while still a research student, he collaborated with M. Krull in constructing a microscope in which a beam of electrons was focused by a magnetic coil, analogous to the way in which lenses focus light in a conventional instrument. This gave a modest X17 magnification but by 1933 he had constructed a far more sophisticated instrument giving X12,000 magnification, six times greater than that of the best optical microscope. Magnifications up to one million times were eventually achieved. In 1986 Ruska shared a Nobel prize with H. Rohrer and G. Binnig, of IBM in Zurich.

Rutherford, Ernest 1871–1937

The founder of modern atomic physics. Born in New Zealand, he was educated at Canterbury College (now University of Canterbury), where he did research on magnetism. In 1895 a scholarship took him to Britain to work under J.J. Thomson at Cambridge, where he did research on the conduction of electricity by gases. In 1898 he was appointed professor of physics at McGill University, Montreal, where he investigated the phenomenon of radioactivity discovered by A.H. Becquerel in 1896. He discovered that two kinds of radiation were involved, which he designated alpha and beta. In 1907 he was appointed professor at Manchester and subsequently at Cambridge (1919–37) as head of the famous Cavendish Laboratory. During these years he attracted a succession of brilliant research workers who laid the foundation of modern atomic physics. His supreme achievement was the concept of the nuclear atom – an atom with a relatively heavy nucleus surrounded by a cloud of much lighter electrons. Niels Bohr explained a final anomaly in this by applying quantum theory (1912). Rutherford was awarded a Nobel prize in 1908 – and was chagrined, as a dedicated physicist, to learn that it was the prize for chemistry.

Salam, Abdus 1926–

Pakistani theoretical physicist. He studied at Government College, Lahore, and Cambridge, UK. After returning briefly to Pakistan he returned to Cambridge as lecturer in mathematics (1954–56) and was then appointed professor of theoretical physics at Imperial College, London. His background made him very conscious of the problems of scientists in the Third World and this led him to found the International Center of Theoretical Physics in Trieste in 1964. Four basic forces are now recognized in nature: the long-familiar forces of gravity and electromagnetism and the much more recently discovered "strong" and "weak" intranuclear forces. Salam made an important advance by formulating a theory linking the electromagnetic and the weak nuclear forces, although these differ by a factor of about a million. This theory was independently put forward also by S. Weinberg (1967) and developed by S.L. Glashow. Salam, Weinberg and Glashow were jointly awarded a Nobel prize in 1979.

Salk, Jonas Edward 1914–

US physician. He graduated in medicine at New York University College of Medicine in 1939 and then held appointments at the University of Michigan (1942), working on influenza virus, and then at the University of Pittsburg (1947) as director of the virus research laboratory. In 1948 J.F. Enders and his team at Harvard had devised easier means of propagating the polio virus and Salk sought to prepare an attenuated strain which could be the basis of a vaccine. In 1952 he began clinical trials and in 1954 the vaccine was released for general use. This was followed by tragedy in 1955, when some children who had been injected developed the disease. More stringent precautions overcame the problem and mass vaccination was resumed. However, around 1960 Salk's vaccine began to be replaced by an oral vaccine developed by A.B. Sabin, given on a lump of sugar. In 1963 Salk became director of the Salk Institute for Biological Studies in San Diego.

Sandage, Allan Rex 1926--

US astronomer and first optically to identify a quasar. After graduating from the University of Illinois and the California Institute of Technology, he joined the staff of the Hale Observatories in 1952, at first as assistant to E.P. Hubble, who had provided experimental evidence for the concept of an expanding Universe. In 1960 Sandage turned his attention to radio astronomy and the mysterious remote bodies known as quasars (quasi-stellar objects). At that time they were known only as cosmic sources of radio waves. Concentrating on quasar 3C48 with a powerful optical telescope he – in collaboration with T. Matthews – discovered it to correspond with a visible material object. Observation of the optical spectrum by the Dutch astronomer M. Schmidt showed a peculiarity in the red band indicating that it was receding at around one-fifth of the velocity of light. Sandage subsequently detected other similar quasars, and he also observed quasars which do not emit radio waves.

Schawlow, Arthur Leonard 1921–

US physicist. He graduated at Toronto University in 1921 and after doing postgraduate research at Columbia University joined Bell Telephone Laboratories (1951–61). Subsequently he was appointed professor of physics at Stanford University. He was particularly interested in microwave radiation and with his brother-in-law C.H. Townes formulated in 1958 the principle of the laser (Light Amplification by Stimulated Emission of Radiation). This was designed to effect for light the high degree of amplification already achieved by Townes for microwaves. The first pratical laser was built by T.H. Maiman in 1960.

Schrieffer, John Robert 1931–

US physicist, noted for research on solid state physics and superconductors. After graduating at the University of Illinois he spent some time in Europe, at the University of Birmingham in the UK and at the Niels Bohr Institute in Copenhagen, Denmark. He worked closely with J. Bardeen and L.N. Cooper to in formulating (1957) the BCS theory of superconductivity, for which all three shared a Nobel prize in 1972. Subsequently Schrieffer held appointments in the University of Pennsylvania (1962–79) and as professor of physics in the University of California, Santa Barbara. His later research included surface physics and ferromagnetism.

Erwin Schrödinger

Frederick Soddy

Roger Sperry

Schrödinger, Erwin 1887–1961

Austrian physicist, founder of wave mechanics. He graduated in Vienna, was professor at Breslau and then Zurich. In 1927 he succeeded Max Planck as professor of physics at the University of Berlin but left for Oxford when Hitler came to power in 1933. He returned to Austria , but in 1938 went to the Institute of Advanced Studies, Dublin. In 1957 he accepted a professorship in Vienna. His great achievement was the Schrödinger Equation (1926) in which he expressed mathematically the simultaneous identification of atoms as both waves and particles, postulated by L. de Broglie in 1924. From this arose quantum mechanics.

Shapley, Harlow 1885–1972

US astronomer, the first to formulate a clear picture of the size and structure of our galaxy. Originally a crime reporter, he soon found astronomy more to his taste. Discovering a disproportionate density of stars in the direction of the constallation Sagittarius he reasoned that this must be the galactic center and that the Sun must be some 50,000 light years from it. He further calculated the overall diameter of the galaxy as about 300,000 light years, though later estimates tripled this.

Sherrington, Charles Scott 1857–1952

British neurophysiologist, often described as the William Harvey of the nervous system. After studying in Cambridge and London, Sherrington did physiological research in Germany before becoming professor of physiology at Liverpool in 1895. From 1913 to 1936 he was Waynflete professor of physiology at Oxford. He specialized early in the physiology of the nervous system, and published over three hundred papers on this subject. In the 1890s he mapped the nervous supply to the musculature but from about 1900 became increasingly interested in the physiology of the brain and spinal cord. His *Integrative Action of the Nervous System* (1904) is one of the great classics of medical literature. He was awarded a Nobel prize in 1932.

Shockley, William Bradford 1910–89

US physicist. A graduate of the California Institute of Technology and MIT, he joined Bell Telephone Laboratories in 1936. During the war he worked with the US navy's Antisubmarine Warfare Operational Research Group and was consultant to the secretary for war (1945). He returned to Bell Telephone after the war but left in 1955 to become an industrial consultant. He was professor of engineering science at Stanford University (1963–75). In 1945 he organized a small group of solid-state physicists, including J. Bardeen and W.H. Brattain, to try to produce a semiconductor device to replace thermionic tubes (valves). In 1948 their invention of the point-contact transistor was announced. This made possible the

miniaturization of a wide range of electronic devices. However, the original point-contact transistor was somewhat limited: it was "noisy" and could be used only for low power inputs. Very shortly, Shockley improved it radically with his junction transistor. Shockley, Bardeen and Brattain were jointly awarded a Nobel prize in 1956.

Sidgwick, Nevil Vincent 1873–1952

British chemist. He graduated at Oxford with first class honors in chemistry and classics. After working in Leipzig he returned to Oxford in 1901 becoming reader (1924) and professor (1935). In 1914 he visited Australia and met Ernest Rutherford, then developing his theories on atomic structure. This led him to attempt to interpret the chemical bonds between atoms in terms of the electronic theory being developed by the physicists, which led to the concept of two different types of bond: the coordinate link, comprising two electrons from one atom, and the covalent link, comprising one electron from each atom. His great *Electronic Theory of Valency* (1927) established his reputation internationally.

Sikorsky, Igor Ivan 1889–1972

Russian-US aeronautical engineer. After graduating in engineering he immediately entered the newly fledged aircraft industry. In 1909 he built a helicopter, which was unsuccessful, but from 1914 had to devote himself to conventional aircraft, building for the Russian government the first four-engined bombers. After the war he emigrated to the USA and set up the Sikorsky Aero Engineering Corporation. This enabled him to resume his interest in helicopters in the 1930s but not until 1939 did he construct a model judged satisfactory: this was the VS30. During World War II he manufactured three variants of this for the US government, producing a total of 400 craft. After the war he designed a series of other helicopters.

Soddy, Frederick 1877–1956

British chemist, pioneer of radiochemistry. After graduating at Oxford he worked in Montreal under Ernest Rutherford, who was investigating radioactivity. They deduced that this must be the consequence of spontaneous atomic disintegration and that in the case of radium a product of decay should be helium gas. In 1903, working with William Ramsay in London, Soddy identified helium in the gaseous emanation of radium bromide. In 1913 Soddy closely related the new concept of atomic number to radioactive decay. Emission of a beta particle, he postulated, increases atomic number by one; of an alpha particle diminishes it by two. In the same year he put forward the idea of atoms of different atomic weights having identical chemical properties: to designate such atoms he coined the word isotope. He was awarded a Nobel prize in 1921.

Sperry, Roger Wolcott 1913–

US psychobiologist. After studying psychology and zoology he did several years of research at Harvard, Yerkes and the National Institutes of Health before being appointed professor of psychobiology at the California Institute of Technology (1954–84). Initially his research was with amphibians which, unlike mammals, can regenerate a severed optic nerve. In them, he was able to demonstrate a specific difference in functions between the two halves of the brain. Turning to the brains of higher animals – primates and humans – he also found a difference in functions between the two hemispheres. In the left hemisphere the processing of verbal messages is generally dominant: in the right, emotional and spatial interpretations. Sperry was awarded a Nobel prize in 1981.

Stanley, Wendell Meredith 1904–71

US biochemist. He graduated in chemistry in 1929, and joined the Rockefeller Institute in Princeton (1931). Learning that J.B. Sumner and J.H. Northrop had prepared enzymes in crystalline form, he sought to do the same thing with viruses. In 1935 he succeeded in crystallizing tobacco mosaic virus (TMV), introducing the concept that a living organism could also be a pure crystalline chemical. During World War II he worked on an influenza virus and prepared a vaccine against it. He shared the Nobel prize for chemistry in 1946 with Northrop and Sumner.

Staudinger, Hermann 1881–1965

German chemist, pioneer of polymer chemistry. After studying in the Universities of Halle, Darmstadt and Munich he became professor of chemistry successively at Karlsrühe (1908), Zurich (1912) and Freiburg (1926–51). About 1920 he became interested in polymers. The conventional view was that these were aggregates of small molecules (monomers) but Staudinger demonstrated that they were in fact giant molecules consisting of thousands of atoms. This opened up new possibilities for the synthesis of many new types of plastics and fibers. Belatedly, he was awarded a Nobel prize in 1953.

Steinberger, Jack 1921–

US particle physicist. He arrived in the USA as a refugee in 1934 and later studied chemistry. During World War II he worked in the Radiation Laboratory at MIT and then took a doctorate in physics at Chicago, doing research on cosmic rays. In 1959 he took a post at Columbia University, moving to CERN in Geneva in 1968. While at Columbia he worked closely with L. Lederman and M. Schwartz: all three shared a Nobel prize in 1988. Steinberger had shown at Chicago that a particle known as a muon decays to give two neutrinos and an electron. They surmised that

▲ Hermann Staudinger

▲ Axel Hugo Theorell

▲ Nikolaas Tinbergen

neutrinos should be of two kinds – muon neutrinos and electron neutrinos, and confirmed this experimentally with the aid of a new high-energy particle accelerator. While at CERN Steinberger continued his work on neutrinos and contributed to the development of a "standard" model for particle physics.

Sumner, James Batcheller 1887–1955

US biochemist. He entered Harvard in 1906 to study engineering but changed to chemistry, graduating in 1910. From 1911 to 1914 he was a research worker there before becoming assistant (later full) professor of biochemistry (1914–38) at Cornell University Medical School, where he was soon appointed director of a laboratory of enzyme chemistry. In 1917, convinced that enzymes were proteins, he set out to isolate one (urease) in pure form. In 1926 he succeeded, producing what he claimed to be a pure crystalline product. This was disputed, however, because it was contrary to the views on enzymes of the eminent German chemist R. Willstätter. Not until J.H. Northrop produced crystalline pepsin in 1930 was Sumner's work validated; he went on to produce a number of other crystalline enzymes. In 1946 he shared a Nobel prize with Northrop and W.M. Stanley.

Swinburne, James 1858–1958

British engineer. After serving an apprenticeship in a locomotive works he was sent by Sir Joseph Swan, pioneer of electric lighting, to establish lamp factories in France and the USA. Returning home, he joined R.E.B. Crompton in his dynamo-manufacturing works (1885–99) before setting up as an independent engineering consultant in London, specializing in electricity generation and electric lighting and taking out more than 100 patents. He appreciated the potential of phenol-formaldehyde resins for electrical components, but in 1907 his patent was anticipated, by a single day, by L.H. Baekeland. Later, however, an accommodation was found and Swinburne became chairman of Bakelite Ltd, retiring in 1948. He was elected to the Royal Society in 1906.

Swinton, Alan Archibald 1863–1930

British electrical engineer, a pioneer of electronic television. After training in engineering in Edinburgh and abroad he joined W.G. Armstrong's engineering works (1882–87) at Newcastle-upon-Tyne and was subsequently engineer to other companies associated with mechanical engineering, electrical supply and transport. His interests included the nascent radio equipment industry and he was president of the Radio Society of Great Britain 1913–21. In this context he became interested in the possibility of television, where in 1907 the Russian physicist Bovis Rosing had proposed a photomechanical system similar to the later system of J.L. Baird but

using a cathode ray tube as receiver. In 1908 Swinton proposed an all-electronic system using cathode ray tubes for both transmission and reception. He elaborated on this system in communications to the scientific journal *Nature* in 1911 and 1920 but never put his ideas into practice.

Tatum, Edward Lawrie 1909–75

US chemist, pioneer of biochemical genetics. After graduating he worked at Stanford before moving on to Yale, latterly as professor of microbiology (1946–48). He then returned to Stanford as professor of biology (1948–56) and biochemistry (1956–57), before joining the Rockefeller Institute. In 1946 he worked with J. Lederberg on bacterial mutants and refuted the then current belief that bacteria have no genes, and thus no sex. In fact, occasional individuals are produced by sexual conjugation. With Lederberg and G.W. Beadle he formulated the "one-gene- one-enzyme" concept, and demonstrated that enzyme action is controlled by genes. Tatum, Lederberg and Beadle shared a Nobel prize in 1958.

Theorell, Axel Hugo Teodor 1903–82

Swedish biochemist. In 1921 he enrolled at the Karolinska Institute, Stockholm to study medicine. On graduating in 1924 he went to Paris and worked briefly at the Pasteur Institute with the bacteriologist L.C.A. Calmette. Returning to Sweden he worked with the Medico-Chemical Institute and in 1936 was appointed director of the new Biochemical Department of the Nobel Medical Institute. His early research (1924–35) was on the constituents of blood – partly in association with The Svedberg, using the ultracentrifuge – but after spending two years with Otto Warburg in Berlin he devoted himself to oxidation enzymes. He succeeded in showing that the so-called "yellow enzyme" was in fact a mixture of a colorless protein moiety and a coenzyme, which proved to be a nucleotide. He received a Nobel prize in 1955.

Thomson, Joseph John 1856–1940

British physicist, discoverer of the electron. After winning a scholarship to Trinity College, Cambridge, in 1876 he graduated in mathematics in 1880. He then joined the staff of the Cavendish Laboratory, Cambridge, latterly as Cavendish Professor (1884–1919). Under his direction it became the world's leading center for research in atomic physics. In 1883 he began to investigate radiation generated by electric discharges through gases at low pressures. The nature of this radiation was uncertain: many physicists believed it to consist of electromagnetic waves, but in 1897 Thomson proved that it was in fact a stream of charged particles. By measuring their deflection in combined electric and magnetic fields he was able to measure the ratio of charge to mass. Subsequently he found the mass to be about one-

thousandth of that of an atom of hydrogen, the lightest element. This was an epoch-making discovery, overturning the accepted belief that atoms were the smallest particles existing in nature. He was awarded a Nobel prize in 1906.

Tinbergen, Nikolaas 1907–88

Dutch-British ethologist. After studying at Leiden, Vienna and Yale Universities he returned to Leiden in 1936, first as lecturer and later (1947) as professor of experimental zoology. He then moved to Oxford University, latterly as professor in animal behavior (1966–74). He had a catholic interest in the behavior of animals both in the wild and in captivity and demonstrated that many species have a stereotyped, rather than random, pattern of behavior. He made a particular study of herring gulls and one of his many books (*The Herring Gull's World*, 1953) is devoted to this species. He also studied human behavior, particularly in the context of autism and aggression. In 1973 he shared a Nobel prize with Karl von Frisch and Konrad Lorenz.

Ting, Samuel Chao Chung 1936–

US physicist. He graduated in physics at the University of Michigan. After appointments at CERN (Geneva) and Columbia University he became professor of physics at MIT in 1969. In 1974 he was working at the Brookhaven National Laboratory, using its very powerful synchrotron accelerator to bombard beryllium with high-speed protons. Among the products of collision was a particle not previously observed, which he named the J particle. Simultaneously, Burton Richter identified the same particle, which he called the psi particle. In 1976 Ting and Richter shared the Nobel Prize for Physics for their discovery of what is now known as the J-psi hadron.

Tiselius, Arne Wilhelm Kaurin 1902–71

Swedish biochemist, remembered for his use of electrophoretic analysis and chromatography for the purification of proteins. After graduating in science and mathematics at Uppsala, he remained there as assistant professor of physical chemistry (1930–37) and of biochemistry (1937–67): latterly he worked in a new Institute of Biochemistry founded in 1946. Initially his research was with The Svedberg on the purification of proteins and other large molecules by ultracentrifugation but he later developed an alternative technique by which such molecules were separated by their different rates of movement in an electric field. He used this technique for many purposes, including the separation of the main blood proteins. After World War II he took a leading role in formulating Swedish scientific policy. In 1947 he was appointed vice-president of the Nobel Foundation and did much to extend its activities. In 1948 he was awarded the Nobel prize for Chemistry.

▲ Shin'ichirō Tomonaga

▲ Konstantin Tsiolkovsky

▲ Alfred Wegener

Tomonaga, Shin'ichirō 1906–79

Japanese theoretical physicist. He graduated at Kyoto Imperial University in 1929. After postgraduate research in Japan he worked for two years (1937–39) with W. Heisenberg. During the war he did research on microwave systems. Afterward he returned to academic work in the Tokyo University of Education, of which he became president in 1963. His research was particularly concerned with quantum electrodynamics. This embraces the quantum mechanical laws governing the interaction of charged particles, particularly electrons, and the electromagnetic field. In the hands of Tomonaga and of J. Schwinger and R. Feynman – with whom he shared a Nobel prize in 1965 – it came to occupy a central position in atomic physics, giving very precise interpretation of events at the atomic level.

Tonegawa, Susumu 1939–

Japanese molecular biologist. He studied science at Kyoto University and did postgraduate research at the University of California, San Diego (1963–69). He then spent ten years in Switzerland at the Basle Institute for Immunology, before returning to the USA in 1981 to teach at MIT. His research has thrown light on a puzzling aspect of the human body's immune system. How does it manage to produce the vast variety of antibodies and antigens, numbered in millions, required to meet every contingency? Tonegawa showed in the 1970s that the cells which produce antibodies (the T-cells) contain about a thousand different pieces of relevant genetic material. The permutations of these make possible at least a billion possible antibodies. For this quantitative explanation of the resources of the immune system Tonegawa was awarded a Nobel prize in 1987.

Townes, Charles Hard 1915–

US physicist. He studied physics at Furman University, Duke University and the California Institute of Technology – he was awarded a doctorate by the latter in 1939. From then until 1947 he worked at Bell Telephone Laboratories on the development of radar bombing systems and later on microwave spectroscopy, which grew out of them. In 1948 he joined the physics faculty of Columbia University and was later Provost of MIT (1961–67) and professor of physics in the University of California, Berkeley. In 1953 he announced the invention of the maser (Microwave Amplification by stimulated Emission of Radiation), a device for amplifying microwave radiation through ammonia gas as an intermediary. In 1958, with A.L. Schawlow, he announced the principle of the laser, by which a similar effect could be achieved with light. In 1964 Townes shared a Nobel prize with the Soviet physicists H.G. Basov and A. M. Prochorov, who had independently devised a form of maser.

Tsiolkovsky, Konstantin 1857–1935

Russian physicist. During his years as a (self-taught) teacher he devoted his spare time to studying the problems of aerial travel, beginning with a metal-skinned dirigible (1892) and an aeroplane with twin propellers (1894). In 1897 he constructed Russia's first experimental wind tunnel. From 1896 he began to explore the possibility of interplanetary travel with rockets and from 1903 to 1917 put forward various plans for the construction of rocket ships, embodying many ideas now widely adopted. However, his work was ignored until the Bolshevik Revolution in 1917. Thereafter he became a pensioned member of the Academy, and was internationally known and respected. In his last years he worked on formulating specifications for rocket fuels.

Twort, Frederick William 1877–1950

British physician. After qualifying he held posts at St Thomas's Hospital, London; the London Hospital; and, as director, the Brown Institution, an animal dispensary (1909). He made many contributions to bacteriology, including identification of the causative agent of Johne's disease, a serious ailment of cattle. His major contribution, however, was made in 1915 when he discovered viruses which infect and destroy bacteria. Felix d'Hérelle in France gave such viruses the generic name bacteriophage. This interaction is, therefore, generally known as the Twort-d'Hérelle phenomenon.

Vine, Frederick John 1939–88

British geologist. After graduating at Cambridge he worked at Princeton University (1965–70) and then returned to the UK as reader, later professor, in the Department of Environmental Sciences, University of East Anglia. In 1963, as a postgraduate research student working with D.H. Matthews, he showed that the magnetism of the ocean bed on either side of the midoceanic ridges alternates in direction as one moves out from the source. This is in accordance with known reversals of the Earth's magnetic field over geological time and gives support to the hypothesis of seafloor spreading advanced by H.H. Hess in 1962.

Vries, Hugo de 1848–1935

Dutch plant geneticist. He studied medicine in Holland and Germany and then taught botany in Amsterdam. With J. von Sachs, he investigated the physiology of water uptake in plants. In the 1870s he prepared for the Prussian Ministry of Agriculture a series of monographs on cultivated plants and this aroused his interest in heredity; he began breeding plants in 1892. He quickly found evidence for the 3:1 ratio discovered by Mendel in the 1850s. When he first came across Mendel's neglected work in 1900 he exerted himself to make it widely known. He took a particular interest in

mutations (sports), plants showing atypical characteristics, and believed that these might cause evolution to proceed more rapidly than Darwinian principles suggested. In the event, however, this proved to be an overestimate.

Walton, Ernest Thomas Sinton 1903–

Irish physicist. A physics graduate from Trinity College, Dublin, he won a scholarship to Cambridge to work in the Cavendish Laboratory under Ernest Rutherford. There, in 1932, he took part with J.D. Cockcroft in a classic experiment in which atoms of lithium and boron were split by bombardment with protons. For this, the two men were jointly awarded a Nobel prize in 1951. Meanwhile Walton had returned to Dublin as fellow of Trinity College (1934–74) and professor of natural and experimental philosophy (1947–74). His later research was on hydrodynamics, nuclear physics and microwaves.

Watson, James Dewey 1928–

US molecular biologist. After graduating, he worked as a virologist and geneticist in the University of Copenhagen (1950–51). He then joined the Medical Research Council Unit in the Cavendish Laboratory, Cambridge, UK. There he collaborated closely with F.H.C. Crick, who was particularly interested in the structure of DNA, by then known to be the carrier of the genetic code. Its structure was known in general chemical terms and there were grounds for supposing that it might be helical, but the precise configuration of the molecule had yet to be established. In collaboration with M.H.F. Wilkins and Rosalind Franklin at King's College, London – already working on DNA – a model of the DNA molecule was constructed in 1953 which was consistent with all the X-ray evidence. It showed that it was a helix – a double helix, in which the two strands were joined by links like the rungs of a ladder. Crick, Watson and Wilkins shared a Nobel prize in 1962.

Watson-Watt, Robert Alexander 1892–1973

British pioneer of radar. After graduating in engineering at Dundee he remained there until joining the Meteorological Office in London in 1915. There he worked on the radio location of thunderstorms: this was based on the detection, with a directional aerial, of the "static" they generate. Later (1921) he was appointed to the government's Radio Research Station and then to the radio division of the Natural Physical Laboratory. In 1935 he put forward a proposal for detecting enemy aircraft by means of pulsed radio signals reflected back from them to a ground base. Field trials quickly proved its feasibility and, with war looming, the British government approved the immediate establishment of a network of radar stations to give early warning of attack. This proved of critical importance during the Battle of

Wilbur Wright (right)

Chen Ning Yang (right)

Britain in 1940. Watson-Watt's genius lay in perceiving that what was needed was not just a detecting device but a coordinated system that could be operated by nonspecialist staff under wartime conditions. After the war he was awarded £50,000, the biggest payment to an individual by the Royal Commission on Awards to Inventors.

Wegener, Alfred Lothar 1880–1930
German meteorologist and geophysicist, formulator of the theory of Continental Drift. After an academic appointment at Marburg, he joined the meteorological research department of the *Deutsche Seewarte* (Marine Observatory) in 1919. In pursuit of his meteorological studies, Wegener joined, in all, four expeditions to Greenland, on the last of which he lost his life. His *Thermodynamik der Atmosphäre* (Thermodynamics of the Atmosphere, 1911) was a standard textbook. He is, however, best known in respect of a secondary interest, deriving from meteorology. Unable to reconcile evidence about climates in the geological past with the existing pattern of the continents, he suggested that over geological time an original supercontinent, Pangaea, had broken up and the fragments had drifted in a sea of molten magma to their present positions. At the time this radical theory found little support, mainly because no appropriate source of power could be discerned, but it is now generally accepted.

Whinfield, John Rex 1901–66
British industrial chemist. Graduating in chemistry at Cambridge, he worked briefly with C.F. Cross and E.J. Bevan – who had invented rayon in 1892 – before joining the Calico Printers Association as research chemist. Stimulated by Carothers' development of nylon for Du Pont, he investigated the fiber-forming potential of other polymers. This led to the discovery of a polyester fiber based on terephthalic acid. This was patented by the CPA in 1941. In 1946 Du Pont acquired the CPA patent for the USA, marketing the fiber as Dacron. In the following year ICI, where Whinfield had completed his research and development work, acquired rights for the rest of the world, selling their product as Terylene.

Whittle, Sir Frank 1907–
British aeronautical engineer. He joined the Royal Air Force in 1923 as an apprentice mechanic, and was sent to Cambridge University to study mechanical sciences. On graduating he was given a commission and after obtaining flying experience transferred to the RAF School of Aeronautical Engineering, Henlow (1932–34). He then did postgraduate engineering research in Cambridge (1934–37). During these years he realized that the future of highspeed flight lay with the gas turbine. In 1935 Power Jets was set up to allow him to develop his ideas and by 1937 he had a prototype

engine running. With war looming he then got massive support from the Air Ministry. Meanwhile H.P. von Ohain's jet-propelled He 178 made its first flight in August 1939. The first British jet-propelled aircraft made its debut in May 1941. It proved to be too late for such aircraft to have a significant effect on the course of the war, but jet propulsion has dominated postwar aviation.

Woodward, Robert Burns 1917–79
US organic chemist. After taking his doctorate at MIT, he spent the rest of his working life at Harvard, where he was appointed professor in 1950. In 1963 he undertook concurrently the directorship of the Woodward Research Institute in Basle. His interest in chemistry was wide, but natural products had a lifelong fascination for him. Probably his best-known success was the elucidation of the structure of strychnine in 1948, confirmed by a vigorous 50-stage synthesis in 1954. He also worked on the chemistry of a number of antibiotics; on steroids and peptides; on chlorophyll and quinine. He was awarded the Nobel Prize for Chemistry in 1965.

Wright, Wilbur 1807–1912 and Orville 1871–1948
US pioneers of heavier-than-air flight. After a modest career in newspaper publishing, they became interested in gliding. In 1900 they built their first glider, similar to those of the Gerrman pioneer Lilienthal but controlled by ailerons (movable flaps) instead of by shifting the center of gravity. Over the next two years they tested over 200 models in a home-made wind tunnel, embodying the results in a glider tested at Kitty Hawk in 1902 in over 1,000 flights. This incorporated a complete control system – ailerons, rudder and flaps. In 1903 they mounted two engines driving propellers in a larger version and on 17 December made four flights, the longest lasting for 59 seconds. By 1905 they succeeded in completing a 39km (24mi) circuit and after a succession of increasingly ambitious flights formed the American Wright Company in 1909.

Yang, Chen Ning 1922–
Chinese-US theoretical physicist who made important contributions to statistical mechanics and symmetry principles. In China he studied at the National Southwest University, Kunming, and at Tsinghua University. In 1945 he went to the USA and did research at Chicago under E. Fermi who greatly influenced him. In 1949 he went to the Institute for Advanced Studies, Princeton, where he was appointed professor in 1955. With T.D. Lee – with whom he shared a Nobel prize in 1957 – he showed that the Law of Conservation of Parity – that there is no difference between a right-handed and a left-handed concept of the Universe – does not hold in the case of weak interactions.

Zeppelin, Ferdinand, Count von 1838–1917
German pioneer of airships. As a soldier (from 1858) he was much impressed by the military use of balloons for observation in the American Civil War (during which he made his own first flight) and the siege of Paris (1870–71). He failed, however, to interest his superiors but when he retired in 1891, as a general, he began experiments on his own. His first airship (LZ1) made its maiden flight in 1900. After mixed success with three later vessels, he formed his own transport and manufacturing company, and up to 1914 carried over 30,000 passengers without accident. From 1908 he received support from the German government. Nearly 90 Zeppelins were built in World War I, but they had limited military success.

Ziegler, Karl 1898–1973
German organic chemist. He graduated in chemistry at Marburg (1920). In 1927 he became professor of chemistry at Heidelberg and in 1936 director of the Chemical Institute Halle-Saale. Finally (1936–69) he was director of the Kaiser Wilhelm (later Max Planck) Institute for Coal Research. His research interests were wide and he made three major contributions to chemistry. First (1923–50), he studied "free radical" compounds of carbon, in which a carbon atom was trivalent instead of the normal quadrivalent. Second (1933–47), he investigated compounds in which the carbon atoms were arranged in large rings. It was, however, his third field of research, spanning nearly all his working life, which was the most productive. In the early days of polythene manufacture it was devoutly believed that the necessary polymerization of ethylene could be effected only at high temperature and pressure. Ziegler devised a process, using a catalyst in the presence of a solvent, which could be conducted at normal pressures and temperatures. G. Natta later adapted this for the manufacture of polypropylene. Ziegler and Natta shared a Nobel prize in 1963.

Zworykin, Vladimir Kosma 1889–1982
Russian-US physicist. An engineering graduate of the University of St Petersburg, he served during World War I as a radio operator. In 1919 he emigrated to the USA and joined the Westinghouse Electric Corporation, moving on to the Radio Corporation of America (RCA) in 1929, but returning to Westinghouse later. He developed (in 1923) the first practical storage camera tube (valve), the iconoscope. In this the image to be televised is projected on to an array of photosensitive cells inside the tube. Each of these holds a charge proportional to the intensity of the light received. The array of cells is then scanned with an electron beam which discharges them in turn, giving signals corresponding to the picture. The electrical signals are then reconstituted to form an image on the screen of a cathode ray tube.

GLOSSARY

Absolute zero
The temperature at which all substances have zero thermal energy.

Accelerator
In particle physics, a research tool used to accelerate subatomic particles to high velocities.

Acid rain
Precipitation, both rain and snow, made acidic in reaction through the chemical pollution of air by waste gases, such as oxides of sulfur and nitrogen, mainly from industry and automobile exhausts.

Alloy
A material of metallic character prepared by combining metals with one another or with nonmetals such as carbon or phosphorus.

Alpha particles
Positively charged helium nuclei emitted from radioactive materials undergoing alpha disintegration.

Anesthesia
The absence of bodily sensation, usually defined with respect to loss of pain sensations. Anesthetics may be local or general (causing a total loss of consciousness).

Antibiotic
Chemical produced by a microorganism and used as a drug to kill or inhibit the growth of other microorganisms.

Antibody
Defensive substance produced by the immune system to neutralize or help destroy a specific foreign substance (antigen).

Atom
Classically one of the minute, indivisible particles of which material objects are composed; in 20th-century science the name given to a relatively stable unit of matter made up of at least two subatomic particles.

Atomic weight
The mean mass of the atoms of an element weighted according to the relative abundance of its naturally occurring isotopes.

Bacteria (singular bacterium)
A large and varied group of microorganisms, classified by their shape and staining ability. They live in many environments; relatively few are harmful to the human body.

Bacteriophage
A virus that attacks bacteria.

Beta particles
Electrons or positrons emitted from radioactive nuclei undergoing beta disintegration.

Big bang
The explosion of unimaginably dense matter widely believed to have been the origin of the Universe.

Biotechnology
The use of microorganisms for industrial purposes.

Bond
The link that holds atoms together in molecules.

Bubble chamber
Device used to observe the paths of subatomic particles, by reducing pressure as the particles pass through so that bubbles form along their paths.

Cancer
A malignant tumor.

CAT scan
Computerized axial tomography, a technique of building up an X-ray image of a section through the body.

Cathode ray tube
An evacuated glass tube containing a cathode and an anode; electrons from the cathode are accelerated through the anode to form a beam that hits a screen. It forms the principal component of an oscilloscope and a television set.

Cepheid variables
Short-period variable stars, whose periods are linked with their real luminosities; the longer the period, the more luminous the star.

Chromosome
A thread of genetic material contained in the cell nucleus and duplicated when the cell divides.

Cloud chamber
A device for making visible the tracks of subatomic particles, by passing them through a saturated vapor; the passage of charged ions through the chamber causes the gas to condense into droplets.

Computer
A device that performs calculations and stores their results, according to a program.

Core
The central mass of the Earth, comprising an inner solid part and an outer liquid part, about 2,470km (1,535mi) in radius, and probably consisting largely of iron.

Crust
The outer layer of the Earth. It consists of two types – continental, which is rich in silica (about 40km, 24mi, thick) and oceanic, which is poor in silica (about 10km, 6mi, thick).

Cyclotron
A particle accelerator in which the particles travel in a spiral path in a strong magnetic field.

Database
A system of centralized information storage to which individual computer users may have access, often by telephonic links.

DNA
Deoxyribonucleic acid; its structure contains the blueprint that contains genetic information.

Ecology
The study of the relationships and the interactions of living organisms, with each other and with the physical world.

Electromagnetic radiation
The form in which energy is transmitted through space or matter. Its spectrum includes radio waves and infrared waves, visible light, ultraviolet, X-rays and gamma rays.

Electron
A subatomic particle of negative charge, commonly in orbit around an atomic nucleus.

Electronics
Science dealing with semiconductors and devices where the motion of electrons is controlled.

Element
Simple substance composed of atoms of the same atomic number.

Energy
One of the fundamental modes of existence, equivalent to and interconvertible with matter.

Enzyme
A protein which is a catalyst of biochemical reactions. There are many different kinds, each kind directly promoting only one or a very limited range of reactions.

Ether
A medium postulated by 19th-century physicists to explain how light could be propagated as a wave motion through otherwise empty space.

Fission
In nuclear physics, the changing of an element into two or more elements of lower atomic weight, with the release of energy.

Frequency
The rate at which a wave motion completes its cycle, measured in hertz (Hz).

Fusion
In nuclear physics, the merging of two nuclei to form a new element of higher atomic weight, resulting in the release of energy.

Galaxy
(A) A distinct star system: Galaxies may contain from a few million to a few million million stars together with differing proportions of interstellar matter (gas and dust). (B) The star-system of which our Sun is a member: it contains about 100,000 million stars.

Gamma rays
High-energy photons emitted from atomic nuclei during radioactive decay.

Gas
One of the three states in which almost all matter can exist. With its freely moving molecules, a gas will fill the available volume.

Genes
The units of inheritance which are transmitted from generation to generation and control the development of an individual.

Genetics
The study of heredity.

Gravitation
The force of attraction between all matter, one of the fundamental forces of nature.

Greenhouse effect
An increase in atmospheric temperature caused by an increased volume of carbon dioxide in the upper atmosphere. This absorbs infrared radiation reflected from the Earth's surface.

Habitat
In ecology, the type of environment that an organism inhabits.

Halflife
The time taken for the activity of a radioactive sample to decrease to half its original value.

Heredity
The passing of genetic characteristics from parents to offspring.

Holography
The creation of three-dimensional images by photographing the subject when illuminated by a split laser beam, and reproducing the image by recreating the beam.

Hormone
Organic substance produced in minute quantity in one part of an organism and transported to other parts where it exerts a specific effect, eg stimulating growth of a specific type of cell.

Immunity
State of resistance to an infection, through the existence of antibodies specific to that pathogen.

Inorganic chemistry
Major branch of chemistry comprising the study of all the elements and their compounds, except carbon compounds containing hydrogen.

Integrated circuit
A structure (often a silicon chip) on which many individual electronic components are assembled.

In vitro
In glass; outside the living body and within an artificial environment such as a testtube.

Ion
An atom or group of atoms that has become electrically charged by the gain or loss of electrons.

Ionization
The formation of ions.

Isotopes
Atoms of an element with the same number of protons in the nucleus but different numbers of neutrons.

IVF
In vitro fertilization.

Laser
Device producing an intense beam of parallel light with a precisely defined wavelength; laser is short for "light amplification by stimulated emission of radiation".

Liquid
One of the three states of matter, taking the shape of the container but of a fixed volume at a particular temperature.

Mantle
The silica-rich layer that constitutes the bulk of the Earth. It lies between the core and the crust.

Maser
A device used as a microwave oscillator or amplifier; "microwave amplification by stimulated emission of radiation".

Mass
A measure of the amount of matter in a body.

Mass spectroscopy
Technique in which electric and magnetic fields are used to deflect moving charged particles differentially according to their mass.

Matter
Material substance, with extension in space and time; the three physical states are solids, liquids and gases. Also regarded as a specialized form of energy.

Microcomputer
A computer with the central processing unit contained on a single silicon chip.

Microorganism
A living organism too small to be seen without a microscope.

Microprocessor
An electronic device that receives, processes, stores and outputs information according to a preprogrammed set of instructions.

Molecular weight
The sum of the atomic weights of all the atoms in a molecule.

Molecule
Entity composed of atoms linked by chemical bonds and acting as a unit; its composition is represented by its molecular formula.

Moho
Abbreviation for the Mohorovičić discontinuity – the boundary between the crust and the mantle.

Neoplasm
A mass of new and abnormal cells; a tumor.

Neuron
A nerve cell.

Neutron
An uncharged subatomic particle.

NMR
Nuclear magnetic resonance; an effect used in imaging the interior of the body using the magnetic resonance of the atoms comprising the elements of the body tissues.

Nuclear physics
Study of the properties and mathematical treatment of the atomic nucleus and subatomic particles.

Nucleus
The core of an atom, containing positively charged protons and electrically neutral neutrons.

Optical fiber
Extruded glass fiber of high purity, used to transmit a light signal; used widely for telephone systems.

Orbital
The mathematical wave function describing the motion of an electron around the nucleus of an atom or several nuclei in a molecule.

Organic chemistry
Major branch of chemistry comprising the study of carbon compounds containing hydrogen.

Paleomagnetism
The properties of the Earth's magnetic field at a time in the past revealed by examining the alignment of magnetic particles in rocks formed at that time.

Periodic table
A table of the elements in order of atomic number, arranged to illustrate periodic similarities and trends in physical and chemical properties.

Photon
The quantum of electromagnetic energy, often thought of as the particle associated with light or other electromagnetic radiation.

Plasma
Almost completely ionized gas, containing equal numbers of free electrons and positive ions.

Plate tectonics
The study of the movement of plates over the surface of the globe. This is the mechanism that moves the continents and opens and closes the oceans.

Polymer
Substance composed of very large molecules built up by repeated linking of small molecules.

Protein
A complex biomolecule, made up of one or more chains of amino acids. Where made of several chains, each of these is known as a polypeptide chain.

Proton
Stable, positively charged subatomic particle found in the nucleus of all atoms.

Pulsar
A star made up chiefly of neutrons. Many neutron stars are emitting radio waves and are rotating rapidly, so that their emissions arrive in pulses: they are, therefore, known as pulsars.

Quantum mechanics theory
Theory of small-scale physical phenomena, such as the motions of electrons and nuclei within atoms.

Quark
Fundamental subatomic particle.

Quasar
A very remote, superluminous object. Quasars are now believed to be the nuclei inside very active galaxies.

Radar
"Radio detection and ranging": a technique that locates the position of distant objects by measuring the time taken for radio waves to travel to them, be reflected and return.

Radiation
The emission and propagation through space of electromagnetic radiation or subatomic particles.

Resistance
The ratio of the voltage applied to a conductor to the current flowing through it.

RNA
Ribonucleic acid; a single-stranded nucleic acid that cooperates with DNA for protein synthesis.

Semiconductor
Substance intermediate between conductor and insulator. Its properties vary with temperature and state of purity. Basis of all integrated circuits.

Solar wind
A stream of low-energy atomic particles continuously sent out in all directions by the Sun.

Spin
Intrinsic angular momentum of a nucleus or subatomic particle arising from its rotation about its axis.

Stereochemistry
The study of the spatial arrangement of atoms in molecules, and of their properties.

Subatomic particles (elementary particles)
Small packets of matter-energy that are constituent of atoms or are produced in nuclear reactions or in interactions between other subatomic particles.

Superconductivity
A condition occurring in many metals, alloys, etc., which have zero electrical resistance at very low temperatures.

Synchrotron
A large accelerator in which the particles are accelerated around a circular path.

Teratogen
An agent that deforms or causes physical defects in an embryo.

Transistor
An electronic device made of semiconductors used in a circuit as an amplifier, rectifier, detector or switch.

Tumor
A growth of excess tissue due to abnormal cell division.

Vitamin
Any of several organic substances, distinguished as vitamins A, B, etc. occurring naturally in minute quantities in many foodstuffs and regarded as essential to normal growth, especially through their activity in conjunction with enzymes in the regulation of metabolism.

X-rays
Invisible electromagnetic radiation with a wavelength between that of ultraviolet radiation and gamma rays.

FURTHER READING

Technology

Baker, W J *A History of the Marconi Company* (London, 1970)

Beaubois, H *Airships: an Illustrated History* (London, 1973)

Braun, W von, Ordway, F and Dooling, D *Space Travel – A History* (London, 1985)

Braun, E and MacDonald, S *Revolution in Miniature: the History and Impact of Semiconductor Electronics* (London, 1978)

Byers, A *Centenary of Service: a History of Electricity in the Home* (London, 1981)

Chant, C *Aviation: an Illustrated History* (London, 1983)

Chant, C (ed) *Science, Technology, and Everyday Life* (London, 1989)

Davies, W J K *Diesel Rail Traction* (London, 1973)

Dummer, G W A *Electronic Inventions and Discoveries 1745–1976* (Oxford, 1978)

Easterling, K *Tomorrow's Materials* (London, 1988)

Ellul, J *The Technological Society* (New York, 1964)

Evans, C F *Making of the Micro: A History of the Computer* (London and New York, 1981)

Feigenbaum, E A and McCorduck, P *The Fifth Generation: Artificial Intelligence and Japan's Computer Challenge to the World* (Reading, Mass, 1983)

Fielding, R (ed) *A Technological History of Motion Pictures and Television* (Berkeley, Calif, 1967)

Furter, W F (ed) *History of Chemical Engineering* (Washington DC, 1980)

Gibbs-Smith, C H *The Aeroplane: an Historical Survey of its Origins and Development* (London, 1960)

Gowing, M *Britain and Atomic Energy 1939–1945* (London, 1964)

Gowing, M *Britain and Atomic Energy 1945–1952* (2 vols) (London, 1974)

Grayson, M *Concise Encyclopedia of Chemical Technology* (Chichester, Sussex, 1985)

Gunston, W *The Illustrated Encyclopedia of the World's Missiles* (New York 1978, London, 1979)

Haber, L F *The Chemical Industry 1900–1930* (Oxford, 1971)

Haber, L F *The Poisonous Cloud: Chemical Warfare in the First World War* (Oxford, 1986)

Hewlett, R G and Anderson, O E *History of the USAEC, Vol I: The New World, 1939–1946* (University Park, Pa, 1962)

Hewlett, R G and Duncan, F *History of the USAEC, Vols II and III: Atomic Shield* (University Park, Pa, 1969)

Hinton, Lord *Heavy Current Electricity in the United Kingdom: History and Development* (Oxford, 1979)

Longstaff, M *Unlocking the Atom: a Hundred Years of Nuclear Energy* (London, 1980)

McNeil, I (ed) *An Encyclopaedia of the History of Technology* (London, 1990)

Medvedev, Z *The Legacy of Chernobyl* (Oxford, 1990)

Mould, R F *Chernobyl: The Real Story* (Oxford, 1988)

Okoshi, T *Three Dimensional Imaging Techniques* (New York, 1976)

Page, R M *The Origin of Radar* (New York, 1962)

Porter, R W *The Versatile Satellite* (New York and Oxford, 1977)

Schlaifer, R and Heron, S D *Development of Aircraft Engines and Fuels* (Boston, Mass, 1950)

Thévenot, R *The History of Refrigeration Throughout the World* (Paris, 1979)

Walker, P (ed) *Chambers Science and Technology Dictionary* (Edinburgh, 1988)

Williams, T I (ed) *A History of Technology: Vols VI and VII* (Oxford, 1978)

Williams, T I *A Short History of 20th Century Technology: c. 1900–c.1950* (Oxford, 1982)

Williams, T I *The Chemical Industry: Past and Present* (Milton Keynes, 1972)

Williams, T I *A History of the British Gas Industry* (Oxford, 1981)

Williams, T I and Withers, S (eds) *A Biographical Dictionary of Scientists* (London, 1982)

Williamson, H F, Andreano, R L, Daum, A R, and Klose, G C *The American Petroleum Industry: the Age of Energy, 1899-1959* (Evanston, Ill, 1970)

Biological and Medical Sciences

Allen, G E *Life Science in the Twentieth Century* (London, 1978)

Bleich, A R *The Story of X-rays from Röntgen to Isotopes* (London, 1960)

Bliss, M *The Discovery of Insulin* (Chicago, Ill, 1982)

Carson, R *Silent Spring* (Boston, Mass and London, 1963)

Carter, R *Breakthrough: the Saga of Jonas Salk* (New York, 1969)

Cartwright, F F *The Development of Modern Surgery* (London, 1967)

Cohen, D L and Segelman, A B *Antibiotics in Historical Perspective* (Rahway, NJ, 1981)

Cope, Z *A History of the Second World War: Surgery* (London, 1953)

Corsi, P and Weindling, P *Information Sources in the History of Science and Medicine* (London, 1983)

Dowling, H F *Fighting Infection: Conquests of the Twentieth Century* (Cambridge, Mass, 1977)

Dunn, L C A *Short History of Genetics* (New York, 1965)

Fine, R *A History of Psychoanalysis* (New York, 1979)

Foster, W D *A History of Medical Bacteriology and Immunology* (London, 1973)

Fox, D M and Lawrence, C *Photographing Medicine: Images and Power in Britain and American Since 1840* (New York, 1988)

Goodfield, J *From the Face of the Earth* (London, 1985)

Gribbin, J *In Search of the Double Helix* (London and New York, 1988)

Hackling, A J *Economic Aspects of Biotechnology* (Cambridge, 1986)

Hester, R E *Understanding our Environment* (London, 1986)

Kennedy, I R *Acid Soil and Acid Rain: the Impact on the Environment of Nitrogen and Sulphur Cycling* (Chichester, Sussex, 1986)

McCollum, E V *A History of Nutrition* (Boston, Mass, 1957)

McGrew, R E *An Encyclopedia of Medical History* (London, 1985)

MacFarlane, G *Alexander Fleming: the Man and the Myth* (Oxford, 1984)

MacFarlane, G *Howard Florey: the Making of a Great Scientist* (Oxford, 1979)

McKeown, T *The Role of Medicine: Dream, Mirage or Nemesis* (London, 1976)

Marlo, G J, Hollingworth, R M, and Durham, W (eds) *Silent Spring Revisited* (Washington DC, 1987)

Marx, J L (ed) *A Revolution in Biotechnology* (Cambridge, 1989)

Nossal, G J V and Coppel, R L *Reshaping Life: Key Issues in Genetic Engineering* (Cambridge, 1990)

Olby, R *The Path to the Double Helix* (London, 1974)

Paul, J R *A History of Poliomyelitis* (New York, 1971)

Razzell, P *The Conquest of Smallpox* (Firle, 1976)

Reed, J *From Private Vice to Public Virtue: the Birth Control Movement and American Society since 1830* (New York, 1978)

Reiser, S J *Medicine and the Reign of Technology* (Cambridge, 1978)

Rhodes, P *An Outline History of Medicine* (London, 1985)

Shyrock, R H *The Development of Modern Medicine: an Interpretation of the Social and Scientific Factors Involved* (Madison, Wisc, 1980)

Singer, C and Underwood, E A *A Short History of Medicine* (London, 1962)

Smith, F B *The Retreat of Tuberculosis, 1850–1950* (London, 1988)

Smith, W D A *Under the Influence* (London, 1982)

Watson, J D *The Double Helix: a Personal Account of the Discovery of the Structure of DNA* (New York, 1968)

Williams, T I *Howard Florey: Penicillin and After* (Oxford and New York, 1984)

Physical sciences

Atkins, P W *Molecules* (New York, 1988)

Augarten, S *Bit by Bit – an Illustrated History of Computers* (New York, 1984)

Barrow, J D and Silk, J *The Left Hand of Creation* (New York, 1983)

Bell, J S *Speakable and Unspeakable in Quantum Mechanics* (Cambridge, 1988)

Calder, N *Einstein's Universe* (London, 1979)

Cline, B L *Men Who Made a New Physics* (Chicago, Ill, 1987)

Close, F, Marten, M and Sutton, C *The Particle Explosion* (Oxford, 1987)

Feynman, R *QED* (Princeton, NJ, 1985)

Graham-Smith, F and Lovell, B *Pathways to the Universe* (Cambridge, 1988)

Hanle, P A and Chamberlain, V D *Space Science Comes of Age* (Washington DC, 1981)

Harwitt, M *Cosmic Discovery* (Brighton, Sussex, 1981)

Henbest, N and Marten, M *The New Astronomy* (Cambridge, 1983)

Hermann, A et al *History of CERN, Vol 1* (Cambridge, 1987)

Hey, T and Walters, P *The Quantum Universe* (Cambridge, 1987)

Jungk, R *Brighter than 1000 Suns: a Personal History of the Atomic Scientists* (San Diego, Calif, 1970)

Marschall, L A *The Supernova Story*

Marton, L *Early History of the Electron Microscope* (San Francisco, Calif, 1968)

Mendelssohn, K *The Quest for Absolute Zero* (London, 1966)

Miller, D et al *Chambers Concise Dictionary of Scientists* (Edinburgh, 1989)

Rhodes, R, *The Making of the Atomic Bomb* (London, 1988)

Rohrlich, F *From Paradox to Reality: Our Basic Concepts of the Physical World* (Cambridge, 1987)

Segrè, E *From X-rays to Quarks* (New York, 1980)

Seymour, R B (ed) *Pioneers in Polymer Science* (Lancaster, Lancs, 1989)

Sherwood, M *New Worlds in Chemistry* (London, 1977)

Walton, D W H (ed) *Antarctic Science* (Cambridge, 1987)

Weedman, D W *Quasar Astronomy* (Cambridge, 1986)

Weinberg, S *The Discovery of Subatomic Particles* (New York, 1983)

Weinberg, S *The First Three Minutes* (London, 1977)

ACKNOWLEDGEMENTS

Picture credits

1 J.P Holland, the US submarine designer: Royal Navy Submarine Museum
2–3 US missiles and rockets: MacQuitty International Collection
4 Particle detector at CERN: SPL/D. Parker
6 Wembley School astronomy club: PF
20–21 The Wright Brothers' first powered flight, 17 December 1903: Smithsonian Institution
54–55 Einstein and the staff of the Yerkes Observatory, Williams Bay, Wisconsin, USA, 1921: University of Chicago/Yerkes Observatory
88–89 The explosion of the *Hindenburg*, 1937: FPG International
122–123 Inoculation against smallpox, Havana, Cuba, 1949: PF
156–157 Edwin Aldrin deploys a solar wind experiment on the Moon, July 1969: NASA
192–193 Dump for low-level industrial nuclear waste: Zefa/Black Star

9 Marine Biological Laboratory, Woods Hole, Mass. 10 Novosti Press Agency 12–13 Chicago Historical Society 15 SPL/St Bartholomew's Hospital 16–17 SPL/David Parker 18–19 Sygma Projects/David Baker 25t Henry Ford Museum, Dearborn, Michigan 25b National Motor Museum, Beaulieu 26 Metropolitan Museum of Art 27 EA 28 Seaver Center, LA County Museum of Natural History 29t Luftschiffbau Zeppelin 29c Royal Navy Submarine Museum 29b IWM 30 The Marconi Company Ltd 31t Foothill Electronics Museum of the Perham Foundation 31b HDC 32–33 Brown Brothers 32 UB 33 Illustrated London News/Sphere 34–35 Smithsonian Institution 34t EA 34b Rank Xerox Ltd 35t Henry Ford Museum, Dearborn, Michigan 35c CP 35b SPL/Heini Schneebeli 37t EA 37b Brown Brothers 38t, 39b Wellcome Institute Library 38b Science Museum Libary 39t Süddeutscher Verlag 40t, 40c Private Collection, London 40b Bridgeman Art Library 41l EA 41tr Museum Boerhaave, Leiden 41cr EA 43, 44tr Jean-Loup Charmet 44tl Los Angeles Herald Examiner 44b Department of Physics, Schuster Laboratory, Manchester University 45t, 45b The Nobel Foundation 46, 46–47 Süddeutscher Verlag 47t, 47c, 47b EA 48 Museum Boerhaave, Leiden 49 Stockholms Stadsmuseum 50–51 International Museum of Photography at George Eastman House 51l Bildarchiv Preussischer Kulturbesitz 51r EA 52 Minnesota Historical Society/John W.G. Dunn 53t SPL/Fermi National Accelerator Laboratory 53b SPL/Dr David Roberts 59 National Archives, Washington, DC 60l Robert Hunt Library 60r, 61t IWM 61b PF 62t, 62b, 62–63 Kobal Collection 63b Science Museum, London 64 HDC 65t Ann Ronan Picture Library 65b, inset 65b David Sarnoff Research Center Archives 66t Archive Center, National Museum of American History 66b PF 67bl Katz Collection 67tl Galerie Loft/Christian Gervais/Phillips 67r Brown Brothers 68–69, 68t British Museum (Natural History) 68b PF 69t HDC 69c SPL/Philippe Plailly 69b Geoscience Features 71 IWM 72t, 72c EA 72–73 The Rockefeller Archive Center 72b PF 74 Greater London Photo Library 74 inset tr, 74 inset br, 75t, 75b RV 76t, 76b Eli Lily and Company 77 Tate Gallery 79 AIP Niels Bohr Library/© Institute International de Physique Solvay 80l, 80b Cavendish Laboratory, Cambridge 80r SPL 81 EA 82–83, 83t The Observatories of the Carnegie Institution of Washington 83b Alabama Space and Rocket Center 84 EA 85 Illustrated London News 86t OCD 86b EA 87t, 87tc Kodak 87c SPL/Eric Gravé 87b SPL/Tektoff-Mervieux, CNRI 93 Lenin Library, Moscow 94 Consolidated Edison Company of New York Inc 95t EA 95b EA 96t Texaco 96–97 Millbrook House Ltd 96b Photograph by British Petroleum Co. Ltd 97 US Army 98t Süddeutscher Verlag 98b EA 99t Willard R. Culver 99 inset Hagley Museum & Library 100t, 100b, 101, 100–101 IWM 102t HDC 102b Archive of the United Technologies Corporation, Hartford, Connecticut 103 EA 104–105 Adler Planetarium, Chicago 104t

Kobal Collection 104bl Donald K. Yeomans, California Institute of Technology 104br Mary Evans Picture Library 105t Kobal Collection 105b Boeing Aerospace Company 107PF 108–109 AVC Department, St Mary's Hospital Medical School, London 108c SPL/St Mary's Hospital Medical School 108b Northern Regional Research Center 109 Library of Congress 110t IWM 110b American Red Cross 111t EA 111b IWM 113t SPL/D. McMullan 113b Estate of Ernst Ruska 114 EA 115t EA 115b SPL/Prof. D. Skobeltzyn 116 Curie Institute Archives 117l SPL/P.I. Dee 117r Cavendish Laboratory, University of Cambridge 118 UKAEA. Harwell Laboratory 119l SPL/US Army 119tr TPS 119b Los Alamos Scientific Laboratory, AIP Niels Bohr Library 120t SPL/C. Powell, P. Fowler and D. Perkins 120b SPL/Patrick Blackett 121r SPL/I. Curie and F. Joliot 121t SPL/US Department of Energy 127 PF 128t AT & T Archives 128b SPL/Dr. J. Burgess 129 The Computer Museum, Boston 130 RCA Archives 131t, 131b PF 132 British Aerospace 133t Rolls Royce Ltd 133b The Popular Mechanics Magazine 134, 134–135 Novosti Press Agency 135 SPL/US Navy 136t HDC 136b EA 137 CP 138–139 AP 138t PF 138c Stanley Miller 139c CP 139t M/Philip Jones-Griffiths 139b Petit Format/John Watney 141 Dr. A. M. Joekes 142–143 March of Dimes Birth Defects Foundation 143 Cold Spring Harbour Laboratory 144t PF 144b Greater London Photo Library 145 The Rockefeller Archive Center 146t EA 146b Professor Bernard John/Australian National University 147t SPL/Science Source 147bl OCD 147br Celltech 149 PF 150t US Navy 150b National Academy of Science, National Research Center 151 USNC/IGY 152 SPL/C. Powell, P. Fowler and D. Perkins 153t SPL/Lawrence Berkeley Laboratory 153b PF 154 The Nobel Foundation 154–155 PF 161 Zefa/Armstrong 162–163, 163r M/Marc Riboud 164 PF 165t Novosti Press Agency 165b Zefa 166 M/René Burri 167t British Aerospace 167b PF 168–169 MacQuitty International Collection 169r HDC 169l SPL/NASA 170t US Geological Survey 170b Novosti Press Agency 171 NASA 172–173 SPL/Dr G.M. Rackham 172t SPL/CNRI 172b SPL/NASA 173t US Navy 173c OSF/Owen Newman 173b SPL/Simon Fraser 175, 176l, 176r TPS 177 Dr Michael E. DeBakey, Baylor College of Medicine, Houston, Texas 178 PF 179t M/David Hurn 179b TPS 181 SPL/David Parker 182t SPL/Lawrence Berkeley Laboratory 182b SPL/CERN 183t CP/D. Channer 183b FSP/Gamma/D. Darr 184, 185t, 185b PF 186–187 SPL/NOAA 187 Alabama Space and Rocket Center 188 National Astronomy and Ionosphere Center, Cornell University 189t Novosti Press Agency 189b NASA 190t, 190b SPL/Peter Ryan/Scripps 191t NASA 191b SPL/Soames Summerhays 197 SPL/US Department of Energy 198t Novosti Press Agency 198b SPL/Agema Infrared Systems 199 Statoil, Norway 200 FSP/Gamma/K. Kurita 201tl SPL/Hank Morgan 201tr Telefocus 201cr Zefa/Bramaz 201b Zefa 202 IMAX Space Technology Inc. 203t, 203b NASA 205 Zefa 206 Network/Sylvester 207 Greenpeace/Morgan 208t WHO 208b FSP 208–209 SPL/Prof. Luc Montagnier, Institut Pasteur/CNRI 210 Novosti Press Agency 211t SPL/Petit Format/CSI 211b SPL/Hank Morgan 212 SPL/P. A. McTurk, University of Leicester and David Parker 213t SPL/D. Parker 213c SPL/Jeremy Burgess 213b Munton and Fison 214–215 FSP/Dryden/Liaison 214 Novosti Press Agency 215tl, 215tr SPL/NASA 215c Art Directors 215b Greenpeace/Perez 217 SPL/Fermi National Accelerator Laboratory 218, 219b SPL/D. Parker 219t SPL/Patrice Loiez, CERN 220t St Andrews University/Professor J.F. Allen 220–221 JET Joint Undertaking 220b OCD 222 NASA 223 Network/Rapho 224b SPL/Max Planck Institute/David Parker 225b NASA/Jet Propulsion Laboratory 224–225 Rex Features 225t SPL/Dr. R. J. Allen and AL 226b Hatfield Polytechnic Observatory 227 SPL/Patrice Loiez/CERN 228l WHO/Best Institute 228c Bell Laboratories 228r EA 229l PF 229c Jocelyn Bell Burnett 229r CP 230l HDC/Bettmann Archive 230c Lawrence Berkeley Laboratory 230r, 231c PF 231l, 231r HDC 232l EA

232c Yale University Art Gallery 232r, 233c HDC 233l Brown Brothers 233r, 234c PF 234l EA 234r TPS 235l NASA 235r AIP Niels Bohr Library/F. D. Rasetti, Segre Collection 236l, 236c HDC 236r Henry E. Huntington Library and Art Gallery 237l Novosti Press Agency 237c, 237r PF 238l Novosti Press Agency 238c HDC 238r PF 239l EA 239c HDC 239r, 240l PF 240c EA 240r, 241r HDC 241l PF 242l AIP Niels Bohr Library/Burndy Library 242c EA 242r PF 243l HDC 243r March of Dimes Birth Defects Foundation 244l Österreichische National Bibliothèque 244c TPS 244r EA 245l HDC 245c PF 245r, 246l EA 246c Novosti Press Agency 246r Ullstein Bilderdienst 247l HDC/Bettman Archive 247r PF

Abbreviations

AP — Associated Press, London
CP — Camera Press, London
EA — Equinox Archive
FSP — Frank Spooner Pictures, London
HDC — Hulton Deutsch Collection, London
IWM — Imperial War Museum, London
M — Magnum, London
NASA — National Aeronautics and Space Administration, USA
NOAA — National Oceanic and Atmospheric Administration, US Dept. of Commerce
OCD — Oxford Chemical Designs Ltd
OSF — Oxford Scientific Films, Long Hanborough, Oxford, UK
PF — Popperfoto, London
RV — Roger-Viollet, Paris
SPL — Science Picture Library, London
TPS — Topham Picture Source, Kent, UK
UB — Ullstein Bilderdienst, Berlin

Abbreviations

t = top, tl = top left, tr = top right, c = center, b = bottom, etc.

Editorial and research assistance

Steven Chapman, Alastair Gray, Jane Higgins, Louise Jones, Christopher Lawrence (Wellcome Centre, London), Rhys Lewis, John Maple, Jack Meadows, Malcolm Oster, Andy Over, Simon Schaffer, Graham Speake, Christine Sutton, Michelle Von Ahn, Katherine Watson, Paul Weindling, Elaine Welsh. Special thanks to Macmillan Educational Company, New York.

Artists

Robert and Rhoda Burns, Alan Hollingbery, Colin Salmon, Dave Smith, Del Tolton

Photographs

Shirley Jamieson

Typesetting

Brian Blackmore, Catherine Boyd; OPUS Ltd

Production

Stephen Elliott, Clive Sparling

Color Origination

J Film Process, Bangkok

Index

Ann Barrett

INDEX